Play of Chance and Purpose

Play of Chance and Purpose

An Invitation to Probability, Statistics, and Stochasticity Using Simulations and Projects

Amrik Sen

CAMBRIDGE
UNIVERSITY PRESS

Shaftesbury Road, Cambridge CB2 8EA, United Kingdom

One Liberty Plaza, 20th Floor, New York, NY 10006, USA

477 Williamstown Road, Port Melbourne, vic 3207, Australia

314 to 321, 3rd Floor, Plot No.3, Splendor Forum, Jasola District Centre, New Delhi 110025, India

103 Penang Road, #05–06/07, Visioncrest Commercial, Singapore 238467

Cambridge University Press is part of Cambridge University Press & Assessment,
a department of the University of Cambridge.

We share the University's mission to contribute to society through the pursuit of
education, learning and research at the highest international levels of excellence.

www.cambridge.org
Information on this title: www.cambridge.org/9781009338394

© Cambridge University Press & Assessment 2025

First published 2025

Printed in India by Avantika Printers Pvt. Ltd.

A catalogue record for this publication is available from the British Library

ISBN 978-1-009-33839-4 Paperback

To Indiya Rusak Sen

Contents

Table and Figures

URL Links to Data in CSV Format

Ch. 4: Table-4.1.csv	https://drupal-s3fs-prod.s3.eu-west-1.amazonaws.com/resources/Table-4.1.csv
Ch. 5: Coke-1.csv	https://drupal-s3fs-prod.s3.eu-west-1.amazonaws.com/resources/Coke-1%20%283%29.csv
Ch. 6: World_Bank_Data.csv	https://drupal-s3fs-prod.s3.eu-west-1.amazonaws.com/resources/World_Bank_Data%20%285%29.csv
Ch. 6: EnergyConsumptionMP_1996-2018.csv	https://drupal-s3fs-prod.s3.eu-west-1.amazonaws.com/resources/EnergyConsumptionMP_1996-2018%20%281%29.csv
Ch. 6: JOBS.csv	https://drupal-s3fs-prod.s3.eu-west-1.amazonaws.com/resources/JOBS.csv

Preface

Our lived experiences are punctuated by events that are sometimes a result of our purposeful intentions and at other times outcomes that happen by pure chance. Even at an abstract level, it is a very human endeavor to deduce meaning from seemingly random observations – an exercise whose primary objective is to derive a causal structure in observed phenomena. In fact, our whole intellectual pursuit that differentiates us from other beings can be understood through our inner urge to discover the very purpose of our existence and the conditions that make this possible. This eternal play[1] between chance episodes and purposeful volition manifests in diverse situations that I have labored to recreate through computer simulations of realistic events. This play has a dual role – first, it binds together the flow of our varied experiences and, second, it offers us a perspective to assimilate our understanding of events happening around us that affect us. In order to appreciate this play of chance and purpose, it is essential that students and readers have a conceptual grounding in the areas of probability, statistics, and stochastic processes. Therefore, several playful computer simulations and projects are interlaced with theoretical foundations and numerical examples – both solved and exercise problems. In this way, the presentation in this book remains true to its spirit of inviting thoughtful readers to the various aspects of this area of study.

[1] Here, *play* may be interpreted as *interplay*.

Historical remark

The advent of a rigorous framework for studying probability and statistics dates back to the eighth century AD and is documented in the works of Al-Khalil, who was an Arab philologist.[2] This branch of mathematics continues to be under development with major contributions from Soviet mathematician Andrey N. Kolmogorov, who developed the modern foundations of probability and statistical theory from a measure-theoretic standpoint in the twentieth century.[3]

[2] Lyle D. Broemeling, An Account of Early Statistical Inference in Arab Cryptology, *The American Statistician* **65**, no. 4 (2011), pp. 255–257, DOI:10.1198/TAS.2011.10191·

[3] A. N. Kolmogorov, *Foundations of the Theory of Probability*, 2nd ed. (Chelsea Publishing Co., 1956), ISBN: 0828400237.

Why should you study probability and statistics?

Both theoretical and applied aspects of probability and statistics are subjects of broad intellectual and practical interest to engineers and scientists alike. This genre of mathematics is widely practiced in diverse areas of engineering and sciences, ranging from artificial intelligence, bioinformatics, wireless communications, aerodynamics, operations research, financial engineering, modern physical and chemical sciences to the social and humanitarian sciences, demography, weather forecasting, and data analytics, among other fields. Over the past century, on the one hand, the scientific community has gradually come to accept the limitations of scientific measurements using precision instruments (owing to the theory of quantum mechanics, and, to be precise, the unpredictability stemming from the consequences of the uncertainty principle). On the other hand, engineers and scientists have been confronted by an explosion of information that has increased the dimensions of variability in data. These challenges have expanded the scope and dependence on probabilistic and statistical techniques to solve real-world problems.[4]

In light of the above, there is a necessity and opportunity to train students in the practical aspects of probability and statistics so that they can readily apply their knowledge to find solutions to engineering and scientific problems. There is an urgent need to introduce the fundamentals and applications of this subject in a less intimidating manner, by not relying excessively on abstract deductions. An alternate approach that allows students to assimilate foundational concepts through projects and real data is often more helpful. This experiential mode of learning not only familiarizes students with applications but also keeps them engrossed in the learning process by making it fun. My objective is to foster the gradual cultivation of abstract thinking in students by first engaging them with a diversity of experiential projects and simulations that enhance the faculties of reflection and imagination.

About the book

This book covers broad areas of probability and statistics, starting with introductory preliminaries, probability distributions, introduction to stochastic processes through discrete and continuous time Markov chains, regression analysis, sampling distributions and their applications in statistical inference and hypothesis testing (including ANOVA), calculation of covariance functions, notion of stationarity and ergodicity of data, an introduction to time series analysis of data (moving average and autoregressive models), and a basic introduction to multivariate statistics and their application through principal component analysis. Almost all topics discussed in the book are presented using both a practical (project-based) and a theoretical framework. The book will be useful for advanced undergraduate-level students, beginning postgraduate-level students, and practitioners from the engineering, basic, and social sciences. The content of this book can be covered over one five-month-long semester or two three-month-long trimesters.

The computational and simulation instructions presented in this book are based on Matlab[5] tutorials and projects. The book includes a helpful appendix on how to get started with Matlab. Additionally, readers can access a Python compendium for the book on the dedicated Cambridge University Press website.

Each chapter features a project-based case study[6] laced together with concepts that are introduced in a systematic manner. Chapters begin with a brief introduction to the case-study (project) and the tools necessary to solve the problem. The concepts are then developed gradually, and their relevance to the project is established systematically. In this way, as readers sift through the various sections of each chapter, they not only learn the concepts but also get an opportunity to apply this knowledge toward solving the case-study project. For example, the chapter

---

[4] Uncertainty breeds risk and thereby rewards!

[5] https://matlab.mathworks.com.

[6] All project-related sections in each chapter is typeset in **bold**.

on "Statistical Experiments" has a case study on "post-disaster reconstruction management". In this project, students learn about some of the main issues that impede post-disaster reconstruction of infrastructure by identifying key bottlenecks such as lack of public funds, shortage of technical manpower to rebuild infrastructure, and lack of community participation, among other factors. A detailed ANOVA hypothesis test is conducted for each of the aforementioned bottleneck areas across six different cities. The tests are based on numerical ratings registered by project managers overlooking the reconstruction projects in each city. This statistical analysis will enable students to identify the bottleneck issues plaguing each city, for which corrective measures are solicited from the authorities. Similar projects are designed for other topics throughout the book. These built-in projects will also enable practitioners to design their own projects and statistical experiments by referring to the design of similar experiments and case studies presented in this book.

Another important feature of the presentation of material in this book is the use of several illustrations of mathematical concepts, figures, pictures, additional references, and footnotes along the margins. This feature is aimed at providing visual aids for the concepts being discussed in the corresponding section of the book. Such visuals often bring the more abstract mathematical formalism to life and enhance the learning experience of the reader.

The book has evolved through several rounds of lectures delivered to undergraduate and postgraduate students from diverse backgrounds in both the United States and India. Students will find it useful as a textbook. Instructors of probability and statistics will find it very useful for both theoretical and laboratory components of their courses. The book is self-sufficient for the most part. This means students with an elementary understanding of probability and statistics (standard XI–XII level) can use this book easily. The only prerequisites are elementary calculus (e.g., single-variable calculus, including differentiation, integration, series, and sequences) and high-school-level familiarity with matrices and algebra. The book will serve as an invitation to students to further explore the fascinating area of probability and statistics through more advanced courses.

A note on the choice of chapter topics

A unique feature of the book becomes clear upon a closer inspection of the table of contents and the sequence of its topics. Typically, classic texts on applied probability and statistics for engineering and science students do not cover topics such as Markov chains, time series analysis, and principal component analysis at the introductory level. Instead, they delve into statistical estimation theory, non-parametric tests, or more theoretical topics from probability theory. However, I feel that there is a tremendous opportunity to introduce students to applied topics like Markov chains (discrete and continuous) immediately after the chapter on conditional probability. This has three advantages:

1. The area of Markov chains has wide applications in a variety of engineering and scientific fields, ranging from queuing theory, weather modeling to financial engineering, natural language processing, AI, and engineering physics. This way we can capture students' attention and foster their appreciation for the subject at a very early stage of the learning process.

2. Two dedicated chapters on Markov chains also retain the alignment with the intended target audience of the book: engineering and science students and/or industry practitioners, who are perhaps more interested in the mathematical modeling aspect of the subject than the estimation-theoretic approaches more suitable for pure statisticians.

3. Markov models, by their construction, are an excellent illustration for appreciating abstract concepts like *conditioning* over disjoint events, evaluating conditional probabilities, and applying the law of total probability.

Likewise, there is an opportunity to introduce techniques to analyze data through time series models and principal component analysis in an introductory text of this nature so that we are able to prepare students interested in pursuing data science as majors. In this way, the need to cater to the target audience has been paramount in my mind while writing this textbook.

Acknowledgments

The book, in its present form, could not have been possible without the unrelenting support of my professional and personal family and friends. I am grateful to the university community for providing me with the right blend of creative and compassionate environment that is conducive to intellectual endeavors such as this one. I want to thank my wife, Olga Rusak, for her patience and diligent proofreading of the material presented here. Several young scientists, mathematicians, and students worked with me at various stages of development of this manuscript – I would like to put on record the contributions of Manraj Singh Ghumman, Vanita Sharma, Sandeep Kumar, Arpit Shukla, Rajat Singla, Abhishek Thakur, Vijay Sahani, Chandan Yadav, Vishal Paudel, and Tejasvi Birdh. I am hugely indebted to the reviewers from all around the world for offering valuable suggestions and constructive critiques that enhanced the quality of the work presented in this book.

I am deeply grateful to Professor Kanchi Gopinath for many insightful discussions on the foundations of artificial intelligence, which highlighted the relevance of numerous topics covered in this book. He also carefully proofread the chapter on multivariate statistics and offered detailed suggestions.

I am indebted to Jennifer Hinneburg of the Mathematisches Forschungsinstitut Oberwolfach; Jeremy Brown of the Wellcome Collection; Professor Ilia Ponomarenko and Victor S. Mikhaylov of the St. Petersburg Department of V.A. Steklov Institute of Mathematics of the Russian Academy of Sciences (PDMI RAS); Michael Sughrue of Omniscient Neurotechnology; and Brian Hayes of bit-player.org for their generous assistance in securing permission to reproduce certain images in this book.

I take this opportunity to express my gratitude to my editors at Cambridge University Press—Vaishali Thapliyal, Qudsiya Ahmed, Vikash Tiwari, Aniruddha De, and others—for trusting me with the authorship of this book and for being patient with me during several missed deadlines. Their generosity immensely contributed to the scholarship embodied in this book.

Finally, I would like to dedicate this book to my eight-year-old daughter, Indiya Rusak Sen, who never shies away from impressing upon me, through all her adventures, the power of imagination and the force of volition that are necessary to create a beautiful and better future for humanity.

Chandigarh, 2025   **Amrik Sen**

1

Thinking in Probability

CHANCE PERMEATES OUR physical and mental universe. While the role of chance in human lives has had a longer history, starting with the more authoritative influence of the nobility, the more rationally sound theory of probability and statistics has come into practice in diverse areas of science and engineering starting from the early to mid-twentieth century.[1] Practical applications of statistical theories proliferated to such an extent in the previous century that the American government-sponsored RAND corporation published a 600-page book that wholly consisted of a random number table and a table of standard normal deviates.[2] One of the primary objectives of this book was to enable a computer-simulated approximate solution of an exact but unsolvable problem by a procedure known as the *Monte Carlo method* devised by Fermi, von Neumann, and Ulam in the 1930s–40s.[3]

Statistical methods are the mainstay of conducting modern scientific experiments. One such experimental paradigm is known as a *randomized control trial,* which is widely used in a variety of fields such as psychology, drug verification, testing the efficacy of vaccines, agricultural sciences, and demography. These statistical experiments require sophisticated sampling techniques in order to nullify experimental biases. With the explosion of information in the modern era, the need to develop advanced and accurate predictive capabilities has grown manifold. This has led to the emergence of modern artificial intelligence (AI) technologies. Further, climate change has become a reality of modern civilization. Accurate prediction of weather and climatic patterns relies on sophisticated AI and statistical techniques. It is impossible to think of a modern economy and social life without the influence and role of chance, and hence without the influence of technological interventions based on statistical principles. We must begin this journey by learning the foundational tenets of probability and statistics.

1.1 Chapter objectives

The chapter objectives are listed as follows.

1. Students will learn the fundamental axioms of probability.

2. Students will apply elementary principles of probability, permutation, and combination to solve simple numerical examples.

3. Students will learn the meaning of random variables and formulate solutions to problems involving random events.

Figure 1.1 A portrait of Shakuni's game of dice from the Indian epic *Mahabharata* composed in the third century BCE (*courtesy:* Wikimedia Commons).

[1] Mauricio Suarez, *Philosophy of Probability and Statistical Modelling* (Cambridge University Press, 2020), DOI: 10.1017/9781108985826.

[2] RAND Corporation, *A Million Random Digits with 100,000 Normal Deviates* (RAND Corporation, 2001 [1955]).

[3] Alfred M. Bock, Randomness and the twentieth century, *The Antioch Review* **27**, no. 1 (1967): 40–61.

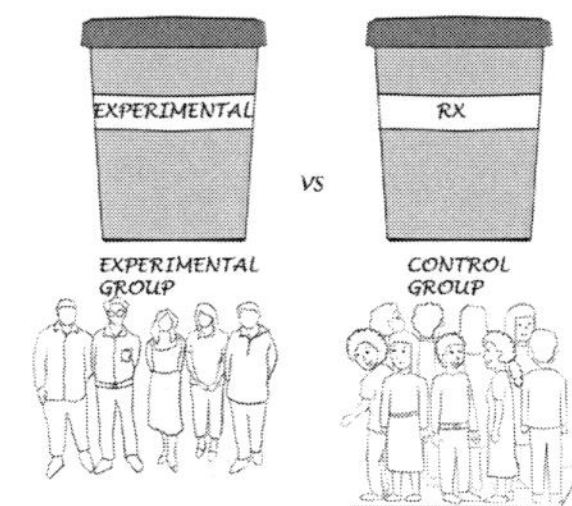

Figure 1.2 Schemata of a *randomized control trial* for evaluating the efficacy of a treatment intervention by a new drug launched in the market.

4. Students will learn to apply Bayes' theorem and the law of total probability to solve complex problems.

5. Students will learn to calculate statistical averages in terms of expectation of random variables.

6. Students will learn to use the techniques of computing probability and expectation of random events to solve a practical simulation project on one-dimensional random walk.

1.2   Chapter project: Random walk on a lonely island

1.2.1   *Prologue: Will Squeaky drift off to the edge and fall off the cliff or keep hopping back and forth forever?*

Figure 1.3 Squeaky is trapped on a lonely island hill with sharp cliffs on both sides.

Our friend Squeaky is trapped somewhere in the middle of a lonely island hill with sharp cliffs on both sides. Squeaky is excited and jumps around in her merry way. At any given instance, she makes a decision to hop to the left or to the right independent of her past moves. Squeaky is unaware of impending danger.

In this project, we will use calculations based on the principles of conditional probability, the law of total probability, and the law of total expectation to predict her fate. In other words, what are the odds that she will bounce around on the island hill, her left-sided moves balancing out her right-sided moves on an average, and never actually trip and fall off on either side? Or will chance play the devil's role and will she eventually drift off to one side and perish? And if the latter turns out to be true, then what is her life expectancy in terms of the total number of hops starting from her first move? Does a certain initial position on the hill give her the best chance to survive the longest?

In addition to our theoretical calculations, we will also build a computer simulation of her actions to corroborate our result. For convenience, we shall assume that the island is one dimensional, i.e., Squeaky's movements are restricted exclusively to lateral directions (left or right). While we build the computer-simulated solution, we will learn to apply a random number generator using a computer software in order to mimic Squeaky's mental choices to hop either to the left or to the right independent of her past moves.

1.3 Deterministic versus probabilistic outcomes

1.3.1 *Deterministic outcomes*

Permutations

Consider an assortment of five differently colored buttons. A simple question may be to find out all the different ways in which we may be able to arrange the five buttons without piling them on top of each other.

This is a classic example of the number of *permutations* of n distinct things. If we consider five empty slots in which to host the individual buttons, and begin with the leftmost slot, then this slot may be occupied by any of the five buttons. So depending on the color we choose, there are *five* different ways of filling the leftmost slot. Subsequently, we are left with four differently colored buttons, and hence the second slot can be filled in *four* different ways. This is followed by the third slot, which can be occupied in *three* distinct manners. The penultimate and ultimate slots can be filled in two and one different ways, respectively. Therefore, the number of *permutations* of n different things is $n! = n \times (n-1) \times \cdots \times 2 \times 1$.

However, if there are r different *types* of $n = n_1 + n_2 + \cdots + n_r$ objects, then there are $\dfrac{n!}{n_1! n_2! n_3! \ldots n_r!}$ different ways of arranging them. Here, the ith type has n_i counts, $i = 1, 2, 3, \ldots, r$.

> For example, if we have nine buttons of which three are of red color, four are of green color, and two are of blue color, then there are $\dfrac{9!}{3!4!2!} = 1260$ ways of arranging them.

Alternatively, we may have n different things, and we may want to know the number of permutations by taking only $r \leq n$ of them at a time. The number of possible ways are

$$P_r^n = n \times (n-1) \times (n-2) \times \cdots \times (n-r+1) = \frac{n!}{(n-r)!}.$$

> For example, let us say there are nine slots and four *differently* colored buttons. There are $\dfrac{9!}{5!} = 3024$ ways of arranging them.

Combinations

In many other situations, the order of arrangement is not so important. In such cases, we may only care about the number of subsets of r items from amongst a total of n items. The number of ways n things can be combined by taking r at a time is given by the formula

$$C_r^n = \binom{n}{r} = \frac{n!}{r!(n-r)!}.$$

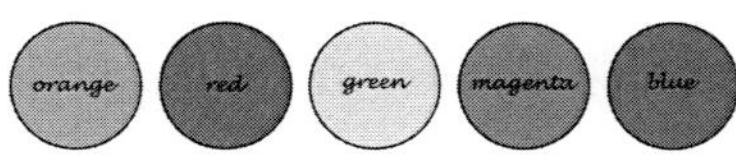

Figure 1.4 Number of different permutations of five differently colored buttons is 5!, i.e., there are $5 \times 4 \times 3 \times 2 \times 1 = 120$ different ways of arranging these five buttons.

For example, if there are nine slots and four *identically* colored buttons that can be placed in any of these slots, then there are $\binom{9}{4} = 126$ different designs/patterns that can emerge upon hosting any four buttons in the nine slots. Obviously, $C_r^n < P_r^n$ when $r > 1$.

1.3.2 *Probabilistic outcomes*

Most importantly, permutations and combinations belong to a class of experiments that have deterministic outcomes. There are a finite and fixed number of ways of arranging or collecting (combining) items. None of the aforementioned examples has a chance outcome. However, we may have to perform experiments whereby the outcome is not certain, at least not in the *a priori* sense. For example, we may ask that in any given arrangement of the five distinctly colored buttons in the five available slots, what is the probability that the first slot is filled by a red-colored button? Implicit in this question is the fact that this particular arrangement of the five buttons is made blindfolded (without the person actually making a conscious decision of placing the red button in the first slot). Under such circumstances, the placement of the red button in the first slot is a matter of chance. The probability of such an outcome is 1/5 because only one out of the possible five differently colored buttons that could have been placed in the first slot is red. We shall devote the rest of this book to the study of random events and statistical experiments that have a probabilistic outcome.

1.3.3 *A note of caution*

Statistical forecasts depend on good and reliable data. Biases in data can skew statistical predictions hugely, as is often noticed in faulty exit poll results. In order to address these biases, statisticians are often concerned with the appropriate design of their experiments.

Moreover, statistical inferences are based on the principles of probability (chance). They explain what outcome is likely to happen. However, in order to understand the rationale behind a particular outcome or the underlying principles responsible for a certain observation, one has to rely on physical theories that fall outside the scope of statistical techniques. Statistical theories shed light on idealized averages[4] of stochastic phenomena. Thus, the reach of statistical inferences may be far removed from individual experiences. A distinctive dimension of reality is its individual aspect, which may not be gleaned from statistical approaches.[5]

1.4   Foundations of probability

1.4.1 *Definition: Probability*

It is the measure of likelihood that an event will occur. For example, we may ask: what are the chances that it will rain today? Most weather prediction websites may give us an answer in terms of a probability measure, say, 75%.

1.4.2 *Definition: Statistics*

It is the branch of mathematics that deals with the collection (sampling), organization, analysis, and interpretation of data, including making inferences and forecasts. It relies on the principles of probability.

[4] Here the word *averages* is used in a broader sense of all statistical moments and not just the mean value.

[5] See Carl Gustav Jung, *The Undiscovered Self* (Routledge Classics, 2021 [1958]), 5.

1.4.3 *Definition: Probability space*

A probability space comprises a triple $(\Omega, \mathfrak{F}, P)$. Here Ω denotes the *sample space,* which is the set of all possible outcomes;[6] $\mathfrak{F}$ denotes the *σ-algebra,* which is a collection of all events of concern to us in a certain statistical experiment and is generated by Ω; and P is the *probability measure,* defined as a function $P : \mathfrak{F} \to [0,1]$. We may think of $\mathfrak{F}$ as an *event space.*

> For example, for the coin tossing experiment illustrated in Figure 1.5, the σ-algebra generated by $\Omega_2 = \{HH, HT, TH, TT\}$ can be taken as the power set of Ω_2,
>
> $$\mathfrak{F} = 2^{\Omega_2} = \{\{\}, \{HH\}, \{TT\}, \{HT\}, ..., \{HT, TT\}, ..., \{HH, HT, TH, TT\}\}.$$
>
> The cardinality of $\mathfrak{F}$ is $2^{|\Omega_2|} = 2^4 = 16$. The σ-algebra defined above is the largest such set. The smallest σ-algebra over Ω_2 is $\{\{\}, \{HH, HT, TH, TT\}\}$. The probability of observing two successive heads is 1/4.

It may be useful to state here that if we have disjoint sets (events) $E_1, E_2, ... \in \mathfrak{F}$, then $\cup_i E_i \in \mathfrak{F}$. A rigorous treatment of σ-algebra will be avoided in this introductory-level text; wherever necessary we will loosely refer to the notion of an event space.

1.4.4 *Axioms of probability*

We begin this short section by asking: *why do we need axioms at all?* In fact, the first foundational axioms of mathematics appeared only as recently as 1879, courtesy Gottlob Frege.[7] Axioms may be regarded as *a priori propositions* whose veracity is accepted universally without requiring their validation by demonstration. The utility of axioms lies in the fact that they enable the deduction of realizable experiences that can be supported by sense perceptions.[8]

> The axioms of probability were formulated by Andrey N. Kolmogorov in 1933.
>
> 1. $P(E) \geq 0$ for all $E \in \mathfrak{F}$ (non-negativity),
>
> 2. $P(\Omega) = 1$ (unitarity), and
>
> 3. $P(\cup_{i=1}^{\infty} E_i) = \Sigma_{i=1}^{\infty} P(E_i)$ for a countable sequence of disjoint events $E_1, E_2, ...$ (σ-additivity).

1.4.5 *Supplementary properties of probability measure*

In addition to the axioms of probability listed above, it is often helpful to consider the following properties of P while performing calculations.

1. Consider $E_1, E_2 \in \mathfrak{F}$; then $P(E_1 \cup E_2) = P(E_1) + P(E_2) - P(E_1 \cap E_2)$. This result may be generalized to n events $E_1, E_2, E_3, ..., E_n$ by induction. This result is known as the *principle of inclusion–exclusion.*

2. If E_1 and E_2 are *independent* events, then $P(E_1 \cap E_2) = P(E_1)P(E_2)$.

3. If A^c stands for the complementary event of A, then $P(A^c) = 1 - P(A)$.

4. The probability of the *impossible* event is zero, i.e., $P(\{\}) = 0$.

5. A distinction must be made between *mutually exclusive* (disjoint) events and *independent* events. E_1 and E_2 are mutually exclusive when $E_1 \cap E_2 = \{\}$.

[6] An *outcome* is the result of a single realization of the model experiment.

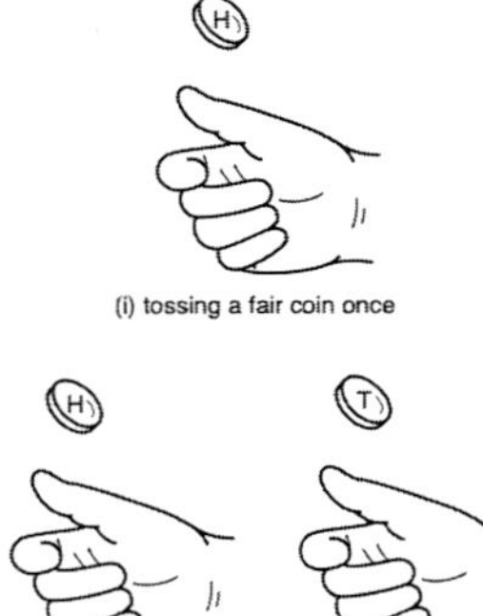

Figure 1.5 The outcome of a toss of a fair coin is heads or tails. Therefore, $\Omega_1 = \{H, T\}$, with the usual abbreviations for *heads* and *tails.* We may conduct an experiment in which we toss the coin twice where $\Omega_2 = \{HH, HT, TH, TT\}$.

[7] https://iep.utm.edu/frege/, accessed June 16, 2023.

[8] Kurt Gödel, Russell's mathematical logic (1944), reprinted in *Philosophy of Mathematics,* ed. Paul Benacerraf and Hilary Putnam, 447–469 (Cambridge University Press, 1983); Kenny Easwaran, The role of axioms in mathematics, *Erkenn* **68** (2008): 381–391.

Figure 1.6 Russian Mathematician A. N. Kolmogorov (*author:* Konrad Jacobs; *source:* Archives of the Mathematisches Forschungsinstitut Oberwolfach).

[9] The prevailing weather pattern in a given locality may be either sunny or rainy because these are *mutually exclusive* weather events in commonly used terminology. The outcomes of tossing a fair coin twice are *independent* events.

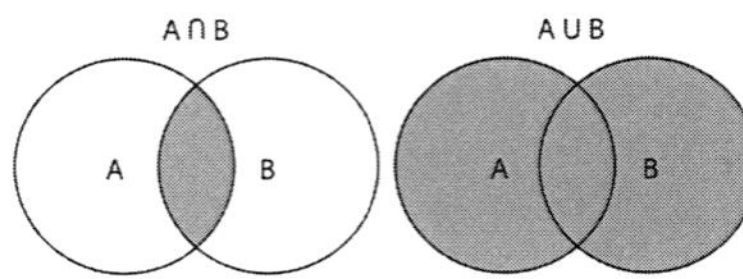

Figure 1.7 Venn diagram showing overlapping events and union of events.

In such a case, $P(E_1 \cap E_2) = P(\{\}) = 0$ and $P(E_1|E_2) = 0$. On the other hand, if two events A_1 and A_2 are independent, then $P(A_1 \cap A_2) = P(A_1)P(A_2)$ and $P(A_1|A_2) = P(A_1)$. Here $A_1|A_2$ refers to the occurrence of the event A_1 with the knowledge that the event A_2 has already happened. In essence, two events are mutually exclusive if they cannot happen concurrently, whereas two independent events may happen concurrently but the outcome of one does not influence the outcome of the other.[9]

The symbols $\cap$ and $\cup$ denote *overlapping* and *union* of events, respectively (see Figure 1.7). Sometimes we will use the notation AB to denote the event $A \cap B$.

Example 1.1   Defining events in probability space

Two dice are thrown. Let E be the event that the sum of the dice is odd, let F be the event that the first die lands on 1, and let G be the event that the sum is 5. Describe the events $EF, E \cup F, FG, EF^c, EFG$.

Solution: $\Omega = \{(1,1)\ (1,2)\ (1,3)\ ....\ (6,4)\ (6,5)\ (6,6)\}$
$EF = \{(1,2),(1,4),(1,6)\}$
$E \cup F = \{(1,1), (1,2), (1,3), (1,4), (1,5), (1,6), ... \}$, where the remaining entries of the set are the fifteen other possible pairs where the second number is odd if the first number is even, and the second number is even if the first number is odd (not 1).
$FG = \{(1,4)\}$
$EF^c = \{$all the 15 possible outcomes where the first die is not 1 and the two dice are not either both even or both odd$\}$
$EF^c = E \cap F^c = \{(2,1)\ (2,3)\ (2,5)\ (3,2)\ (3,4)\ (3,6)\ ....\ (6,1)\ (6,3)\ (6,5)\}$
$EFG = E \cap F \cap G = \{(1,4)\} = F \cap G.$

Example 1.2   Rolling two dice concurrently

Consider an experiment comprising throws of two independent dice. The sample set is the Cartesian product comprising ordered pairs,

$$\Omega = \{1,2,3,4,5,6\} \times \{1,2,3,4,5,6\} = \{(1,1),(1,2),...,(2,1),(2,2),...,(6,5),(6,6)\}.$$

We may be interested in knowing the odds that the sum of the outcomes from each die is greater than or equal to ten. In this case, the event space is

$$\mathscr{E} = \{(4,6),(5,5),(5,6),(6,4),(6,5),(6,6)\},$$

and hence the required probability is $\dfrac{|\mathscr{E}|}{|\Omega|} = \dfrac{6}{36}.$

Example 1.3   Probability of a complementary event

A, B, and C are three mutually exclusive and exhaustive events of a random experiment such that $P(B) = \dfrac{3}{2}P(A)$ and $P(C) = \dfrac{1}{2}P(B)$. What is the probability of non-occurrence of event A?

Solution: $P(C) = \frac{1}{2} \times \frac{3}{2} P(A) = \frac{3}{4} P(A)$

The second axiom of probability (unitarity) demands

$$P(A) + P(B) + P(C) = 1.$$

$$P(A) + \frac{3}{2} P(A) + \frac{3}{4} P(A) = 1$$

$$P(A)\left(1 + \frac{3}{2} + \frac{3}{4}\right) = 1$$

$$P(A) = \frac{4}{13}$$

$$P(A^c) = 1 - \frac{4}{13} = \frac{9}{13}.$$

Example 1.4   Probabilities of composite events originating from rolling two dice

If two dice are thrown, what is the probability that their sum is (a) greater than 9, (b) neither 7 nor 11?

Solution: The cardinality of the sample space is $n(\Omega) \equiv |\Omega| = 36$.

Let S_i be the event when the sum of the outcomes of the two dice equals i.

$$\text{a)} \, P(\text{sum is greater than } 9) = P(S_{10}) + P(S_{11}) + P(S_{12})$$

$$= \frac{3}{36} + \frac{2}{36} + \frac{1}{36}$$

$$= \frac{1}{6}.$$

$$\text{b) Event } A := S_7, \quad P(A) = \frac{1}{6}$$

$$\text{Event } B := S_{11} \quad P(B) = \frac{1}{18}$$

Since the event $(A^c \cap B^c)$ is equivalent to the event $(A \cup B)^c$, we have

$$P(A^c \cap B^c) = P(\Omega) - P(A \cup B)$$

$$= 1 - (P(A) + P(B))$$

$$= 1 - \left(\frac{1}{6} + \frac{1}{18}\right)$$

$$= \frac{7}{9}.$$

[10] *Notation*: By convention, a random variable is denoted by an uppercase letter, such as X.

1.5 Random variable

Consider a probability space $(\Omega, \mathfrak{F}, P)$. A *random variable*[10] is a *measurable* function, $X : \Omega \to \mathbb{R}$, which maps each outcome in the sample space to a real number, i.e.,

$$\{\omega \in \Omega;\ X(\omega) \le x\} \in \mathfrak{F}, \quad x \in \mathbb{R}.$$

A random variable may be *discrete* or *continuous* depending on whether it takes on discrete values or a continuum of values. Next, we will discuss some concrete examples.

Example 1.5   Coin tossing experiments

Consider a simple experiment of tossing a fair coin. $\Omega_1 = \{H, T\}$. $X(H) = 1$, $X(T) = 0$ is a re-labelling of every outcome in Ω_1 to a measurable space (often taken as $\mathbb{R}$).

In the case of an experiment where we toss the coin twice, the outcomes are extracted from $\Omega_2 = \{HH, HT, TH, TT\}$. We may wish to know the number of heads observed in a given realization. Therefore, $X(HH) = 2$, $X(HT) = 1$, $X(TH) = 1$, $X(TT) = 0$.

Example 1.6   Indicator random variable

Often, in calculations, it is convenient to define an *indicator random variable**
as follows:

$$\mathbb{1}_A(\omega) \equiv \mathbb{I}_{\{\omega \in A\}} = \begin{cases} 1, & \omega \in A, \\ 0, & \omega \notin A. \end{cases} \tag{1.1}$$

*It is also known as Bernoulli random variable.

In the second experiment described above, where we toss a fair coin twice, we may be interested in an outcome where we observe at least one head. We may define $A = \{HH, HT, TH\}$ and use the indicator random variable to represent the events where we observe at least one head. In section 1.8.5 of this chapter, we will use the indicator random variable to solve a problem encountered by a hiring manager of a company.

Example 1.7   Defining random variables for events

Consider a fair coin being tossed thrice. Consider the number of heads obtained after three tosses. Find the sample space, and therefore, define a random variable. Also find the probabilities associated with each value of the random variable.

Solution: Sample space, $\Omega = \{HHH, HHT, HTH, THH, HTT, THT, TTH, TTT\}$. Let X be the number of heads obtained after three tosses.

$$X(\omega) = \begin{cases} 3, & \text{if } \omega \in \{HHH\}, \\ 2, & \text{if } \omega \in \{HHT, HTH, THH\}, \\ 1, & \text{if } \omega \in \{HTT, THT, TTH\}, \\ 0, & \text{if } \omega \in \{TTT\}. \end{cases}$$

$$P(X=0) = \frac{1}{8}, \; P(X=1) = \frac{3}{8}, \; P(X=2) = \frac{3}{8}, \; P(X=3) = \frac{1}{8}.$$

Example 1.8   Defining an indicator random variable for a stochastic event

Consider an unbiased die being rolled once, where the outcome of interest is one which displays a prime number. Find the sample space, the collection of events we are interested in, and define an indicator random variable to represent the event when a prime number appears. Also find the probabilities associated for each value of the random variable.

Solution: Sample space, $\Omega = \{\{1\}, \{2\}, \{3\}, \{4\}, \{5\}, \{6\}\}$.

Let $A = \{\{2\}, \{3\}, \{5\}\}$.

Then, the indicator random variable is

$$\mathbb{1}_A(\omega) = \begin{cases} 1, & \text{if } \omega \in \{\{2\}, \{3\}, \{5\}\}, \\ 0, & \text{if } \omega \in \{\{1\}, \{4\}, \{6\}\}. \end{cases}$$

$$P(X=0) = \frac{3}{6} = \frac{1}{2}, \; P(X=1) = \frac{3}{6} = \frac{1}{2}.$$

The examples discussed above are discrete-type random variables. Some examples of continuous random variables are listed below.[11]

1. Time duration between successive arrivals of buses in a station (this random time interval follows an exponential distribution).

2. Distribution of wealth in a society follows a Pareto distribution, unraveling the fact that a high proportion of wealth is held by a small fraction of people in a society.

3. Scores obtained by students in an engineering class may follow a bell-shaped curve (see Figure 1.8).

1.6 Conditional probability

The occurrence of certain events may depend on the occurrence of other events. In fact, their likelihood of happening may be boosted (or diminished) by the outcomes of the preceding events. For example, the chances of rain are certainly higher on a cloudy day than on a day with clear skies. In this simple example, knowledge of the prevailing weather (cloudy/sunny) can greatly enhance our ability to predict

[11] We will discuss different types of discrete and continuous random variables (and their probability distributions) in more detail in the next chapter.

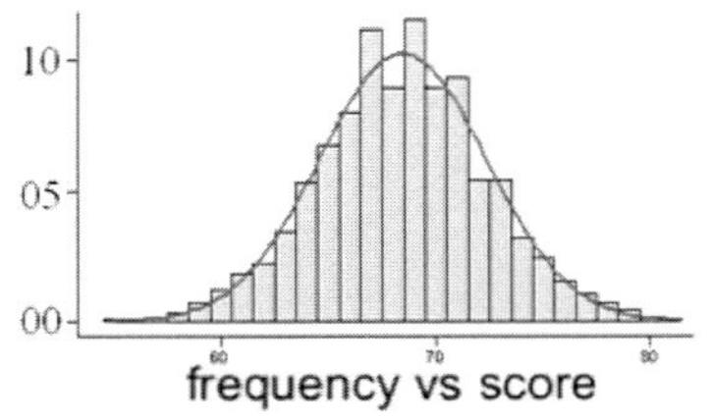

Figure 1.8 Scores obtained by students in a class may follow a Gaussian distribution. The scores can take on a *continuum* of values between the lowest score and the highest score.

Figure 1.9 Likelihood of occurrence of an event (rain) may depend on another event (sunny/cloudy).

the chance of rain. This underscores the importance of calculating *conditional probability*, where we may want to know the chance of occurrence of a certain event conditioned upon our knowledge of a preceding event.

Consider two events A and B. The conditional probability of event A given the occurrence of event B is given by the following relation:

$$P(A|B) = \frac{P(A \cap B)}{P(B)}. \tag{1.2}$$

If A and B are independent events, then $P(A \cap B) = P(A)P(B) \Rightarrow P(A|B) = P(A)$. Let us understand this concept by considering another simple example.

Example 1.9   Elementary idea of conditional probability

Let A be the event that we make the following observation on two successive tosses of a fair coin: "heads" followed by "tails." Let B be the event that in any two successive tosses of a fair coin, the outcome of the first toss is "heads." Let us evaluate the conditional probability $P(A|B)$ using two different approaches.

Method 1: Given the knowledge of the event B, the only possible way that the event A can happen is if the outcome of the second event turns out to be "tails."

$$\text{Therefore, } P(A|B) = \text{Prob(second toss turns up as "tails")} = \frac{1}{2}.$$

Method 2: An alternative approach would be to use the formula (1.2).

$$P(A|B) = \frac{P(A \cap B)}{P(B)} = \frac{1/2 \times 1/2}{1/2} = \frac{1}{2}.$$

Here, it is essential to explain the calculation of $P(A \cap B)$. First, we shall analyze the meaning of the event $A \cap B$. $A \cap B$ stands for the event that is *common* to both A and B, i.e., it is that special event when each of event A and event B is guaranteed to have happened. A little introspection may reveal that this event must be the appearance of "heads" in the first toss and "tails" in the second toss, which happens with a probability $1/2 \times 1/2 = 1/4$.[*]

---

[*]It may help to reflect if the event B may be a candidate for the event $A \cap B$. It turns out that the occurrence of event B does not guarantee the event A as there is a possibility that the second toss may turn out to be "heads." You may proceed by a careful elimination process to comprehend the special event $A \cap B$.

Example 1.10   Simple concurrent events

Suppose that a bag contains six pink balls and two grey balls and we draw two balls randomly from the bag without replacement. If at each draw, each ball in the bag is equally likely to be drawn, what is the probability that both balls that are drawn are pink?

Solution: Let R_{1p} be the event that the first ball drawn is pink and let R_{2p} be the event that the second ball drawn is pink.

$P\left(R_{1p}\right)=\dfrac{6}{8}$. Given that the first ball selected is pink, there remain five pink balls and two grey balls. Therefore, $P\left(R_{2p}\mid R_{1p}\right)=\dfrac{5}{7}$.

$$P\left(R_{1p}\cap R_{2p}\right)=P\left(R_{1p}\right)P\left(R_{2p}\mid R_{1p}\right)=\frac{6}{8}\times\frac{5}{7}=\frac{30}{56}=\frac{15}{28}.$$

Example 1.11   Estimating chance of a defect in a supply line

A box contains 2000 components of which 5% are defective, a second box contains 500 components of which 40% are defective, and two other boxes contain 1000 components each with 10% defective components. We select one of the boxes at random and remove a single component from it. What is the probability that the component is defective?

Solution: Let B_i be the event that denotes the selection of the ith box and let A be the event that the selected component is defective. Then the required probability is $P(A)$.

$$
\begin{aligned}
P(A) &= P\left(A\cap B_1\right)+P\left(A\cap B_2\right)+P\left(A\cap B_3\right)+P\left(A\cap B_4\right)\\
&= P\left(B_1\right)P\left(A\mid B_1\right)+P\left(B_2\right)P\left(A\mid B_2\right)+P\left(B_3\right)P\left(A\mid B_3\right)+P\left(B_4\right)P\left(A\mid B_4\right)\\
&= \frac{1}{4}\times\frac{5}{100}+\frac{1}{4}\times\frac{40}{100}+\frac{1}{4}\times\frac{10}{100}+\frac{1}{4}\times\frac{10}{100}\\
&= \frac{65}{400}\\
P(A) &= \frac{13}{80}.
\end{aligned}
$$

1.6.1   *Law of total probability*

The sample space Ω may be partitioned into k disjoint sets (events), namely E_i, where $i = 1, 2, ..., k$. The probability of a certain event $A\subset\Omega$ can then be computed by the weighted sum of the conditional probabilities, $P\left(A\mid E_i\right)$, where the weights are given by the probability of the partitioning events $P\left(E_i\right)$. This is the *law of total probability*, stated succinctly as follows.

$$P(A)=\sum_{i=1}^{k}P\left(A\mid E_i\right)P\left(E_i\right).\tag{1.3}$$

1.6.2   *Bayes' theorem*

Bayes' theorem helps us to compute the *posterior* probability $P(A|B)$ by using the concepts of conditional probability and the law of total probability as follows.

$$P\left(A\mid B\right)=\frac{P\left(B\mid A\right)P\left(A\right)}{P\left(B\right)}.\tag{1.4}$$

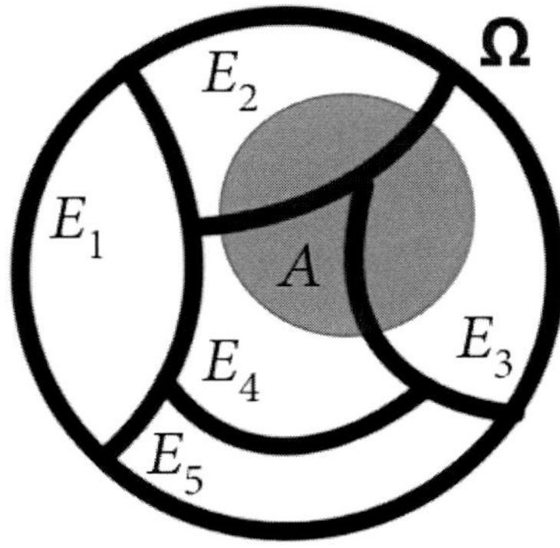

Figure 1.10 Here, the sample space Ω is partitioned into $E_1, E_2, ..., E_5$ in order to facilitate the computation of the probability of the event A in terms of the conditional probabilities.

$P(A)$ and $P(B)$ are known as *prior* probabilities. The prior probabilities may be computed using the law of total probability. The formula in equation 1.4 can be deduced by using the definition of conditional probability:

$$P(A|B) = \frac{P(A \cap B)}{P(B)} = \frac{P(B \cap A)}{P(B)} = \frac{P(B|A)P(A)}{P(B)}.$$

Bayesian statistics is a model for capturing epistemological uncertainty within the framework of probability. The prior probabilities constitute our original belief sets that are conditioned (over time) by the diversity of our experiences (data) and manifest as posterior probabilities. These posterior probabilities constitute our refined and conditioned beliefs that form the basis of inferential decisions.[12] Let us understand the essence of this framework through a simple example.

[12] To put this in context of our formalism, $P(A)$ constitutes our original belief sets, and the posterior probabilities $P(A|B)$ are the updated beliefs that are attained by the process of conditioning over experiences and data (represented here by the event(s) B). This update is made possible through the *likelihood* model $P(B|A)$.
Note: A detailed discussion on the likelihood function, used in estimation theory, is beyond the scope of this text.

Example 1.12   Bayes' theorem and law of total probability

A factory unit uses three automatic bolt threading machines (rollers), each accounting for 20%, 30% and 50% of the factory output of ready-to-use bolts for the aerospace industry. The precision rating (number of non-defective parts produced per 100) of each of the rollers is 95%, 97%, and 99%, respectively. If a part is picked up at random from the production line and found to be defective, what is the probability that it was produced by the second machine?

Solution: Let us begin by defining the relevant events: A_i is the event that a randomly picked bolt is manufactured by the ith machine, $i = 1, 2, 3$; B is the event that a randomly chosen part is defective. Based on the information provided, we glean that the prior probabilities are $P(A_1) = 0.2$, $P(A_2) = 0.3$, $P(A_3) = 0.5$. Further, $P(B|A_1) = 0.05$, $P(B|A_2) = 0.03$, $P(B|A_3) = 0.01$. We are asked to find $P(A_2|B)$. Using Bayes' theorem,

$$P(A_2|B) = \frac{P(B|A_2)P(A_2)}{P(B)} = \frac{(0.03)(0.3)}{\sum_{i=1}^{3} P(B|A_i)P(A_i)}$$

$$= \frac{0.009}{(0.05)(0.2) + (0.03)(0.3) + (0.01)(0.5)}$$

$$= \frac{9}{24} = 0.3750. \tag{1.5}$$

Figure 1.11 A quality control engineer who understands the nuances of Bayesian statistics and its implications.

Here we have considered the case of a small factory that has only three operational rolling machines. In a more realistic setting, we may expect that the factory quality control engineer may have to deal with a large pool of machines producing bolts *en masse*. Her prior belief set may hint to her that there is a 30% chance this defective bolt came from the second machine because the second machine has a production rate of 30% of the total output. However, the extra information gleaned from randomly picking a part and noticing it to be defective has led to an *update* in her belief system that manifests in terms of the posterior probability $P(A_2|B) = 0.375$. The update from 30% to 37.5% is a significant change of 25% that is likely to draw a greater attention of the engineer to the operational fitness of the second machine. Bayesian inference, thus, enables an enhancement of predictive knowledge of a phenomenon by synthesizing information and data from experiences.

Example 1.13   Diagnosis of disease

The chance a doctor D will diagnose a disease X correctly is 60%. The chance that a patient will die by his treatment after correct diagnosis is 40%, and the chance of death by wrong diagnosis is 70%. A patient of doctor D who had disease X died. What is the chance that his disease was diagnosed correctly?

Solution: Let B_1 be the event that disease X is diagnosed correctly by doctor D. Let B_2 be the event that disease X is not diagnosed correctly by doctor D.

Let A be the event that the patient who had disease X dies.

$$P(B_1 \mid A) = \frac{P(B_1)P(A \mid B_1)}{P(B_1)P(A \mid B_1) + P(B_2)P(A \mid B_2)}$$

$$= \frac{\dfrac{60}{100} \times \dfrac{40}{100}}{\dfrac{60}{100} \times \dfrac{40}{100} + \dfrac{40}{100} \times \dfrac{70}{100}}$$

$$= \frac{24}{24 + 28}$$

$$= \frac{6}{13}.$$

Example 1.14   Academic leadership and curriculum matters

In late 2022, there are three candidates for the position of director at a university. Mr. X, Mr. Y, and Mr. Z have chances of getting appointed to the post in the ratio 4:2:3. If Mr. X is selected, he could introduce a new syllabus in the university with a probability 0.3. The probabilities for Mr. Y and Mr. Z doing the same, if selected, are 0.5 and 0.8, respectively.

(i) What is the probability that there will be a new syllabus in 2023?

(ii) If there is a new syllabus in 2023, what is the probability that Mr. Z is the newly appointed director?

Solution: Let A be the event where a new syllabus is introduced in 2023. Let X, Y, and Z denote the events where Mr. X, Mr. Y, and Mr. Z are the directors, respectively.

1. $P(A) = P(X)P(A \mid X) + P(Y)P(A \mid Y) + P(Z)P(A \mid Z)$

$$= \frac{4}{9} \times 0.3 + \frac{2}{9} \times 0.5 + \frac{3}{9} \times 0.8$$

$$= \frac{12 + 10 + 24}{90}$$

$$= \frac{46}{90}$$

$$= \frac{23}{45}.$$

$$2.\ P(Z\,|\,A) = \frac{P(Z)P(A\,|\,Z)}{P(A)}$$

$$= \frac{\dfrac{3}{9} \times 0.8}{\dfrac{46}{90}}$$

$$= \frac{24}{46}$$

$$= \frac{12}{23}.$$

1.7   Chapter project: Random walk on a lonely island

1.7.1   *Interlude: Analytical calculations and computer simulations to predict the fate of Squeaky*

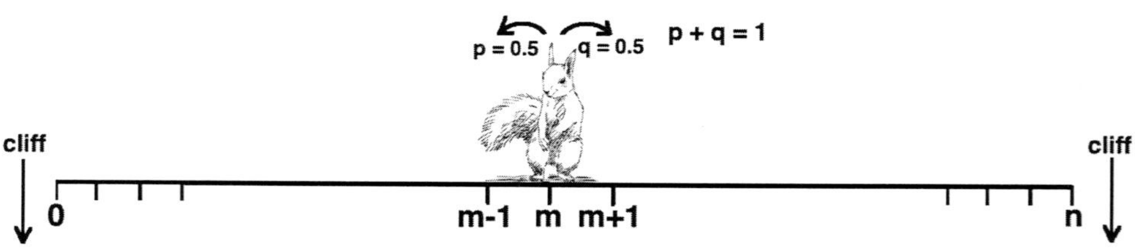

Figure 1.12 Schematic portrait of Squeaky's hopping adventure on the one-dimensional island hill. At any given instance, she jumps to the left with probability 1/2 and jumps to the right with probability 1/2.

Consider the schematic diagram of Squeaky's hopping adventure on the one-dimensional island hill as shown in Figure 1.12. In order to make the calculations tractable, we may consider dividing the island into discrete grid points that can host Squeaky. The grid points run from location 0 to location n. At a certain time, let us consider that Squeaky is at location m, and she makes a choice to jump to her left to location $(m + 1)$ with probability $q = 0.5$ and to jump to her right with probability $p = 0.5$. The probabilities p and q are assigned the value 0.5 because we have assumed that she does not have any inherent bias or preference in choosing between left- and right-sided moves. In this example, we will take her decision instances to jump either to the left or the right as the time stamps, i.e., in any given time point so defined, she does not decide to stay where she is.

Before we attempt the theoretical calculations to predict her fate, let us consider the case phenomenologically with the help of a spanning diagram as shown in Figure 1.13. Since it is quite obvious from this diagram that the decision paths span the entire breadth of the one-dimensional island, our hunch is that Squeaky will make it all the way to the edge and trip. Let us see if the calculations below validate our intuition.

Let us define the following events. W is the event that Squeaky falls in the pit to our left. We would like to compute the following probability.

$$P_m = P_m\ (\text{left pit}) = \text{Probability of the event } W \text{ when Squeaky starts at } X_0 = m,\ (1.6)$$

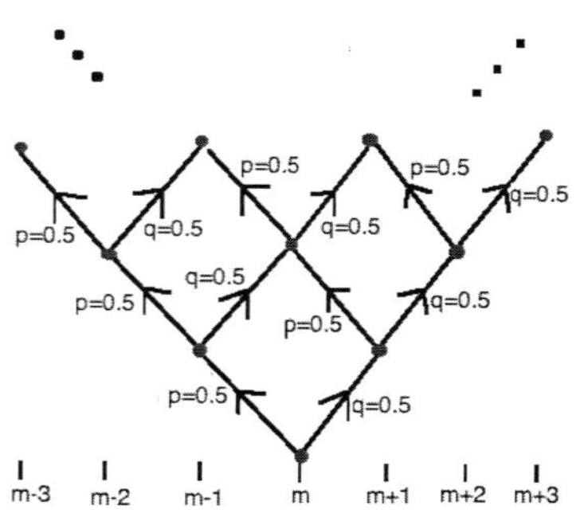

Figure 1.13 Spanning tree showing the possible decision paths that Squeaky could opt for starting from the location m.

with $P_0 = 1$, $P_n = 0$ for obvious reasons. Further, let E be the event that the *first hop* is to the *left*. We will use the law of total probability and *condition* our computation upon this event as follows:

$$P_m = P\left(W \text{ and } E \mid X_0 = m\right) + P\left(W \text{ and } E^c \mid X_0 = m\right)$$

E and E^c partition the choice space

$$= P\left(W \mid E \text{ and } X_0 = m\right) P\left(E \mid X_0 = m\right)$$

Law of total probability

$$+ P\left(W \mid E^c \text{ and } X_0 = m\right) P\left(E^c \mid X_0 = m\right)$$

$$= P\left(W \mid X_1 = m-1\right) \times \frac{1}{2} + P\left(W \mid X_1 = m+1\right) \times \frac{1}{2}$$

$$= \frac{1}{2} P\left(W \mid X_0 = m-1\right) + \frac{1}{2} P\left(W \mid X_0 = m+1\right)$$

Independent hops

$$P_m = \frac{1}{2} P_{m-1} + \frac{1}{2} P_{m+1}. \tag{1.7}$$

Equation 1.7 is a *recurrence* relation whose solution can be readily computed.

Questions

1. Solve the recurrence relation 1.7 for $P_m = P_m$ (left pit).

2. Use an argument based on symmetry to deduce the solution for $P'_m = P_m$ (right pit).

3. Compute $P_m + P'_m$ and thereafter comment on the fate of Squeaky based on your theoretical calculations.

4. Build a computer simulation of Squeaky's exploration on the island hill and comment whether the results of the simulation corroborate with your theoretical calculations about Squeaky's fate. In order to develop the simulation, you may refer to the pseudocode provided below and turn it into a computer-executable code using a programming language of your choice.

Software Implementation

Pseudocode of the random walk algorithm

```
INPUT: grid_length, start_pos.

initialise curr_pos = start_pos;
initialise num_hops = 1;
while (curr_pos > 0 && curr_pos < grid_length)
    toss = rand(1);
    if (toss < 0.5)
```

```
        curr_pos = curr_pos - 1;
    elseif (toss >= 0.5)
        curr_pos = curr_pos + 1;
    end
    plot curr_pos and record graphic frame;
    num_hops = num_hops + 1;
end


OUTPUT: num_hops, play recorded animation.
```

In case you prefer to use Matlab, some useful commands to build your code may be `rand, stem, getframe, movie`.

We will return to the random walk expedition undertaken by Squeaky later in this chapter. Let us now get introduced to some new concepts on computing statistical averages.

1.8 Expected values of random variables

The most common entity of interest while conducting an experiment is perhaps its *average* output. Since we are largely interested in *statistical* experiments, we may expect the average output in terms of a *statistical average* due to the *random* or *stochastic* nature of the underlying process/model. What this means is the following: *if we were to repeat the same experiment many times over, each time recording the output of the model (or process), and subsequently, take the average of all the recorded outputs, then this ensemble average may be regarded as the mean behavior of the experimental model (or process).*

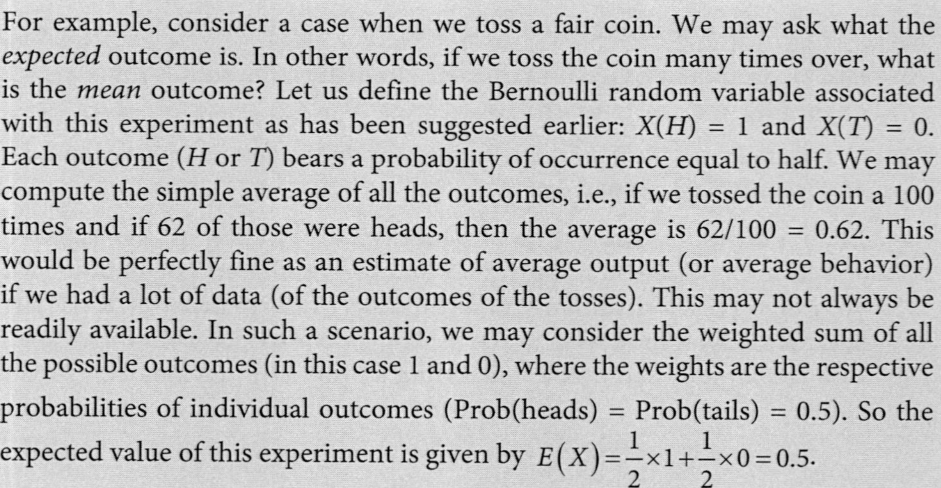

For example, consider a case when we toss a fair coin. We may ask what the *expected* outcome is. In other words, if we toss the coin many times over, what is the *mean* outcome? Let us define the Bernoulli random variable associated with this experiment as has been suggested earlier: $X(H) = 1$ and $X(T) = 0$. Each outcome (H or T) bears a probability of occurrence equal to half. We may compute the simple average of all the outcomes, i.e., if we tossed the coin a 100 times and if 62 of those were heads, then the average is $62/100 = 0.62$. This would be perfectly fine as an estimate of average output (or average behavior) if we had a lot of data (of the outcomes of the tosses). This may not always be readily available. In such a scenario, we may consider the weighted sum of all the possible outcomes (in this case 1 and 0), where the weights are the respective probabilities of individual outcomes (Prob(heads) = Prob(tails) = 0.5). So the expected value of this experiment is given by $E(X) = \frac{1}{2} \times 1 + \frac{1}{2} \times 0 = 0.5$.

In fact, had we conducted our coin tossing experiment many more times than 100 (as was done above), then the average would be observed to converge to the value 0.5. As a matter of fact, we have just stated a very important result of probability theory known as the *law of large numbers.*[13] We will revisit this law in greater detail in the next chapter after we introduce the notion of probability distributions.

It is also essential to state that the expected value of a statistical experiment may take on a value that is not equal to the elements of the range of the random variable.

Figure 1.14 Let's flip some coins or do some math! Oh well, the law of large numbers will prevail!

[13] There are two variants of this law, namely, the *weak law of large numbers* and the *strong law of large numbers*, depending on the nature of convergence of the *sample mean* to the *expected value.*

In the above example, X takes on values 0 and 1 but $E(X) = 0.5$, which does not belong to the range of X. This observation must be noted in conjunction with our comment earlier in section 1.3.3 that statistics deals with idealized averages that are far removed from individual experiences (outcomes).

The expected value of a random variable may thus be generalized as follows.

$$\mu_X = E(X) = \sum_{x \,\in\, \text{range}(X)} x P(X = x). \tag{1.8}$$

When it is understood, we will simply write μ and omit the subscript used to denote the relevant random variable. The formulation in equation 1.8 is simply a generalization of the explanation in the preceding paragraph where range(X) = {1, 0}. The expected value is the *first statistical moment* and the variation in the outcomes is given by the variance, which is the *second statistical moment* and is defined as follows.

$$\sigma^2 = Var(X) = E\left((X - \mu)^2\right) = \sum_{x \,\in\, \text{range}(X)} (x - \mu)^2 \, P(X = x). \tag{1.9}$$

The variance is related to the expectation in yet another useful manner.

$$\text{Var}(X) = E(X^2) - \mu_X^2. \tag{1.10}$$

We will revisit the calculations of expectation and variance again with more rigor in the next chapter on probability distributions. Here we will simply state some useful results and study a few examples.

1. $E(cX) = cE(X)$, where c is a constant.
2. $E(X + c) = E(X) + c$, where c is again a constant.
3. $E(X + Y) = E(X) + E(Y)$.
4. $Var(cX) = c^2 Var(X)$, where c is a constant.
5. $Var(X + c) = Var(X) + 0 = Var(X)$, where c is a constant.
6. $Var(aX \pm bY) = a^2 Var(X) + b^2 Var(Y) \pm 2ab Cov(X, Y)$, where a, b are constants. $Cov(X, Y) = E(X - \mu_X)(Y - \mu_Y)$ is the *covariance* of X and Y.

> ### Example 1.15 *Expectation of an indicator random variable*
>
> A very useful result we will employ in a subsequent example is the expected value of an indicator random variable defined in equation 1.1. By definition 1.8,
>
> $$E(\mathbb{1}_A) = 1 \times P(A) + 0 \times P(A^c) = P(A).$$

> ### Example 1.16 *Success rate in an infinite series of independent trials*
>
> What is the expectation and variance of the number of failures preceding the first success in an infinite series of independent trials with probability p of success in each trial?

Solution: $\Omega = \{S, FS, FFS, FFFS, FFFFS, \ldots\}$, where S = Success and F = Failure. Let X be the number of failures.

$$E(X) = \sum_{x=0}^{\infty} x P\{X = x\}$$

$$E(X) = (1-p)p + 2(1-p)^2 p + 3(1-p)^3 p + \cdots$$
$$(1-p)E(X) = (1-p)^2 p + 2(1-p)^3 p + \cdots$$

We subtract the last two equations and get:

$$(1-1+p)E(X) = (1-p)p + (1-p)^2 p + (1-p)^3 p +$$
$$pE(X) = \frac{p(1-p)}{1-(1-p)}$$
$$pE(X) = 1-p$$
$$E(X) = \frac{1-p}{p}.$$

Now we calculate the variance:

$$Var(X) = E(X^2) - (E(X))^2$$

$$E(X^2) = \sum_{x=0}^{\infty} x^2 P\{X = x\}$$

$$E(X^2) = (1-p)p + 2^2(1-p)^2 p + 3^2(1-p)^3 p + 4^2(1-p)^4 p + \cdots$$
$$E(X^2) = (1-p)p + 4(1-p)^2 p + 9(1-p)^3 p + 16(1-p)^4 p + \cdots$$
$$(1-p)E(X^2) = (1-p)^2 p + 4(1-p)^3 p + 9(1-p)^4 p + \cdots$$

We subtract the last two equations and get:

$$(1-1+p)E(X^2) = (1-p)p + 3(1-p)^2 p + 5(1-p)^3 p + 7(1-p)^3 p + \cdots$$
$$pE(X^2) = (1-p)p + 3(1-p)^2 p + 5(1-p)^3 p + 7(1-p)^3 p + \cdots$$
$$(1-p)pE(X^2) = (1-p)^2 p + 3(1-p)^3 p + 5(1-p)^4 p + \cdots$$

Again, we subtract the last two equations and get:

$$\left(p-p+p^2\right)E\left(X^2\right)=(1-p)p+2\left[(1-p)^2p+(1-p)^3p+(1-p)^3p+\cdots\right]$$

$$p^2E\left(X^2\right)=(1-p)p+2\left[\frac{(1-p)^2p}{1-(1-p)}\right]$$

$$p^2E\left(X^2\right)=(1-p)p+2(1-p)^2$$

$$p^2E\left(X^2\right)=(1-p)(p+2-2p)$$

$$p^2E\left(X^2\right)=(1-p)(2-p)$$

$$E\left(X^2\right)=\frac{(1-p)(2-p)}{p^2}.$$

Finally, we obtain the variance:

$$Var(X)=\frac{(1-p)(2-p)}{p^2}-\left[\frac{1-p}{p}\right]^2$$

$$Var(X)=\frac{(1-p)(2-p)}{p^2}-\frac{(1-p)^2}{p^2}$$

$$Var(X)=\frac{(1-p)\left[(2-p)-(1-p)\right]}{p^2}$$

$$Var(X)=\frac{1-p}{p^2}.$$

Example 1.17 *Expectation and variance of a mean subtracted normalized random variable*

Suppose X is a random variable that takes the values 1, 2, 3, and 4 with probabilities $\frac{1}{2}, \frac{1}{4}, \frac{1}{8}$, and $\frac{1}{8}$, respectively. Let a new random variable Y be defined as $Y=\frac{X-\mu_X}{\sigma_X}$, where μ_X is the mean and σ_X^2 is the variance of X. Use the properties of expectation to find the expectation and variance of Y. Is this true for all random variables X?

Solution: Take $\mu_X=\mu$ and $\sigma_X=\sigma$.

$$E(Y)=E\left(\frac{X-\mu}{\sigma}\right)$$

$$=E\left(\frac{X}{\sigma}-\frac{\mu}{\sigma}\right)$$

$$=\frac{1}{\sigma}E(X)-\frac{\mu}{\sigma}$$

$$=\frac{\mu}{\sigma}-\frac{\mu}{\sigma}$$

$$=0.$$

Since $Var(c) = 0$ and $Cov(X, c) = 0$ for all constants c,

$$Var(Y) = Var\left(\frac{X - \mu}{\sigma}\right)$$

$$= Var\left(\frac{X}{\sigma} - \frac{\mu}{\sigma}\right)$$

$$= \frac{1}{\sigma^2} Var(X) + 0$$

$$= \frac{\sigma^2}{\sigma^2}$$

$$= 1.$$

Hence, this result is true for all X.

Example 1.18   Variance of sum of two random variables

Prove that $Var(aX \pm bY) = a^2 Var(X) + b^2 Var(Y) \pm 2ab Cov(X, Y)$, where $Cov(X, Y) = E(X - \mu_X)(Y - \mu_Y)$.

Solution: Note that $E(aX \pm bY) = aE(X) \pm bE(Y) = a\mu_X \pm b\mu_Y$.

$$Var(aX \pm bY) = E\left(aX \pm bY - (a\mu_X \pm b\mu_Y)\right)^2$$

$$= E\left(a(X - \mu_X) \pm b(Y - \mu_Y)\right)^2$$

$$= E\left(a^2(X - \mu_X)^2 + b^2(Y - \mu_Y)^2 \pm 2ab(X - \mu_X)(Y - \mu_Y)\right)$$

$$= a^2 E(X - \mu_X)^2 + b^2 E(Y - \mu_Y)^2 \pm 2ab E(X - \mu_X)(Y - \mu_Y)$$

$$= a^2 Var(X) + b^2 Var(Y) \pm 2ab Cov(X, Y).$$

Example 1.19   Expected number of new recruits per n hiring interviews

Let us consider that a hiring manager has the responsibility of conducting interviews of n candidates for the post of a peon over a certain period of time. The candidates appear for the interviews in a random fashion, i.e., from the perspective of the hiring manager; prior to the interview, there is an equal probability among candidates to be the most suitable candidate. The hires are made on a rolling basis in the sense that whenever he encounters a better candidate than the existing one, he hires that person and keeps him in the job until a better candidate is found. How many hires are made in this process? Can we give an estimate of the cost associated with this firing–recruiting process?

Figure 1.15 Homer's probability of getting hired is $\dfrac{1}{513}$ because he is the 513th candidate. How he wishes he had applied earlier! Had he been the first candidate to be interviewed, he would have most certainly been recruited because $E(\mathbb{1}_1) = p_1 = 1$. Moral of the story: *Act fast, do not procrastinate!*

Consider an indicator random variable $\mathbb{1}_i = \begin{cases} 1, & \text{when the } i^{th} \text{ candidate is hired,} \\ 0, & \text{when the } i^{th} \text{ candidate is not hired.} \end{cases}$

For the ith candidate to be hired, the preceding $(i-1)$ candidates must not have been better than this candidate. But since each of these i candidates had an equal chance to be hired, the probability that the ith candidate is hired is $p_i = \dfrac{1}{i}$. Thus, the total number of hires is given by

$$X = \sum_{i=1}^{n} \mathbb{1}_i. \tag{1.11}$$

We compute the expected value of X, and by linearity of expectation, we have

$$E(X) = E\left(\sum_{i=1}^{n} \mathbb{1}_i\right) = \sum_{i=1}^{n} E(\mathbb{1}_i) \ = \ \sum_{i=1}^{n} p_i$$

Because $E(\mathbb{1}_A) = P(A)$ as per section 1.8.

$$= \sum_{i=1}^{n} \frac{1}{i}$$
$$\approx \log n + \mathcal{O}(1), \text{ for very large } n \tag{1.12}$$

Euler–Mascheroni result

This means that for every n interviews conducted by the hiring manager, approximately $\log n$ of them get hired on an average. The cost of the recruitment process is $\mathcal{O}\left(c_H \log n\right)$, where c_H is the hiring cost factor.

Example 1.20   *Analysis of sorting algorithm*

One of the essential features while analyzing the cost of sorting an array is the number of existing inversions* in the array. We will denote an inversion by $\mathfrak{J}$. Let us define an indicator random variable $\mathbb{1}_{A[i]>A[j]}$ (when $1 \leq i < j \leq n$) to analyze the average number of inversions in an array of length n. Let X denote the total number of inversions in the array, $X = \sum_{i=1}^{n-1} \sum_{j=i+1}^{n} \mathbb{1}_{A[i]>A[j]}$. Further, $P(\mathfrak{J}) = \dfrac{1}{2}$ because given any two distinct random numbers, the probability that one is bigger than the other is half. This entails $E\left(\mathbb{1}_{A[i]>A[j]}\right) = P(\mathfrak{J}) = 1/2$ for all $i < j$. Therefore,

$$E(X) = \sum_{i=1}^{n-1}\sum_{j=i+1}^{n}\frac{1}{2} = \frac{1}{2}\sum_{i=1}^{n-1}(n-i) = \frac{1}{2}\sum_{i=1}^{n-1}n - \frac{1}{2}\sum_{i=1}^{n-1}i \ = \ \frac{1}{2}\left(n(n-1) - \frac{n(n-1)}{2}\right)$$

$$= \frac{n(n-1)}{4}. \tag{1.13}$$

The expected number of inversions in an array of size n is $\mathcal{O}\left(n^2\right)$.

---

*An inversion in an array between a pair of entries is a condition when i < j but A[i] > A[j].

1.8.1   *Law of total expectation*

Just like we could compute the probability of an event by conditioning over the partitioning events E_i and calculating the weighted sum, we can perform a very similar calculation for computing the expected value. This is known as the *law of total expectation*.

$$E(\mathbb{1}_A) = \sum_{i=1}^{k} E\left(\mathbb{1}_A \mid \mathbb{1}_{E_i}\right) P(E_i). \tag{1.14}$$

We will use the law of total expectation to estimate the life expectancy of Squeaky in terms of the average number of hops till the end.

A generalization of the law of total expectation is known as the *law of iterated expectations* (LIE).[14]

$$E(X) = E\left(E(X \mid Y)\right) = \sum_{y} E(X \mid Y = y) P(Y = y). \tag{1.15}$$

Here X and Y belong to the same probability space. $E\left(E(X \mid Y)\right)$ must be understood as $E_Y\left(E_X(X \mid Y)\right)$ to make the order of the expectation operator with respect to the random variables, X and Y, explicitly clear.

Example 1.21   *Average number of mangoes eaten per week in a population of engineers*

Let us consider a dietary survey of 1000 engineers working in a factory. The population of the engineers has males and females. The number of young males is 300 and the number of old males is 500. The corresponding figures for female engineers are 50 and 150. The survey reveals that among the males, the younger folks eat 4 mangoes per week while the older folks eat 6 mangoes per week.* The corresponding figures for the women engineers are 0 and 4 per week. The question is to find the average number of mangoes eaten per week by any person from the whole population. Let M be the number of mangoes eaten by a factory engineer per week. We proceed by first calculating the chance of encountering a male engineer, $P_m = \dfrac{800}{1000}$. Likewise, the corresponding estimate for a female engineer is $P_f = \dfrac{200}{1000}$. Here the subscripts m and f refer to males and females, respectively; and the subscripts y and o refer to young and old, respectively.** Further, $P_{y|m} = \dfrac{300}{300+500} = 3/8$. $P_{o|m} = 1 - P_{y|m} = 5/8$. Therefore, computing the expected value of mangoes eaten per week by the male and female engineers as a weighted sum of the mangoes eaten by the respective age-groups,*** we have the following estimates:

$$E(M \mid m) = M_{y|m} \times P_{y|m} + M_{o|m} \times P_{o|m} = 4 \times \frac{3}{8} + 6 \times \frac{5}{8} = \frac{21}{4} \tag{1.16}$$

for the male engineers. Similarly,

$$E(M \mid f) = M_{y|f} \times P_{y|f} + M_{o|f} \times P_{o|f} = 0 \times \frac{1}{4} + 4 \times \frac{3}{4} = 3 \tag{1.17}$$

for the female engineers.

Now, using the *law of iterated expectation*, we have

$$E(M) = E\left(E(M \mid \text{gender})\right) = E(M \mid m) P_m + E(M \mid f) P_f = \frac{21}{4}\frac{800}{1000} + 3\frac{200}{1000} = 4.8. \tag{1.18}$$

---

[14] The law of iterated expectation implies the law of total probability as follows. Consider an event A and a random variable Y. Then,

$$
\begin{aligned}
P(A) &= E(\mathbb{1}_A) \\
&= E(E(\mathbb{1}_A \mid Y)) \\
&\quad\text{Law of iterated expectation} \\
&= \sum_{y} E(\mathbb{1}_A \mid Y = y) P(Y = y) \\
P(A) &= \sum_{y} P(A \mid Y = y) P(Y = y) \\
&\quad\text{Law of total probability}
\end{aligned}
$$

Figure 1.16 Should the mango vendor bring more mangoes to the factory during lunch breaks?

So, on an average, the number of mangoes eaten by an engineer from the whole population in the factory is 4.8 per week.

---

*For simplicity, let us assume that within the same age group, the answers in the survey are consistent and identical.

**This is one of the very rare instances in the book where we have used both uppercase (M) and lowercase letters (y, o, m, f) to denote random variables.

***See explanation given in the introductory paragraphs of section 1.8.

1.8.2   *Law of total variance*

The *law of total variance*[15] is another useful result that we will simply state here as follows.

[15] There is a similar law of total covariance, which we will not discuss in this text.

$$Var(Y) = E(Var(Y|X)) + Var(E(Y|X)). \tag{1.19}$$

1.9   Chapter project: Random walk on a lonely island

1.9.1   *Epilogue: Life expectancy of Squeaky*

Given that we have predicted Squeaky's fate, our next question of interest is: *what is her life expectancy in terms of the number of hops from the beginning till the end?* Let D be the number of hops till the end. We will use the law of total expectation and once again *condition* upon the event E as follows:

$$
\begin{aligned}
E_m &= E\left(D\,|\,X_0 = m\right)\\
&= E\left(D\,|\,E \text{ and } X_0 = m\right)P\left(E\,|\,X_0 = m\right) + E\left(D\,|\,E^c \text{ and } X_0 = m\right)P\left(E^c\,|\,X_0 = m\right)\\
&= \frac{1}{2}E\left(D\,|\,X_1 = m-1\right) + \frac{1}{2}E\left(D\,|\,X_1 = m+1\right)\\
&\overset{\text{reset chain}}{=} \frac{1}{2}\left(1 + E\left(D\,|\,X_0 = m-1\right)\right) + \frac{1}{2}\left(1 + E\left(D\,|\,X_0 = m-1\right)\right)\\
E_m &= 1 + \frac{1}{2}E_{m-1} + \frac{1}{2}E_{m+1}.
\end{aligned}
\tag{1.20}
$$

Equation 1.20 is a *non-homogeneous* recurrence relation.

Questions

1. Explain the appearance of the numeral 1 in the recurrence relation 1.20.

2. Solve the non-homogeneous recurrence relation 1.20 to estimate the life expectancy of Squeaky E_m.

3. Find the starting location of Squeaky's hopping expedition to maximize her life span.

4. Compare the estimate of E_m from the computer simulation you developed earlier in the chapter with the theoretical estimate of E_m above. Comment on the origin of any discrepancy you observe in the comparison.

1.10 Selected bibliography

Feller, William. *An Introduction to Probability Theory and Its Applications*, Vol. 1 (3rd edition). John Wiley & Sons, Inc., 1968.

Gupta, S. C. and V. K. Gupta. *Fundamentals of Mathematical Statistics* (11th edition). Sultan Chand and Sons, 2017.

Kolmogorov, Andrey N. *Foundations of the Theory of Probability* (2nd English translation). Chelsea Publishing Co., 1956.

Montgomery, Douglas C. and George C. Runger. *Applied Statistics and Probability for Engineers* (6th edition). Wiley, 2014.

Renyi, Alfred. *Probability Theory* (Illustrated edition). Dover Publications Inc., 2007.

Rosenthal, Jeffrey S. *A First Look at Rigorous Probability Theory* (2nd edition). World Scientific, 2006.

Ross, Sheldon M. *Introduction to Probability and Statistics for Engineers and Scientists* (6th edition). Academic Press (Elsevier), 2021.

Stark, Henry and John W. Woods. *Probability and Random Processes with Applications to Signal Processing* (4th edition). Pearson Education, 2011.

1.11 Exercise problems

1. (***Combinatorics***) Out of a population of 10 digits running from 0 through 9, what is the probability that 5 consecutive random digits are all different?

2. (***Occupancy problems: Bose–Einstein, Fermi–Dirac, and Maxwell–Boltzmann statistics***) This model relates to placing randomly r *indistinguishable* balls (particles) into n cells (quantum states). Consider the *occupancy numbers* r_1, r_2, ..., r_n [16] satisfying $\sum_{i=1}^{n} r_i = r$. Two distributions of the balls are distinguishable only if the n tuples $(r_1, r_2, ..., r_n)$ are not identical.

   2.I Bose–Einstein statistics: This is a model for photons, nuclei, and atoms containing an even number of particles.

   (a) Find an expression for the number of distinguishable distributions, $A_{r,n}$.

   (b) Find an expression for the number of distinguishable distributions in which no cells are empty.

   (c) What is the probability of each distribution in (a)?

   (d) Given $n = 5$ quantum states (cells) and $r = 3$ indistinguishable particles, what is the probability of the distribution $\left(*|_-|*|*|_-\right)$ in Bose–Einstein statistics? Here $*$ represents a particle, an empty orbital is denoted by $_-$, and the barrier between two successive orbitals is denoted by the symbol $|$.

   2.II Fermi–Dirac statistics: This is a model for electrons, neutrons, and protons. This model assumes that (i) it is impossible for two or more

[16] Occupancy number r_k stands for the number of balls in the kth cell.

identical particles to be in the same quantum state,[17] i.e., $r \leq n$, and (ii) all distinguishable distributions have equal probabilities.

(a) How many such distributions are possible?

(b) What is the probability of the distribution $\left(*|\_|*|*|\_\right)$ in Fermi–Dirac statistics?

2.III <u>Maxwell–Boltzmann statistics</u>: This is a model for material particles distributed over various energy states in thermal equilibrium in classical mechanics. This model does not apply to quantum particles. In this model, the number of ways we can place r distinguishable particles in n cells is certainly $\underbrace{n \times n \times \cdots \times n}_{r \text{ times}} = n^r$ when sampling *with replacement* is permissible.[18]

(a) Find the number of ways in which a population of r particles can be partitioned into n cells such that $r_1 + r_2 + \cdots r_n = r$.

(b) Given occupancy numbers $r_1, \ldots, r_n$, what is the probability of this distribution in Maxwell–Boltzmann statistics?

(c) What is the probability of the distribution $\left(*|\_|*|*|\_\right)$ in Maxwell–Boltzmann statistics?

3. (***Quadratic equation with stochastic coefficients***) Each coefficient of a quadratic equation $ax^2 + bx + c = 0$ is determined by the throw of a regular die. Find the probability that the equation will have at least one real root.

4. (***Medical diagnosis of prostate cancer***) Prostate cancer is the most common type of cancer found in males. As an indicator of whether a male has prostate cancer, doctors often perform a test that measures the level of the PSA protein (prostate specific antigen) that is produced only by the prostate gland. The test is highly unreliable even though there is a strong correlation between high PSA value and incidence of cancer. Indeed, the probability that a non-cancerous man will have an elevated PSA level is approximately 0.115, with this probability increasing to approximately 0.273 if the man does have cancer. If, based on other factors, a physician is 81% certain that a male has prostate cancer, what is the conditional probability that he has the cancer given that

(a) the test indicates a high PSA level;

(b) the test does not indicate a high PSA value?

Re-estimate your probabilities if the physician initially believes there is a 29% chance the man has prostate cancer.

5. (***Spacecraft control system***)[19] The flight control computer on a spacecraft employs four independent flight computers that work in parallel – a much required redundancy built into the system. During every critical flight path decision, the computers "vote" to decide on the most important step. For example, in case of a "roll" decision, the probability that a computer will make an error is 0.0002. Let X denote the number of computers that vote for an erroneous roll movement. Compute $E(X)$ and $Var(X)$.

6. (***A game of dart***) In a game of dart, a participant is given three attempts to hit a target. On each try, she either scores a hit, H, or a miss, M. The game requires that the player must alternate which hand she uses in successive

---

[17] Pauli's exclusion principle.

[18] To illustrate this further, consider that the n cells are in a *bag*; we randomly select a cell from this bag and place one of the r particles in it. Consequently, we put this cell back in the bag and continue sampling, accounting for the fact that the same cell (as was chosen from the bag before) may be sampled again to be filled by yet another particle. This is akin to sampling with replacement. This way of placing r particles in n cells (with replacement) is similar to throwing an n-sided dice r times.

[19] Gene D. Carlow, Architecture of the space shuttle primary avionics software system, *Communications of the ACM* **27**, no. 9 (1984).

attempts. That is, if she makes her first attempt with her right hand, she must use her left hand for the second attempt and her right hand for the third. Her chance of scoring a hit with her right hand is 0.7 and with her left hand is 0.4. Assume that the results of successive attempts are independent and that she wins the game if she scores at least two hits in a row. If she makes her first attempt with her right hand, what is the probability that she wins the game?

7. (***Random sums***) Let $X_1, X_2, \ldots$ be i.i.d.[20] random variables and let $E(X_1) = \mu_X$. Let N be a non-negative integer-valued random variable that is independent of the sequence of Xs and let $E(N) = \mu_N$. Define the random sum S to be $S = X_1 + X_2 + \ldots + X_N$, where $S = 0$ if $N = 0$.[21] What is $E(S)$? *Hint: Use the law of iterated expectation, i.e., condition your computation on $N = n$.*

[20] Independent and identically distributed random variables.

[21] Notice that S is the sum of a random number of terms.

8. (***Population growth model with random progeny and no deaths***) Consider a population of tumor cells where each cell has a random number of progeny. Consider that the number of progeny of the proliferating cells is i.i.d. with mean μ. Suppose the process starts with one cell in generation zero. For simplicity, let us assume there are no deaths. What is the expected total number of tumor cells in n generations? For $n \to \infty$, find a condition for arresting the rate of growth of tumor cells.

Hint: In the formulation of the question above, consider $T_k \equiv S =$ number of cells in generation k. $T_k = X_1 + X_2 + \ldots + X_{T_{k-1}}$. Here X_1, X_2, etc., ... are the numbers of progeny of the first, second, etc., ... cells in generation k − 1. This gives $E(T_k) = \mu E(T_{k-1})$. The final answer is $\sum_{k=0}^{n} E(T_k)$.

9. (***Matching probability***) Consider n letters that are designated for n envelopes. However, the letters and envelopes are not in order (they are randomly arranged) and hence the letters may not go in the correct envelop. Let A_k denote an event when a match occurs in the kth place. What is the probability of the event A_k? What happens as $n \to \infty$?

10. (***Ordering of events and their probability***) Consider a probability measure P defined on an appropriate probability space $(\Omega, \mathfrak{F}, P)$. Let E_1, E_2 be events from the event space $\mathfrak{F}$ such that $E_1 \subset E_2$. Deduce a relationship between $P(E_1)$ and $P(E_2)$.

2

Probability Distributions

DISTRIBUTIONS ARE GENERALIZATIONS of mathematical functions from a purely technical standpoint. But perhaps it is most pertinent to begin by asking a more utilitarian question. Why should we study distributions? Specifically, why should we study *probability* distributions? One of the motivations stems from a practical limitation of experimental measurements that is underlined by the uncertainty principle postulated by Werner Heisenberg (see Figure 2.1). The very fabric of reality and the structure of scientific laws that govern our ability to understand physical phenomena demand a probabilistic (statistical) approach. Our inability to make infinite-precision measurements of data necessitates the consideration of averages over many measurements, and under similar conditions, as a more reliable strategy to affix experimental values to unknowns with reasonable accuracy. Typically, the operation of *averaging* is equivalent to performing an integral of the form $\bar{u} := \int_{\mathcal{D}} u(x)\,dx$ over the domain $\mathcal{D}$. The function $u(x)$ is a mathematical representation of a quantity of interest whose average we are interested in calculating over $\mathcal{D}$. This simple expression for the average is computed by integrating *uniformly* over $\mathcal{D}$. More generally, the average may be computed by using an appropriate weight function $f(x)$ and integrating as $\int_{\mathcal{D}} u(x) f(x)\,dx$. Here $f(x)$ serves as the distribution function over $\mathcal{D}$ that appropriately weighs $u(x)$ during the averaging process.

As an example, consider a large collection of particles moving in one-dimensional space (the particles are free to move either to the left or to the right with any speed). The velocity of the ith particle is given by $u^{(i)}$. Let us suppose that we want to find the average velocity of the flow. One option is to find an ensemble average of the velocities of many such particles. This is equivalent to computing $\bar{u} := \int_{\mathcal{D}} u(x)\,dx$, if the velocity profile of the overall flow, $u(x)$, is known to us over the domain $\mathcal{D}$.

Alternatively, if the velocity distribution function of the particles $f(u)$ is known, then we can compute $\bar{u} \equiv \langle U \rangle = \int_{\mathcal{U}} u f(u)\,du$, where $\mathcal{U}$ is the appropriate domain in the velocity space and $f(u)$ prescribes the probability distribution of U.[1] Two interesting cases are noteworthy here: (i) if the aforementioned particles follow a Brownian motion (see Figure 2.2), then $f(u)$ is the Maxwell–Boltzmann (symmetric and bell-shaped) distribution and $\bar{u} = 0$ as expected and (ii) if the flow is unidirectional and every particle moves with constant speed u_0, then clearly $\bar{u} = u_0$. The latter result may be expressed as a weighted integral by using the Dirac delta distribution as

Figure 2.1 A cartoon capturing the essence of the Heisenberg's uncertainty principle: $\Delta x \Delta p \geq \dfrac{\hbar}{2}$, where Δx is the uncertainty in position, Δp is the uncertainty in momentum (or velocity), and $\hbar$ is the Planck's constant.

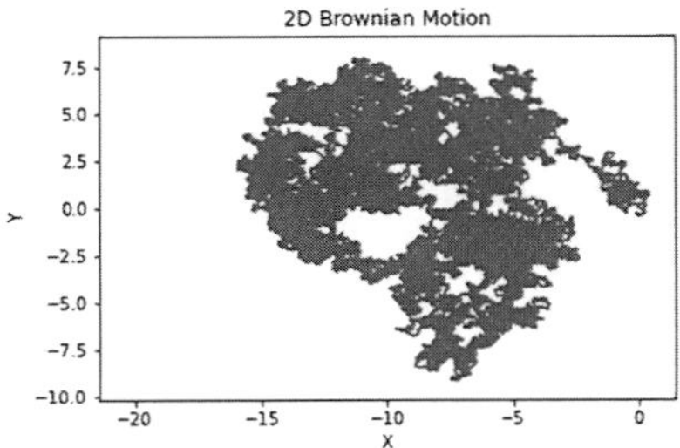

Figure 2.2 Signature of a two-dimensional Brownian motion. Like in the one-dimensional case, the expected value of the two-dimensional random velocity vector $\langle U \rangle = 0$.

[1] $\langle U \rangle \equiv E(U)$ is the expected value of the random velocity variable U and is analogous to the ensemble average.

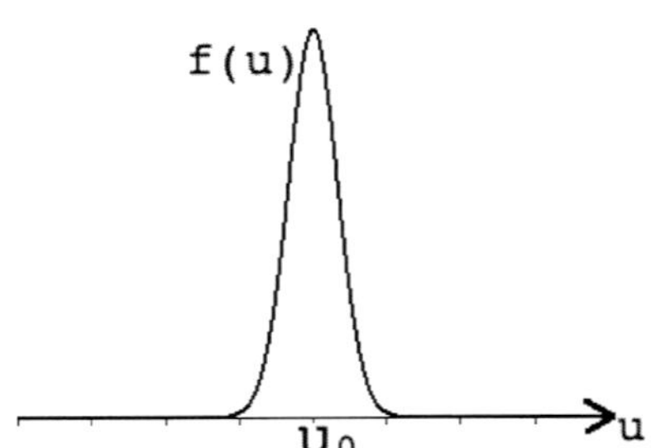

Figure 2.3 Distribution of velocity of particles in a uniform flow shows that a majority of the particles move with a velocity u_0 although there may be some small deviations.

---

[2] In this chapter, we will encounter terms such as $E(X^r)$, $r \in \mathbb{I}^+$, that are known as the moments of the random variable X.

follows: $\overline{u} = \int_{\mathfrak{u}} u \delta (u - u_0) du = u_0$. In order to facilitate a deeper appreciation of the practical utility of distribution functions, let us dwell a little more on this example of a collection of particles moving with velocity u_0. A gas in thermal equilibrium comprises a collection of such particles. Let us say we want to deduce the notion of temperature of the gas. This may be done by associating the concept of temperature to the average kinetic energy of the particles. If all the particles have an identical velocity u_0, then the average kinetic energy is simply $\frac{1}{2} m u_0^2$. However, even in a steady and uniform flow, it is unlikely that every particle constituting the flow will move identically with velocity u_0. Even though the majority of the particles may have the velocity u_0, there are likely going to be exceptional particles with slightly higher or lower velocities. So, instead of applying the Dirac delta distribution, a more appropriate weight function may be $f(u)$, with a distribution profile akin to the one shown in Figure 2.3. So in order to estimate the average kinetic energy of the particles, the distribution $f(u)$ must be employed to probe the kinetic energy law $\psi(u) = \frac{1}{2} m u^2$ across the spectrum of particle velocities. Thence, the average kinetic energy is $\int_{-\infty}^{\infty} \psi(u) f(u) du$, which now has a clear mathematical interpretation in the sense of the distribution $f(u)$.

Thus, statistical averages (and *moments*)[2] rely on the probability distribution functions of the relevant variables and are commonly used in developing many statistical models of practical importance, such as the kinetic theory of gases. We must now begin a formal study of such distributions that lie at the heart of all statistical analyses.

2.1 Chapter objectives

The chapter objectives are listed as follows.

1. Students will learn the notion of probability mass function, probability density function, and cumulative distribution function.

2. Students will learn to compute expected values and higher order statistical moments using probability distribution functions.

3. Students will learn the concept of moment generating function and use it to deduce the distribution of a random variable.

4. Students will learn the concept of joint probability distribution of multiple random variables. They will learn to deduce the marginal probability distribution from the joint probability distribution.

5. Students will study different types of discrete and continuous random variables. They will learn how to appropriately characterize a given phenomenon using one or many of these named probability models.

6. Students will learn to design and analyze an application project from the actuarial sciences (e.g., insurance model) by using a certain compound probability distribution model.

2.2   Chapter project: Predicting insurance claim aggregates during a policy period

2.2.1   *Epilogue: Modeling insurance claims using a compound probability distribution*

A certain insurance company is interested in predicting the total aggregate of all claims made during a fixed policy period from a portfolio of insurance products. Such an exercise will enable the company to make an assessment of its financial risks while charting out product launch schedules for the upcoming financial year.

A consultant to the company designs the following mathematical model to accomplish this task. Consider that the firm expects a certain number (N_j) of claims from amongst its clients during a fixed period j. Since there is no reason for this number N_j to be deterministically computable,[3] it is reasonable to assume N_j to be a random variable. Now there are N_j number of these claims: each claim amount is independent of the other and is also independent of N_j. This is also reasonable because each claim is made by a different client acting independently of the other. Further, each claim amount is also a random number that possibly corresponds to a common probability distribution. Let the claim amount by the ith client be denoted by X_i. X_i corresponds to a probability distribution function $F_X(x)$. The aggregate claim for the policy period j under consideration is also a random quantity $Y_j = X_1 + X_2 + \cdots + X_{N_j} = \sum_{i=1}^{N_j} X_i$ that obeys a compound probability distribution. Based on this model, a quantity of interest to the insurance firm is $E(Y_j)$, which you as the consultant will have to estimate in this project.

Moreover, consider there are four policy periods in a given financial year. The total premium collected at the beginning of the year by the insurance firm is $\$\,m$. Let λ_j be the rate at which claims are received per policy period j. Now consider $Z = \sum_{k=1}^{4} Y_k$ as the aggregate claim at the end of the fourth policy period (year end). The company incurs a loss if $Z > \$\,m$. In this project, you will simulate a certain compound stochastic process in Matlab and compute the associated risk for the insurance firm in terms of a probability $P(Z > \$\,m)$. Concurrently, you will learn about a composite stochastic model known as the *compound Poisson process* that is used by insurance companies to assess their risks.

Before we urge the readers to work on this aforementioned project, let us first learn some of the essential and fundamental elements of probability distributions.

2.3   Geometrical interpretation of integration with respect to a distribution function

We may recall from our elementary calculus course that the *Riemann integral* $\int u(x)\,dx$ with respect to the independent variable x may be interpreted as the *area under the curve* $u(x)$ (see Figure 2.5). This Riemann integral can be approximated by the Riemann sum $\sum_{x_i \in \mathfrak{D}} u(x_i^*)\Delta x_i$, where $x_i^* \in [x_{i-1}, x_i]$, $\forall i = 1, 2, \cdots, n$ and the x_i s are

Figure 2.4 Claims are made to an insurance manager at the end of a policy period.

[3] A multitude of external factors may determine the value of N_j. The complex interrelationship between these factors may further enhance the uncertainty in knowing what the exact value of N_j might be.

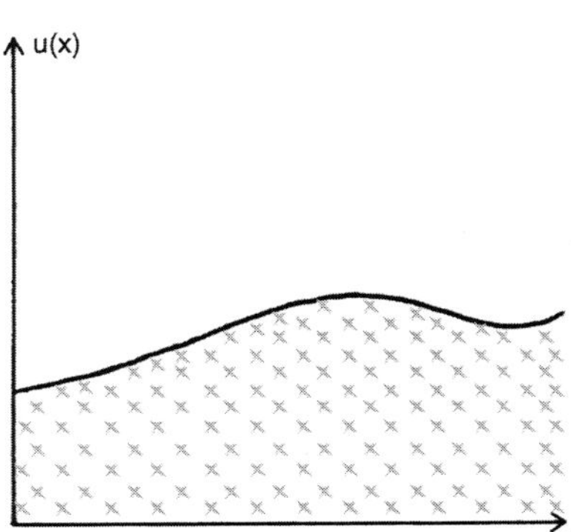

Figure 2.5 Profile of the function $u(x)$ along x. The shaded area under the curve $u(x)$ is given by $\int u(x)\,dx$.

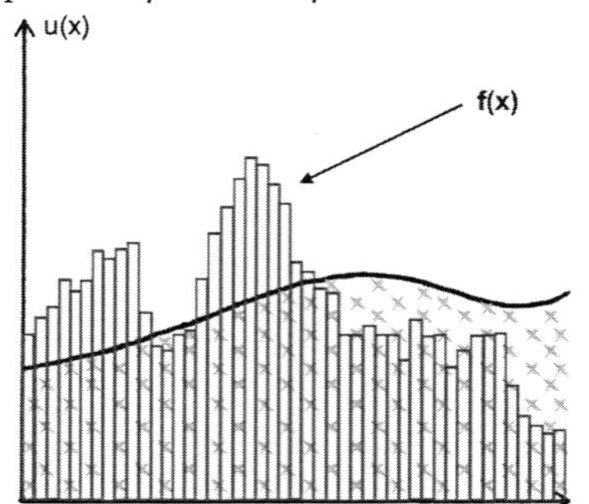

Figure 2.6 Profile of the weight function $f(x)$ demonstrates the relative importance of the observables x in $\mathcal{D}$.

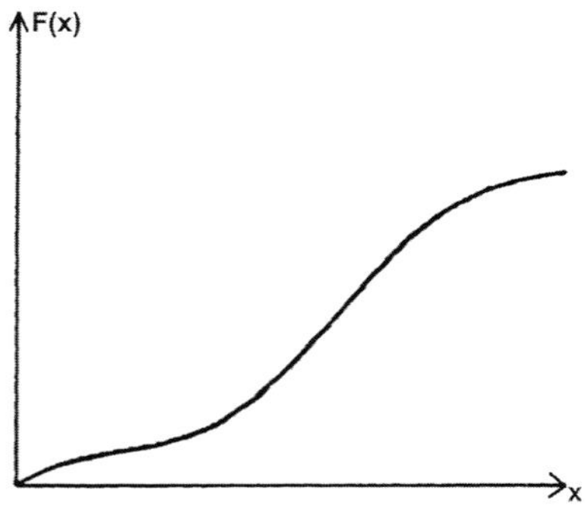

Figure 2.7 Distribution profile of the observables x is prescribed by some function $F(x)$.

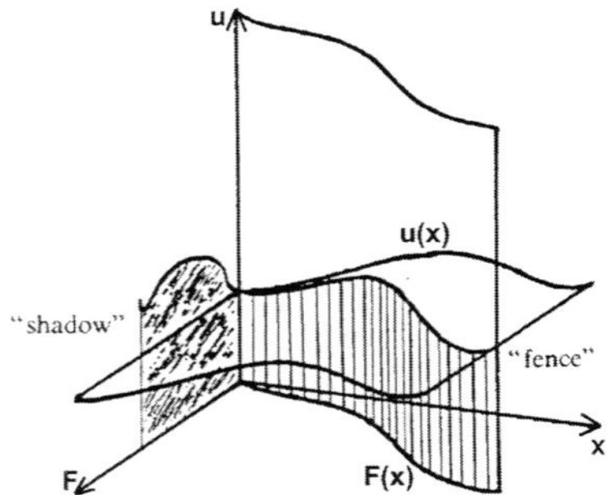

Figure 2.8 Geometrical meaning of an integral with respect to a distribution function. This figure is borrowed from the work of Gregory Bullock referenced below.

generally equally spaced n nodes (grid points) in the domain of integration $\mathcal{D}$ along the x axis. Simply put, this sum is an aggregate of areas of thin rectangular strips of height $u(x_i^*)$ and width Δx_i. The Riemann sum is equal to the integral $\int_{\mathcal{D}} u(x)dx$ in the limit $\Delta x_i \to 0$ (for all i) if and when this limit exists. This computation relies on the fact that all *observable x* values[4] are equally important and equally likely to be encountered in a practical situation. Hence, the function $u(x_i^*)$ is multiplied by the same unit scalar for all values of x_i^* in the summand. However, this need not always be the case as has been explained in the example discussed in the introductory paragraphs of this chapter. Consider that the *importance factors* of the different x_i^* values are illustrated by the profile $f(x)$ defined over all the x_i^*s in $\mathcal{D}$ (see Figure 2.6). This relative importance of certain x_i^*s over the others necessitates scaling the $u(x_i^*)$s appropriately by the weight factors $f(x_i^*)$s. The accurate representation of this case entails that we now have a modified integral of the form $\int_{\mathcal{D}} u(x)f(x)dx$, which may be approximated by the sum $\sum_{x_i \in \mathcal{D}} f(x_i^*)\, u(x_i^*)\Delta x_i$.

It is important to note that the profile of the importance factors of the observables as prescribed by the distribution function $f(x)$ will be different for different practical applications. In fact, much of this chapter and the subsequent models that we will discuss in this book will highlight this fact emphatically. Further, since the averaging procedure of the function $u(x)$ is being performed under the *guidance* of the distribution function $f(x)$, it is more appropriate to underscore this fact by using a modified notation for the integral, namely $\int_{\mathcal{D}} u(x)dF(x)$. In the context of probability theory, there is a unique relation between the *cumulative distribution function $F(x)$* and the *probability density function $f(x)$* that we will address in one of the subsequent sections of this chapter $\left(f(x) = \dfrac{dF(x)}{dx} \right)$. Thus far in our discussion here we have been referring to the probability density function $f(x)$ as the distribution function in the general sense of computing averages of functions. Going forward in subsequent sections of this chapter, and in the following chapters, we will make this distinction explicit in most scenarios.

Before we dive further into the conceptual elements of probability distributions, it may be useful to shed light on the geometrical meaning of the integral $\int_{\mathcal{D}} u(x)dF(x)$, which is widely known as the *Riemann–Stieltjes integral*. For this purpose, we will consider a distribution profile of the observables prescribed by some function $F(x)$ as shown in Figure 2.7. Since the function being averaged ($u(x)$) has an *independent* existence compared to the distribution profile of the observables ($F(x)$), each of x, u, and f can be represented along an independent axis in a three–dimensional representational space. Further, since $F(x)$ and $u(x)$ have independent origin and existence, the profile of a *sheet* traced by $u(x)$ along x, and protruding out of the u–x plane, will resemble hills and valleys along the x axis but look flat (straight) along the F direction. Thus, the height of this undulating sheet is prescribed by $u(x)$. If we were to consider another surface that cuts through this sheet, which emanates out of the u–x plane, under the *guidance* of the curve traced by $F(x)$, then a *fence*-type surface will emerge whose height is given by $u(x)$. This is shown in Figure 2.8. Clearly the area of the projection of this *fence* on the u–x plane gives the familiar area under the curve $u(x)$ that can be computed by the integral

---

[4] We are now using a terminology for x that will serve as a bridge from calculus to probability vocabulary.

$\int_{\mathfrak{D}} u(x)\,dx$. The projection of this *fence* on the u–F plane, on the other hand, is denoted by the shaded shadow region whose area is given by the *Riemann–Stieltjes integral* $\int_{\mathfrak{D}} u(x)\,dF(x)$. For the special case $F(x) = x$ in $\mathfrak{D}$, the Riemann–Stieltjes integral (integration with respect to the distribution $F(x)$) becomes identical to the more familiar Riemann integral $\int_{\mathfrak{D}} u(x)\,dx$. This aforementioned explanation is based on the discussion reported in the article published in the *American Mathematical Monthly* by Gregory L. Bullock.[5]

2.4 Discrete versus continuous probability distributions

In the previous chapter, under the section on random variables, we have encountered two different types, namely, discrete and continuous random variables. In the case of the former type, the random variables take on distinct values. A simple example is the outcome of tossing a fair coin – it is a *head* or a *tail* with a designated random value of 1 or 0, each with a probability equal to $\frac{1}{2}$. In the continuous case, the outcomes are such that the random variable may take on a continuum of values over a range prescribed by the sample space Ω. A classic example is the test score obtained by a student who is registered in a course; this score may be any real number between 0 and 100. It may be considered a random number because without any specific information about this student's performance in the test or his/her talent/skill level in the subject or the nature of the test itself, it may be hard to make a definitive prediction of the student's score. Under the circumstances, it may be prudent to deduce the score by sampling without bias from the bell-shaped Gaussian distribution (see Figure 1.8).

In this section, we will formally define the probability distribution profile for discrete and continuous random variables. Consequently, these definitions will be helpful to calculate statistical moments (means, variances, etc.) of the respective random variable and thereby make forecasts of important events. These calculations become very useful when we do not have access to any sample data but instead have some understanding of the underlying stochastic phenomenon which allows us to identify, with reasonable accuracy, the relevant probability/stochastic model and utilize its probability distribution profile.

2.4.1 *Definition: Probability mass function*

For a discrete random variable, each possible observable $x_i \in \Omega$ has a certain probability of occurrence $p_i := P(X = x_i)$ which we can think of as a *probability mass*. Obviously, $\sum_{x_i \in \Omega} P(X = x_i) = 1$ due to the second axiom of probability (axiom of unitarity), and this serves as a conservation law. We will use the notation $f_X(x_i) \equiv p_i$ to denote the probability mass function (p.m.f.) of a discrete random variable. The profile of $f_X(x_i)$ presents a visual depiction of how the probability masses are interspersed over the sample space Ω (see Figure 2.9).

[5] Gregory L. Bullock, A geometric interpretation of the Riemann–Stieltjes integral, The *American Mathematical Monthly* **95**, no. 5 (1988): 448–455.

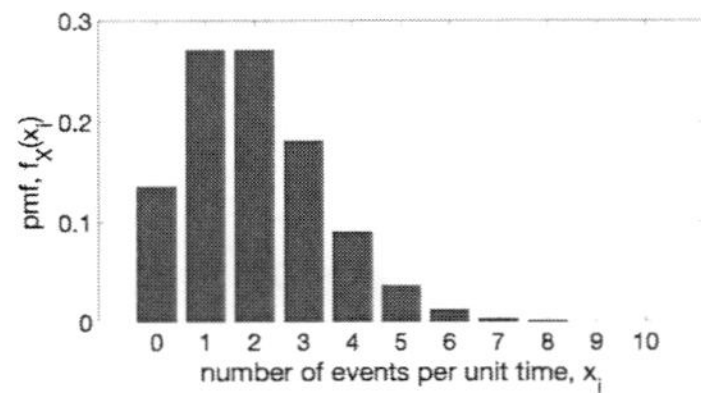

Figure 2.9 Probability mass function $f_X(x_i)$ of a certain discrete random variable. It may be verified that $\sum_{x_i \in \Omega} f_X(x_i) = 1$.

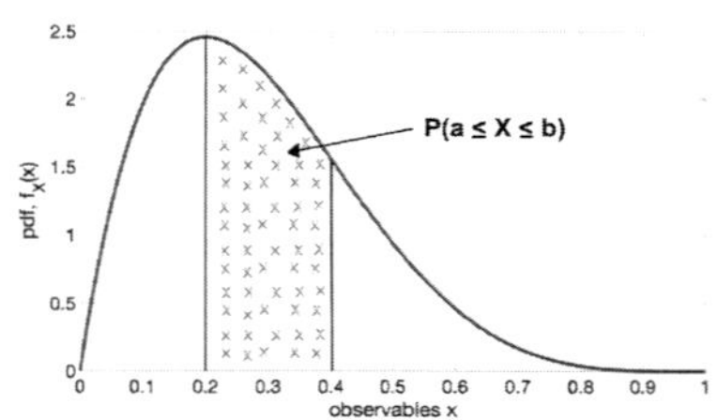

Figure 2.10 Probability density function $f_X(x)$ of a certain continuous random variable.

2.4.2 *Definition: Probability density function*

In the case of a continuous random variable X, the probability mass is spread continuously over the range of the observables. Therefore, it is appropriate to use the notion of a *density function* $f_X(x)$ instead of probability mass. The unitarity axiom of probability enforces the following normalization of the probability density function (p.d.f.): $\int_{x \in \Omega} f_X(x)\,dx = 1$. It follows that $P(a \le X \le b) = \int_a^b f_X(x)\,dx$ represents the area under the curve f between a and b (see Figure 2.10).

Example 2.1 Distribution profile of a discrete random variable

A random variable X has the following p.m.f. $p(x)$:

Value of X	0	1	2	3	4	5	6	7
$p(x) = P\{X = x\}$	0	k	$2k$	$2k$	$3k$	k^2	$2k^2$	$7k^2 + k$

   i. Find the value of k where $k \in \mathbb{R}$ is a constant.

  ii. Evaluate $P\{X < 6\}$ and $P\{0 < X < 5\}$.

Solution:

i.
$$\sum_i p(x_i) = 1$$
$$k + 2k + 2k + 3k + k^2 + 2k^2 + 7k^2 + k = 1$$
$$10k^2 + 9k - 1 = 0$$
$$k = \frac{1}{10}.$$

ii.
$$P(X < 6) = 1 - P(X \ge 6)$$
$$= 1 - \left[P(X = 6) + P(X = 7) \right]$$
$$= 1 - \frac{19}{100}$$
$$= \frac{81}{100}.$$
$$P(0 < X < 5) = 1 - \left[P\{X = 5\} + P\{X = 6\} + P\{X = 7\} \right]$$
$$= 1 - \frac{20}{100}$$
$$= \frac{4}{5}.$$

Example 2.2 Probability mass function of a discrete random variable

A random variable X has the following p.m.f.:

$$p(i) = \frac{c\lambda^i}{i!}, \qquad i = 0, 1, 2, 3, \dots \qquad c, \lambda \text{ are constants}$$

i. Find the value of c.

ii. Evaluate $P\{X = 0\}$.

Solution:

i.
$$\sum p_i = 1$$
$$c\sum \frac{\lambda^i}{i!} = 1$$
$$ce^{-\lambda} = 1$$
$$c = e^{-\lambda}.$$

ii.
$$p(0) = \frac{e^{-\lambda}\lambda^0}{0!} = e^{-\lambda}.$$

Example 2.3 *Probability density function of a continuous random variable*

The p.d.f. of a continuous random variable X is given by

$$f(x) = \begin{cases} ax, & 0 \le x < 1, \\ a, & 1 \le x < 2, \\ -ax + 3a, & 2 \le x \le 3, \\ 0, & \text{otherwise.} \end{cases}$$

Find the value of a.

Solution:

$$\int_{-\infty}^{\infty} f(x)\,dx = 1$$

$$\int_0^1 ax\,dx + \int_1^2 a\,dx + \int_2^3 (-ax + 3a)\,dx = 1$$

$$\frac{a}{2} + a - \frac{5a}{2} + 3a = 1$$

$$2a = 1$$

$$a = \frac{1}{2}.$$

Example 2.4 *Sculpting an electric kettle*

The diameter X of an electric kettle is assumed to be a continuous random variable with p.d.f.

$$f(x) = \begin{cases} 6x(1-x), & 0 \le x \le 1, \\ 0, & \text{otherwise.} \end{cases}$$

i.   Check that the above function is indeed a p.d.f.

ii. Compute $P\left\{X \le \frac{1}{2} \middle| \frac{1}{3} \le X \le \frac{2}{3}\right\}$.

Figure 2.11 Contour of an old metallic electric kettle that has taken many dents on its body over the years.

iii. Determine k such that $P\{X < k\} = P\{X > k\}$.

Solution:

i. $LHS = \displaystyle\int_{-\infty}^{\infty} f(x)\, dx = \int_{0}^{1}\left(6x - 6x^2\right) dx = \left(\dfrac{6x^2}{2} - \dfrac{6x^3}{3}\right)_{0}^{1} = \left(3x^2 - 2x^3\right)_{0}^{1} = 1 = RHS.$

ii. $P\left\{X \le \dfrac{1}{2}\,\middle|\,\dfrac{1}{3} \le X \le \dfrac{2}{3}\right\} = \dfrac{P\left[\left\{X \le \dfrac{1}{2}\right\} \cap \left\{\dfrac{1}{3} \le X \le \dfrac{2}{3}\right\}\right]}{P\left\{\dfrac{1}{3} \le X \le \dfrac{2}{3}\right\}}$

$= \dfrac{P\left\{\dfrac{1}{3} \le X \le \dfrac{1}{2}\right\}}{P\left\{\dfrac{1}{3} \le X \le \dfrac{2}{3}\right\}} = \dfrac{\left(3x^2 - 2x^3\right)_{\frac{1}{3}}^{\frac{1}{2}}}{\left(3x^2 - 2x^3\right)_{\frac{1}{3}}^{\frac{2}{3}}}$

$= \dfrac{\left(\dfrac{3}{4} - \dfrac{2}{8}\right) - \left(\dfrac{3}{9} - \dfrac{2}{27}\right)}{\left(\dfrac{12}{9} - \dfrac{16}{27}\right) - \left(\dfrac{3}{9} - \dfrac{2}{27}\right)} = \dfrac{\left(\dfrac{3}{4} - \dfrac{1}{4}\right) - \left(\dfrac{9}{27} - \dfrac{2}{27}\right)}{\left(\dfrac{36}{27} - \dfrac{16}{27}\right) - \left(\dfrac{9}{27} - \dfrac{2}{27}\right)}$

$= \dfrac{\dfrac{1}{2} - \dfrac{7}{27}}{\dfrac{20}{27} - \dfrac{7}{27}} = \dfrac{27 - 14}{2 \times 13} = \dfrac{13}{26}.$

$P\left\{X \le \dfrac{1}{2}\,\middle|\,\dfrac{1}{3} \le X \le \dfrac{2}{3}\right\} = \dfrac{1}{2}.$

iii. k is positive and $k < 1$, otherwise we will get a nonsensical relation like $0 = 1$.

Therefore, $0 < k < 1$, such that $P\{X < k\} = P\{X > k\}$.

$$P\{X < k\} = P\{X > k\}$$

$$\int_{0}^{k} 6x(1-x)\, dx = \int_{k}^{1} 6x(1-x)\, dx$$

$$3k^2 - 2k^3 = 3\left(1^2 - k^2\right) - 2\left(1^3 - k^3\right)$$

$$3k^2 - 2k^3 = 3 - 3k^2 - 2 + 2k^3$$

$$4k^3 - 6k^2 + 1 = 0$$

$$(2k - 1)\left(2k^2 - 2k - 1\right) = 0$$

$$k = \dfrac{1}{2},\, \dfrac{1 \pm \sqrt{3}}{2}.$$

But since we considered $0 < k < 1$, therefore $k = \dfrac{1}{2}$.

2.4.3   *Definition: Cumulative distribution function*

The cumulative distribution function (c.d.f.) $F_X : \mathbb{R} \to [0,1]$ is defined as $F_X(x) \equiv$ $F(x) := P(X \le x)$, $x \in \mathbb{R}$. It follows that $P(a \le X \le b) = \int_a^b f(x)\,dx = F(b) - F(a)$. The c.d.f. F_X has the following properties:

 i. $\lim_{y \downarrow -\infty} F(y) = 0$,

 ii. $\lim_{y \uparrow \infty} F(y) = 1$,

 iii. $\lim_{y \downarrow x} F(y) = F(x)$, $\forall x \in \mathbb{R}$ (i.e., F_X is right-continuous).

The first two properties imply that F is always a non-decreasing function. It must be noted that the definition of the c.d.f. as stated above is valid for both discrete and continuous random variables. Let us consider the case of a continuous random variable. Clearly $F(x) = \int_{-\infty}^{x} f(\zeta)\,d\zeta$ for a continuous real-valued function f (p.d.f.), where F is uniformly continuous and differentiable, therefore $\dfrac{dF(x)}{dx} = f(x) \Rightarrow dF(x) = f(x)\,dx$, which is a direct consequence of the *fundamental theorem of calculus.*

There are two main interpretations of the distribution function $F_X(x)$ that are noteworthy to mention here.

 I. $F_X(x)$ prescribes the *distribution of probability mass* on the real line. Concomitantly, $F(b) - F(a)$ is the mass concentrated in the interval $[a,b]$. For the discrete case, locations of concentrated point mass on the real line (x_i) are points of discontinuity of F_X with jumps proportional to $p_i \equiv F_X(x_i + \varepsilon) - F_X(x_i - \varepsilon)$ where ε is infinitesimally small.[6] There are a finite or a countable number of such jumps, and F_X is continuous everywhere else.

 II. $F_X(x)$ encompasses the accumulation of probability masses (or density) up to x. Therefore, it is *additive*, non-negative, and has a unit maximum value. Thus, the c.d.f. F qualifies as a *measure* (*F-measure*). In section 2.3, we have commented on this aspect of interpreting the linear functional $F(u)$ defined by $F(u) = \int_{\mathcal{D}} u(x)\,dF(x)$ as an integral of a measurable function $u(x)$ over $\mathcal{D}$ with respect to the F-measure $F(x)$.[7]

[6] A distribution with only concentrated point masses is a discrete distribution and one without is a continuous distribution.

[7] The F-measure corresponds to the *Jordan–Peano* measure (Jordan content) that extends the notion of *size* to more complicated geometry.

Example 2.5   Cumulative distribution function of a discrete random variable

A random variable X has the following p.m.f. $p(x)$:

Value of X	0	1	2	3	4	5	6	7
$p(x) = P\{X = x\}$	0	k	$2k$	$2k$	$3k$	k^2	$2k^2$	$7k^2 + k$

Find the (cumulative) distribution function of X.

Solution:

Value of X	0	1	2	3	4	5	6	7
$p(X) = $ $P\{X = x\}$	0	$\dfrac{1}{10}$	$\dfrac{2}{10}$	$\dfrac{2}{10}$	$\dfrac{3}{10}$	$\dfrac{1}{10^2}$	$\dfrac{2}{10^2}$	$\dfrac{7}{10^2}+\dfrac{1}{10}$
$F(x) = $ $P\{X \le x\}$	0	$\dfrac{1}{10}$	$\dfrac{3}{10}$	$\dfrac{5}{10}$	$\dfrac{8}{10}$	$\dfrac{8}{10}+\dfrac{1}{10^2}=\dfrac{81}{10^2}$	$\dfrac{8}{10}+\dfrac{3}{10^2}=\dfrac{83}{10^2}$	$\dfrac{9}{10}+\dfrac{10}{10^2}=1$

Example 2.6   Convergence of c.d.f. to unity

A random variable X has the following p.m.f.

$$p(i) = \frac{c\lambda^i}{i!}, \text{ where } i = 0, 1, 2, 3, \ldots \quad c, \lambda \text{ are constants}$$

Verify that the c.d.f. converges to 1 as x tends to ∞.

Solution:

$$\lim_{x \to \infty} F(x) = P\{X \le \infty\}$$

$$= \sum_{i=0}^{\infty} p_i$$

$$= \sum_{i=0}^{\infty} \frac{e^{-\lambda} \lambda^i}{i!}$$

$$= e^{-\lambda} \sum_{i=0}^{\infty} \frac{\lambda^i}{i!}$$

$$= e^{-\lambda} e^{\lambda}$$

$$= 1.$$

Example 2.7   Computing probability of events from the c.d.f.

Let X be a random variable with c.d.f. $F(x)$. Compute each of the following probabilities in terms of $F(x)$:

$P\{X \le a\}$,

$P\{X > a\}$,

$P\{a < x \le b\}$,

$P\{X < a\}, P\{X = a\}$,

$P\{a \le X < b\}$,

$P\{a \le X \le b\}$,

$P\{a < X < b\}$ and $P\{X \ge a\}$.

Figure 2.12 The staircase of chance from nullity to unity: $F_X(x) \to 1$ as $x \to \infty$.

Solution:

$$P\{X \le a\} = F(a).$$

$$P\{X > a\} = 1 - F(a).$$

$$P\{a < X \le b\} = F(b) - F(a).$$

$$P\{X < a\} = F(a^-) = \lim_{t \to a^-} F(t).$$

$$P\{X = a\} = F(a) - F(a^-).$$

$$P\{a \le X < b\} = F(b^-) - F(a^-).$$

$$P\{a \le X \le b\} = F(b) - F(a^-).$$

$$P\{a < X < b\} = F(b^-) - F(a).$$

$$P\{X \ge a\} = 1 - F(a^-).$$

Example 2.8 *Cumulative distribution of a continuous random variable*

Let the p.d.f. of a random variable X be given as follows:

$$f(x) = \begin{cases} x, & 0 < x \le 1, \\ 2 - x, & 1 < x \le 2, \\ 0, & \text{otherwise.} \end{cases}$$

Compute the c.d.f. of this random variable.

Solution:

$$F(x) = \int_{-\infty}^{x} f(t)dt = \int_{0}^{x} f(t)dt.$$

$$\text{Therefore, } F(x) = \begin{cases} 0, & x \le 0 \\ \int_{0}^{x} x \ dx, & 0 < x \le 1 \\ \int_{0}^{1} x \ dx + \int_{1}^{x} (2-x)dx, & 1 < x \le 2 \\ \int_{0}^{1} x \ dx + \int_{1}^{2} (2-x)dx + \int_{2}^{x} 0 \ dx, & x \ge 2 \end{cases}$$

$$\Rightarrow F(x) = \begin{cases} 0, & x \le 0 \\ \dfrac{x^2}{2}, & 0 < x \le 1 \\ \dfrac{1}{2} + 2(x-1) - \dfrac{x^2 - 1^2}{2}, & 1 < x \le 2 \\ \dfrac{1}{2} + 2(2-1) - \dfrac{2^2 - 1^2}{2} + 0, & x \ge 2 \end{cases}$$

$$\Rightarrow F(x) = \begin{cases} 0, & x \le 0 \\ \dfrac{x^2}{2}, & 0 < x \le 1 \\ -\dfrac{x^2}{2} + 2x - 1, & 1 < x \le 2 \\ 1, & x \ge 2 \end{cases}$$

Figure 2.13 Pafnuty Lvovich Chebyshev (1821–1894) was a prominent Russian mathematician and professor of algebra, number theory, and probability at St. Petersburg University (*courtesy*: St. Petersburg Department of V. A. Steklov Institute of Mathematics of the Russian Academy of Sciences, PDMI RAS).

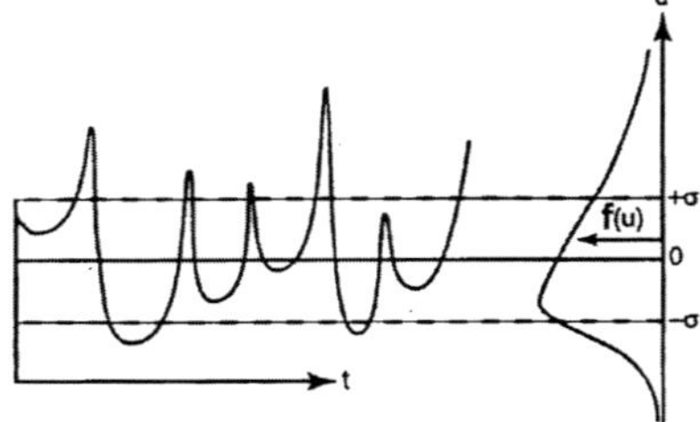

Figure 2.14 Time series data $u(t)$ with positive skewness ($\hat{\mu}_3 > 0$).

[8] The non-normalized version is known as the third moment, $\mu_3 := E\left((X - \mu)^3\right)$.

2.4.4 *Statistical moments and their significance*

Earlier in section 1.8, we have encountered the notion of expected value and variance of a random variable. There we had defined the expectation and variance as a weighted aggregate of the observables and mean squared deviation, respectively. The weight factors were taken to be the probability masses. In this section, we will extend the definitions in terms of the distribution functions. In what follows here, we will consider X as the random variable (discrete or continuous) and the observables $x \in \Omega$. The statistical moments, defined below, determine the shape of the distribution function and hence characterize data. Renowned Russian mathematician Pafnuty Lvovich Chebyshev was the first to systematically define and use statistical moments of random variables during the mid-nineteenth century.

i.   **Mean** (μ or $E(X)$) is the *first statistical moment*.

$$E(X) = \sum_{x \in \Omega} x P(X = x) \qquad \text{(discrete case)}, \qquad (2.1)$$

$$E(X) = \int_{x \in \Omega} x f(x)\, dx = \int_{x \in \Omega} x\, dF(x) \quad \text{(continuous case)}. \qquad (2.2)$$

ii.   **Variance** (σ^2 or $Var(X)$) is the *second statistical moment*.

$$Var(X) = E\left((X - \mu)^2\right) = \sum_{x \in \Omega} (x - \mu)^2 P(X = x) \quad \text{(discrete case)}, \qquad (2.3)$$

$$Var(X) = \int_{x \in \Omega} (x - \mu)^2 f(x)\, dx \qquad \text{(continuous case)}. \qquad (2.4)$$

Equivalently, $\boxed{Var(X) = E(X^2) - (E(X))^2}$, a result that is a consequence of algebraically unraveling the expression $E((X - \mu)^2)$.

iii. **Skewness** ($\hat{\mu}_3$) is the *third standardized moment*.

$$\hat{\mu}_3 = E\left(\left(\frac{X - \mu}{\sigma}\right)^3\right) = \frac{E((X - \mu)^3)}{(Var(X))^{3/2}} \qquad (2.5)$$

$\hat{\mu}_3$ measures the degree of asymmetry of the probability density function. A p.d.f. that is symmetric about the mean has zero skewness. All higher order odd moments of such a symmetric p.d.f. will also be identically zero. Data $u(t)$ with positive skewness is characterized by a p.d.f. with a longer tail for $X - \mu > 0$ than for $X - \mu < 0$ (here X represents the random variable that is sampled over time). Hence, a positive skewness means that deviation $X - \mu$ is more likely to take on large positive values than large negative values. For instance, a time series data with long periods of small negative values and a few instances of large positive values, with zero temporal mean, has positive skewness (see Figure 2.14).[8]

iv. **Kurtosis** ($\hat{\mu}_4$) is the *fourth standardized moment*.

$$\hat{\mu}_4 = E\left(\left(\frac{X - \mu}{\sigma}\right)^4\right) = \frac{E((X - \mu)^4)}{(Var(X))^2} \qquad (2.6)$$

A p.d.f. with longer tails will have a larger kurtosis than a p.d.f. with narrower tails. A time series data $u(t)$ with most measurements clustered around the mean has low kurtosis. A time series dominated by intermittent extreme events has high kurtosis.[9]

2.4.5 Discrete and continuous probability distribution models: construction and applications

Models of the real-world processes must account for the element of uncertainty that is inherently omnipresent. Therefore, there is a strong case for designing and using probabilistic models that suitably capture random phenomena. In this section, we will study many important probabilistic models of both discrete and continuous processes.

Discrete probability models

i. **Bernoulli distribution:** This is a binary probability model with only two possible outcomes. Examples of this model are the outcomes of tossing a fair coin, success or failure of a projectile in hitting its target, etc. Let us consider that the random variable X can take one of two possible values 1 or 0 with probability p and $1 - p$. The p.m.f. is defined below.

$$X \sim Bernoulli(p).$$

$$f_X(x) = \begin{cases} p, & \text{when } x = 1, \\ 1 - p, & \text{when } x = 0. \end{cases} \tag{2.7}$$

The expected value and variance of a Bernoulli random variable are calculated as follows:

$$E(X) = \left((1 \times p) + (0 \times (1 - p))\right) = p. \tag{2.8}$$

$$Var(X) = \sum_{x=\{0,1\}} (x - E(X))^2 f_X(x) = (0 - p)^2 (1 - p) + (1 - p)^2 p$$

$$= p(1 - p). \tag{2.9}$$

Example 2.9 Fixing wooden planks to a wall by a battery of nail guns

A battery of ten nail guns are fired simultaneously to affix wooden planks to a wall. The probability that a nail gun successfully drives a nail into a wooden plank is 0.95, (hit rate). The plank attaches to the wall if at least seven of the ten nails are driven through the plank into the wall successfully. Do you think this battery of guns can successfully affix planks to a wall on an average?

Let $X_i \sim Bernoulli(p = 0.95)$, where $i = 1, 2, 3, \ldots, 10$. $X_i = 1$ represents an event that a nail is successfully driven through the plank into the wall. $X_i = 0$ represents a compromised nail. Consider the random variable $Y = \sum_{i=1}^{10} X_i$ that captures the total number of successful hits by the battery.

[9]The non-normalized version is known as the fourth moment, $\mu_4 := E\left((X - \mu)^4\right)$.

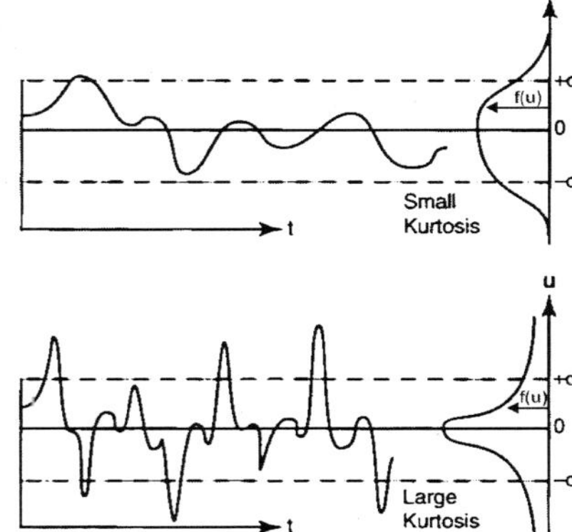

Figure 2.15 Time series data $u(t)$ with small (top) and large (bottom) kurtosis $\hat{\mu}_4$. Large values of $\hat{\mu}_4$ correspond to data with intermittent extreme events.

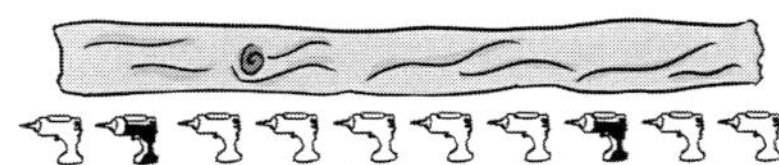

Figure 2.16 A battery of nail guns is used to fix a wooden plank to a wall. The failure rate of each gun $(1 - p)$ = 0.05 is small enough to ensure that the battery successfully fixes planks to the wall on an average.

$$E(Y) = E\left(\sum_{i=1}^{10} X_i\right) = \sum_{i=1}^{10} E(X_i) = 10p = 9.5. \tag{2.10}$$

So the average hit rate of the battery is 9.5, which is greater than 7. So, on an average, the battery of guns can be successfully used to affix planks to a wall.

ii. **Binomial distribution:** The above example demonstrates that a sequence of Bernoulli trials can be used to model a binomial random process. Here, we are interested in accounting for a random number (X) of successes (each with probability p) in n independent Bernoulli trials.

$$X \sim Bin(n, p).$$

$$f_X(k) = \begin{cases} {}^nC_k \; p^k (1-p)^{n-k}, & \text{for } k = 0, 1, 2, 3, \ldots, n, \\ 0, & \text{otherwise.} \end{cases} \tag{2.11}$$

Taking a cue from the calculation shown in equation (2.10), we can estimate the expected value and variance of a binomial random variable as follows:

$$E(X) = np. \tag{2.12}$$

$$Var(X) = np(1 - p). \tag{2.13}$$

Example 2.10   Statistical moments of a binomial random variable

Let X be a binomial random variable with parameters (n, p) with p.m.f.

$$p(x) = P\{X = x\} = \begin{cases} {}^nC_x p^x q^{n-x}, & x = 0, 1, 2, \cdots, n \\ 0, & \text{otherwise,} \end{cases}$$

where $0 \leq p \leq 1$, $q = 1 - p$, and p is the probability of success. Compute the following moments:

i. $E(X)$, ii. $E(X^2)$, iii. $Var(X)$, iv. $E(X^3)$.

Hint: Consider $X^2 = X(X-1) + X$ and $X^3 = X(X-1)(X-2) + 3X(X-1) + X$.

Solution: We will use the following fact:

$$\sum_{x=0}^{n} p(x) = \sum_{x=0}^{n} {}^nC_x p^x q^{n-x} = (p+q)^n = 1.$$

$${}^nC_x = \frac{n}{x} \cdot {}^{n-1}C_{x-1} = \frac{n}{x} \cdot \frac{n-1}{x-1} \cdot {}^{n-2}C_{x-2} = \ldots$$

i. $E(X) = \sum_{x=0}^{n} x p(x) = \sum_{x=1}^{n} x \ ^{n}C_{x} p^{x} q^{n-x}$

$$= \sum_{x=0}^{n} n \cdot \ ^{n-1}C_{x-1} p^{x} q^{n-x}$$

$$= np \sum_{x=0}^{n} \ ^{n-1}C_{x-1} p^{x-1} q^{(n-1)-(x-1)}$$

$$= np(p+q)^{n-1}.$$

$E(X) = np.$

ii. $E(X)(X-1) = \sum_{x=1}^{n} x(x-1) p(x) = \sum_{x=1}^{n} x(x-1) \ ^{n}C_{x} p^{x} q^{n-x}$

$$= \sum_{x=1}^{n} n(n-1)^{n-2}C_{x-2} p^{x} q^{n-x}$$

$$= n(n-1) p^{2} \sum_{x=1}^{n} \ ^{n-2}C_{x-2} p^{x-2} q^{(n-2)-(x-2)}.$$

$E(X)(X-1) = n(n-1) p^{2} (p+q)^{n-2} = n(n-1) p^{2}.$

Using the hint above, we get

$$E(X^{2}) = E(X)(X-1) + E(X)$$
$$= n(n-1) p^{2} + np$$
$$= n^{2} p^{2} - np^{2} + np$$
$$= n^{2} p^{2} + np(1-p).$$

$E(X^{2}) = n^{2} p^{2} + npq.$

iii. $Var(X) = E(X^{2}) - (E(X))^{2}$

$$= n^{2} p^{2} + npq - (np)^{2}$$

$Var(X) = npq.$

iv. Following similar steps,

$$E(X)(X-1)(X-2) = n(n-1)(n-2) p^{3}.$$

$$E(X^{3}) = E(X)(X-1)(X-2) + 3E(X)(X-1) + E(X),$$

$$E(X^{3}) = n(n-1)(n-2) p^{3} + 3n(n-1) p^{2} + np.$$

Example 2.11 *Free cola tasting and smart marketing campaign*

Boca-cola is a newly emerging cola brand in the market that faces stiff competition from a very popular and old cola brand, namely, Moca-cola. Both colas look the same. The Boca-cola company made a smart

Figure 2.17 Boca-cola versus Moca-cola contest: two colas in a blind taste campaign.

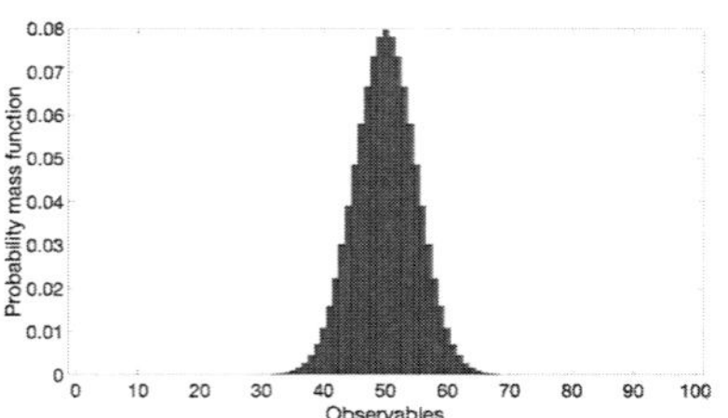

Figure 2.18 The p.m.f. of $Y \sim Bin\,(100, 0.5)$.

move of inviting 100 sworn customers of Moca-cola to a free cola tasting campaign during the innings interval of the world cricket championship final held in the city of Borgo Verde. Two anonymous samples of cola (one from each brand) were offered to each of the hundred tasters, and they were asked to vote their preference. The Boca-cola company made profitable use of the binomial distribution, which has most of its mass concentrated within three standard deviations of the mean. The following calculations demonstrate how they made a compelling case for a surge in Boca-cola sales even in the face of stiff competition from their bigger rival Moca-cola.

The campaign was successful on the premise that a typical Moca-cola drinker will not be able to tell the difference between the two colas during a blind test. Hence, they are equally likely to prefer one cola over the other (i.e., probability that a random participant in the blind cola tasting test prefers Boca-cola [or Moca-cola] is one-half, $p = 0.5$). Consider a Bernoulli random variable X which takes a value 1 (when Boca-cola is preferred) or 0 (when Moca-cola is preferred). Let $Y = \sum_{i=1}^{100} X_i$ denote the number of instances when a sworn Moca-cola drinker preferred the Boca-cola during the blind test. Clearly, $Y \sim Bin\left(n = 100, p = 0.5\right)$. The following lines of Matlab code are used to construct the relevant p.m.f. of Y (see Figure 2.18).

```
n = 100; p = 0.5;
x = 0:n;
y = binopdf(x,n,0.5);
figure, bar(x,y,1);
xlabel('Observables');
ylabel('Probability mass function');
set(gca,'FontSize',60);
```

The mean and variance of Y are calculated as $E(Y) = np = 100 \times 0.5 = 50$ and $\sigma^2 = Var\left(Y\right) = np\left(1 - p\right) = 25 \;\Rightarrow\; \sigma = 5$. It is clear from the profile of the distribution of Y in Figure 2.18 that almost all the probability mass is concentrated between $Y = 35$ and $Y = 65$, i.e., within three standard deviation of the mean. Therefore, $P\left(Y \geq 35\right) \approx 1$, which signifies that more than 35% of sworn Moca-cola drinkers prefer the newly launched Boca-cola.* A 35% switch by sworn clients of a competing brand was indeed a compelling advertisement campaign!

---

*We can check using the Matlab command »sum(y(35:end)) that $P(Y \geq 35) = 0.9996 \approx 1$.

iii. ***Geometric distribution***: Consider a sequence of Bernoulli trials with probability of success equal to p. Let X be the number of failures before the first success. The p.m.f. for such a random variable is defined as follows:

Geometric distribution of type-0

$$X \sim geom_0(p)$$

$$P(X = x) = \begin{cases} (1-p)^x\, p, & \text{for } x = 0, 1, 2, 3, \ldots, \\ 0, & \text{otherwise.} \end{cases} \tag{2.14}$$

$$E(X) = \frac{1-p}{p}. \tag{2.15}$$

$$Var(X) = \frac{1-p}{p^2}. \tag{2.16}$$

If Y is the random variable that counts the number of Bernoulli trials until first success, the p.m.f. of Y is slightly different from the case above as mentioned below.

Geometric distribution of type-1

$$Y \sim geom_1(p)$$

$$P(Y = y) = \begin{cases} (1-p)^{y-1}\, p, & \text{for } y = 1, 2, 3, \ldots, \\ 0, & \text{otherwise.} \end{cases} \tag{2.17}$$

$$E(Y) = \frac{1}{p}. \tag{2.18}$$

$$Var(Y) = \frac{1-p}{p^2}. \tag{2.19}$$

Figure 2.19

Cunning fox, while passing by,
Thought to taste some grapes on high,
So he leaps, and leaps again,
For they say – there is no gain without pain!

The story of the fox and the grapes, adapted from *Aesopica,* is an old anthem to learn new math tricks with the geometric distribution.

Example 2.12   Derivation of $E(X) = \dfrac{1}{p}$ ***for*** $X \sim geom_1(p)$

We will begin with the definition of expectation.

$$E(X) = \sum_{x=1,2,\ldots} x P(X = x)$$

$$= \sum_{x=1,2,\ldots} x (1-p)^{x-1}\, p$$

$$= p \sum_x x (1-p)^{x-1} =: pS. \tag{2.20}$$

Consider $S_1 = \displaystyle\sum_{x=1,2,\ldots} (1-p)^x = \dfrac{1}{1-(1-p)}$, which is a convergent infinite geometric series. $\dfrac{dS_1}{dp} = \displaystyle\sum_{x=1,2,\ldots} x (1-p)^{x-1}(-1) = -\dfrac{1}{p^2}$. Therefore, using equation (2.20), $S = \dfrac{1}{p^2}$ and consequently $E(X) = \dfrac{1}{p}$.

Likewise, using $Var(X) = E(X^2) - (E(X))^2$, writing $E(X^2) = E(X(X-1)) + E(X)$, and following similar steps as shown above, we can deduce the expression for $Var(X) = \dfrac{1-p}{p^2}$.

The distribution function (c.d.f.) of the geometric distribution of type-1 is $F_X(k) := P(X \le k) = 1 - P(X > k)$. $P(X > k)$ can be calculated as follows:

$$
\begin{aligned}
P(X > k) &= P(X = k+1) + P(X = k+2) + \cdots \\
&= (1-p)^k p + (1-p)^{k+1} p + \cdots \\
&= (1-p)^k p \left(1 + (1-p) + (1-p)^2 + + \cdots\right) \\
&= (1-p)^k p \frac{1}{1-(1-p)} \\
&= (1-p)^k.
\end{aligned}
\tag{2.21}
$$

Therefore, $F_X(k) = 1 - (1-p)^k$.

Example 2.13   Memoryless property of geometric distribution

Any random variable X has a memoryless property if, for any n, $m \ge 0$, we have $P(X > n + m | X > m) = P(X > n)$. We will demonstrate that this is certainly true of the geometric random variable $X \sim geom_1(p)$.

$$
P(X > n+m \mid X > m) = \frac{P(\{X > n+m\} \cap \{X > m\})}{P(X > m)}
$$

Definition of conditional probability

$$
= \frac{P(X > n+m)}{P(X > m)}
$$

$$
= \frac{(1-p)^{n+m}}{(1-p)^m}
$$

$$
= (1-p)^n
$$

$$
= P(X > n).
\tag{2.22}
$$

Using equation (2.21)

Practically, what this means is the following. *Suppose we are about to start flipping a fair coin for which the probability of observing a head is p. Then the probability distribution of $X :=$ "number of flips until the first head is observed" is $geom_1(p)$. Now suppose that the first m flips are tails. Then the probability distribution of the new random variable $Y :=$ "number of additional flips until the first head is observed" is still $geom_1(p)$. It is as if the knowledge of the outcomes of the first m flips has been lost from memory!* The geometric distribution is the only known discrete probability distribution with this property.

Figure 2.20 Bummer spent the whole day in class wondering whether or not he flushed the toilet before he left his apartment! Does our friend Bummer have a cognitive state that follows the geometrical distribution? (*Courtesy*: Michael Tran, *Daily Bruin*).

iv. ***Poisson distribution:*** The number of occurrences of an event (e.g., arrival of busses at a stand and calls to a telephone operator) in a fixed interval of time can be random. In order to derive an expression for the distribution of such a random variable, we make two fundamental assumptions about the random instantiations (henceforth referred to as "arrivals").

(a) *Homogeneity*: The arrival rate λ is constant with respect to time. The expected number of arrivals in a given interval of time Δt is $\lambda\Delta t$. This is also known as *weak stationarity,* as will be discussed in one of the latter chapters in this book.

(b) *Independence*: The number of arrivals in any two disjoint intervals of time is independent of each other.

Let N_t be the number of arrivals in an interval $\left[\tau, \tau+t\right]$ for any $\tau > 0$. We seek to know the distribution of N_t. Homogeneity implies that $E(N_t) = \lambda t$. Next, we proceed with constructing n canonical intervals of time t/n in such a way that as $n \to \infty$, M_j is a Bernoulli random variable representing the number of arrivals (0 or 1) in the interval $I_{j,n} := \left[(j-1)\dfrac{t}{n}, j\dfrac{t}{n}\right]$ for any $j \in \mathbb{I}$. By definition,

$$E\left(M_j\right) = 0\left(1 - p_j\right) + 1\left(p_j\right) = p_j = E\left(N_{\frac{t}{n}}\right) = \lambda\frac{t}{n}, \text{ where } p_j \text{ is the probability that } M_j$$

$= 1$ and $(1 - p_j)$ is the probability that $M_j = 0$ in the interval $I_{j,n}$.[10] Consequently,

$$N_t = \sum_{j=1}^{n} M_j \sim Bin(n, p),$$

where $p \equiv p_j = \lambda\dfrac{t}{n}$.[11] Therefore,

$$P\left(N_t = k\right) = \binom{n}{k}\left(\frac{\lambda t}{n}\right)^k \left(1 - \frac{\lambda t}{n}\right)^{n-k}, \tag{2.23}$$

for $k = 0, 1, 2, ..., n$. Here $\binom{n}{k}$ refers to ^nC_k. Explicitly, we have not yet considered the assumption $n \to \infty$ in the mathematical expressions above. We hope that after taking this limit, the distribution function will stabilize. We will consider the limit of each of these terms separately and collate the results thereafter.

$$\lim_{n\to\infty}\binom{n}{k}\frac{1}{n^k} = \lim_{n\to\infty}\frac{n}{n}\frac{n-1}{n}\cdots\frac{n-k+1}{n}\frac{1}{k!} = \frac{1}{k!}, \tag{2.24}$$

$$\lim_{n\to\infty}\left(1 - \frac{\lambda t}{n}\right)^n = e^{-\lambda t}, \tag{2.25}$$

Result from elementary calculus

and certainly,

$$\lim_{n\to\infty}\left(1 - \frac{\lambda t}{n}\right)^{-k} = 1. \tag{2.26}$$

Now combining the results from equations (2.24), (2.25), and (2.26), we have

[10] This is a direct consequence of the homogeneity axiom: $E\left(N_{\Delta t}\right) = \lambda\left(\Delta t\right)$.

[11] Here we have used the fact that the number of arrivals (successes) N_t in n Bernoulli trials is $Bin(n,p)$, where p is the probability of one arrival (success) in each canonical interval $\dfrac{t}{n}$.

Figure 2.21 The number of calls received per hour by a telephone operator follows a Poisson distribution.

$$\lim_{n \to \infty} P\left(N_t = k\right) = \frac{\left(\lambda t\right)^k}{k!} e^{-\lambda t}.$$

(2.27)

By carefully inspecting the term on the right-hand side (RHS) of equation (2.27), we notice that the following result holds.

$$e^{-\lambda t} \sum_{k=0}^{\infty} \frac{\left(\lambda t\right)^k}{k!} = e^{-\lambda t} e^{\lambda t} = 1.$$

(2.28)

Results (2.27) and (2.28) entail that we have indeed chanced upon a legitimate probability distribution (the *Poisson distribution*) that complies with the unitary axiom of probability over the sample space $\Omega = \{0, 1, 2, \cdots\}$. In the expression on the right-hand side of equation (2.27), we have only one parameter λt. This chain of thought motivates the definition of the Poisson distribution with parameter $\mu > 0$[12] to model the counting process of a random number of arrivals in fixed intervals of time.

$$X \sim Poisson\left(\mu\right)$$

$$P\left(X = k\right) = \frac{\mu^k}{k!} e^{-\mu} \quad \text{for } k = 0, 1, 2, \ldots$$

(2.29)

$$E(X) = \mu.$$

(2.30)

$$Var(X) = \mu.$$

(2.31)

The mean can be deduced from the fact that in the preceding paragraph $N_t \sim Bin\left(n, \frac{\lambda t}{n}\right)$ and hence $E\left(N_t\right) = n \frac{\lambda t}{n} = \lambda t, \forall n$. Further, $\lim_{n \to \infty} Var(N_t) = \lim_{n \to \infty} n \frac{\lambda t}{n} \left(1 - \frac{\lambda t}{n}\right) = \lambda t$. Therefore, for $X \sim Poisson\left(\mu\right)$, we expect that $E\left(X\right) = \mu$ and $Var\left(X\right) = \mu$. In fact,

$$
\begin{aligned}
E\left(X\right) &= \sum_{k=0}^{\infty} k \frac{\mu^k}{k!} e^{-\mu} \\
&= e^{-\mu} \sum_{k=1}^{\infty} \frac{\mu^k}{\left(k-1\right)!} \\
&= \mu e^{-\mu} \sum_{k=1}^{\infty} \frac{\mu^{k-1}}{\left(k-1\right)!} \\
&= \mu e^{-\mu} \sum_{j=0}^{\infty} \frac{\mu^j}{j!} \\
&= \mu e^{-\mu} e^{\mu} \\
&= \mu.
\end{aligned}
$$

(2.32)

The calculation for $Var(X)$ follows a similar approach and is left to the reader as a self-exercise.

---

[12] You may think of μ as λt.

Example 2.14   *Risk of loss incurred by Carepal from insurance payouts*

A watered-down version of the chapter project is considered in this example. A workers' insurance company named Carepal has introduced a new insurance policy for factory workers to cover certain types of injuries sustained at work. Since a worker may be inflicted by a diverse type of injuries in the factory, and that the insurance policy may not be applicable for all types of injuries as per the coverage plan, only a certain number out of the total claims get re-reimbursed by Carepal. Further, the insurance scheme allows only a standard payment of ₹1,00,000 for all claims approved by Carepal. The annual premium for the policy is ₹15 for each insured person. Based on claims data available with the company, it was found that on an average, a total of about 100 claims per year get approved for similar schemes. There are about 1,000,000 policy holders of this scheme. What is the risk (calculated in terms of a probability) that this factory workers' scheme will yield an annual loss for Carepal?*

The insurance scheme is designed in such a manner that the probability of successful approval of any given claim is very small. Of course, not all claims made by the workers will get approved by Carepal. Thus, from the perspective of a policy period, there are several probability events being played out at once where only those claims that will be eventually approved by Carepal may be regarded as a *successful instantiation* of a claim (event). Let X denote the total number of claims that will be approved by Carepal during one policy period of a single year. Consequently, X may be regarded as a Poisson random variable with rate $\mu = 100$ such that $E(X) = 100$ and $Var(X) = 100$. Carepal will incur an annual loss if the aggregate of all claims payouts turns out to be greater than the total revenue generated by selling this insurance scheme to 1,000,000 customers. Let us use Matlab to compute the critical number of approved claims that will determine whether Carepal incurs loss from this insurance scheme.

```
total_customers = 1000000;
premium = 15;
std_pay_per_claim = 100000;
total_payout_for_loss = 15000000;
mu = 100;
income = premium*total_customers;
Number_payouts = total_payout_for_loss/std_pay_per_claim;
```

Here the value of `Number_payouts` turns out to be 150, which is the critical number of approved claims above which the company will incur a loss. Our next objective is to estimate the probability that more than 150 claims will be approved by Carepal in the given policy year.**

$$P(X > \text{Number\_payouts}) = 1 - \sum_{k=0}^{\text{Number\_payouts}} e^{-\mu} \frac{\mu^k}{k!}$$

Figure 2.22 The saga of your insurance claims may have been scripted by the French Mathematician Siméon Denis Poisson (*courtesy*: Wellcome Collection).

is calculated as follows:

```
k = [0:1:Number_payouts];
prob_mass_of_payouts = exp(-mu)*(mu.^k)./(factorial(k));
Risk_of_loss = 1 - sum(prob_mass_of_payouts)
```

It turns out that `Risk_of_loss = 1.2331e-06`, i.e., Carepal's probability of incurring a loss from this scheme is incredibly minuscule and the said insurance scheme is risk free by and large.

*Here we are interested in knowing the risk of incurring an annual loss stemming from this particular insurance scheme alone.

**Note that Carepal cannot simply enforce a hard stop on its claims approval process in order to arrest a likely annual loss in its business because such an intervention would severely dent its reputation and credibility as a trusted insurance company among factory workers. Further, it cannot raise its premium arbitrarily because such a measure would almost certainly play into the hands of its competitors.

Poisson heuristic: Recall that the construction of the Poisson distribution, as elucidated above, demands that the probability of an event happening (arrival) is small. Specifically, it is proportional to $\dfrac{\lambda t}{n}$, where $n \to \infty$, which makes $\dfrac{\lambda t}{n} \to 0$. Further, recall that the number of arrivals in any two disjoint time intervals was assumed to be independent. It turns out that even if we relax the criterion of independence and consider *weakly dependent Bernoulli trials*, the Poisson distribution (more specifically the *Poisson heuristic*) may still provide a reasonable estimate. We will demonstrate this point with the help of an example below.

Figure 2.23 If *love is blind* then there is perhaps a case for *blind dates* and some Poisson heuristics.

Example 2.15   *Your luck with the blind dating app RATATOON*

RATATOON is a new mobile application for blind dates. It is hosted on several cloud-based servers. Every month, up to 20 randomly picked clients (10 men and 10 women) from the RATATOON provincial database can register on one of their cloud servers. The intelligence integrated within the software application ensures that hugely different profiles (for instance, men and women more than 20 years apart in age) are not usually hosted on the same server in order to avoid absurd match-ups. Each server allows its customers to post a preferred date every month when they may be up for a blind date. The app locks in a potential blind date if the preferred dates of any two clients (one man and one woman) are within one day of each other, and it is left up to the pair if they are willing to make this minor adjustment in their chosen dates. After a date is locked in, the individual pairs may decide whether to proceed with the date or not, but chances are that they will give it a shot. What is the probability that two or more clients are matched up for a date by a RATATOON server every month? Further, what is the probability that at least 5% of the possible pair-ups (combinations) are matched up for blind dates by RATATOON?

Had RATATOON paired up potential dates only if the preferred dates of each man and woman in the pair matched identically, then there is a precise and direct way of computing the required probability. However, by allowing up to one day difference in their preferred dates, the probability in question becomes very difficult to calculate exactly. In this case, the Poisson heuristic provides a reasonably good approximation of the asked probability.

We begin by identifying the total number of all possible pairs of men and women from a pool of 10 men and 10 women. A combination is represented as (M_j, F_j), where M stands for a man and F stands for a woman, and $j = 1, 2, ..., 10$ tags each of the 10 men and women registered in the same server. This computation may be undertaken by populating each of the two slots that consitute a pair with the names of men and women from the chosen pool. The first slot may be filled up in 10 different ways (corresponding to the 10 different men) and the second slot may be filled up in 10 different ways for similar reasons. Thus, there are a total of $n = 10 \times 10 = 100$ trial combinations, out of which some combinations may lead to a match-up if the preferred dates of the individuals are within a day of each other (the latter are the successful trials). It is not too difficult to find that each of the n trials has the same success probability $p = \dfrac{3}{30}$.*

Each day of a month may be mapped to an empty slot of a 30-slot array. For the ith combination (M_i, F_i), let us suppose that F_i picks the slot k. Now, if M_i picks any of the 30 slots but $(k-1)$, k, or $(k+1)$, then it will be a *failed* match-up. This failure can happen with a probability $p_c = \dfrac{27}{30}$ and this estimate will be true of any failed match-up. Thus, the probability of success for any combination (M, F) is $p = 1 - p_c = 1 - \dfrac{27}{30} = \dfrac{3}{30} = 0.1$.** Let X be a random variable that denotes the number of successful trials (match-ups). Then the probability that two or more clients are matched-up for a date by RATATOON is the same as $P(X \geq 1)$.

A careful reflection of the situation reveals that the aforementioned Bernoulli trials are *weakly dependent* (as opposed to being independent). This is because it is extremely unlikely that preferred dates coming from a diverse population will be clustered together. Consider that F_1 prefers the 14th of the month and M_1 picks the 13th. (M_1, F_1) turns out to be a match-up because each of their preferences belongs to the set $([12, 14] \cap [13, 15])$. Additionally, consider (M_1, F_i) for $i = 3, 5$ are also match-ups due to similar date preferences (e.g., $[12, 14]$). But with every such match-up for M_1, it becomes less probable that (M_1, F_j), where $j \neq 1, 3, 5$, will also be a match-up because it is unlikely that all women from the pool of 10 would have preferred dates between the 12th and 14th of the month (unless of course we are talking about special months like February when many single women [and men] may prefer dates around a special day, on the 14th of February, which is Valentine's day). Thus, in some sense, the outcome of two different trial combinations may bear some element of dependency. This dependence of trial outcomes may be considered *weak* because of a relatively large number (30) of available preferred dates.***

Therefore, $X \approx Poisson(\mu)$, where $\mu = np = 100 \times 0.1 = 10$. Consequently,

$$P(X \geq 1) \approx 1 - P(X = 0) = 1 - e^{-\mu} = 0.9999546. \tag{2.33}$$

Further,

$$P(X \geq 5) \approx 1 - \sum_{k=0}^{k=4} P(X = k) = 0.97074731. \tag{2.34}$$

*For simplicity, we have considered only 30 day months.

**For preferred dates on the start and the end of the month, we have considered that the possible match-up dates can be 30th, 1st, 2nd and 29th, 30th, 1st.

***The estimate would be better if the available number of preferred dates was larger, for example, instead of considering match-ups every month, one could consider match-ups every three months.

In order to establish the veracity of the calculations and results of the above example, we will design a simple computer experiment whereby 10 male and 10 female bots will randomly pick their preferred dates for a blind date on a given month. Consequently, match-ups will be generated by the computer by considering those pairs that have preferred to go on a blind date within a day of each other. This experiment is repeated 100,000 times in order to estimate the probability that at least a certain percentage (q) of pair-ups are matched-up for a blind date. This experimental result is then compared with the theoretical estimations prescribed by the above example.

```
%%%%%%%%%%%%%%%%%%%%%%%%%%%%%%%%% START of CODE %%%%%%%%%%%%%%%%%%%%%%%%%%%%%%%%%%%%%%%%
%%%%%%%%%%%%%%% start of parameters %%%%%%%%%%%%%%%
nm = 10; % num of males
nf = 10; % num of females
n=nm*nf; % total number of possible pair-ups %(combinations)
total_period = 30; % duration of monthly cycle of blind match-ups
match_interval = 3; % this ensures that a match-up happens when the preferred
% dates of a male and female pair are within a day of each other
match_percent = 5; % minimum target match-ups in percentage
req_matchups = ceil(0.01*match_percent*n);% 0.0x*n means atleast x% must be
% matchups
kmax = req_matchups - 1; % the running index for the Poisson heuristic
nmax = 100000; % number of times the computer experiment is conducted
%%%%%%%%%%%%%% end of parameters %%%%%%%%%%%%%%%%%
match_cnt = 0;
for i=1:nmax
   % RATATOON
   males = ceil(total_period*rand(1,nm)); % list of date preferences by
                              % by males
   females = ceil(total_period*rand(1,nf));% list of date preferences by
                              % by females
   [men,women] = meshgrid(males, females);
   pairs = [men(:) women(:)]; % all pair-ups are stored in matrix form
   diff_pairs = mod(abs(pairs(:,1) - pairs(:,2)),total_period); % difference
                                    % between dates preferred by
                                    % the male and the female in a
                                    % pair
   Ans = [diff_pairs pairs]; % concatenating diff_pairs and pairs in one matrix
   matchups = find(Ans(:,1)<=1); % matchups happen if diff_pairs is within
                       % one unit
```

```
    num_matchups = length(matchups);
    prob_sim_matchups = num_matchups/(n); % this is not really necessary but
                                          % just for fun to compare with
                                          % p
    if num_matchups >=  req_matchups % this condition  calculates the required
        match_cnt = match_cnt + 1;   % probability over multiple repetition
    else                             % of the experiment
        continue;
    end
end
prob_sim_final = match_cnt/nmax;     % experimental estimate of the required
                                     % probability


% poisson heuristic approximation
%%%%%%%%%%%%%%%%%%%%%%%%%%%%%%%%%
p=(1 - ((total_period - match_interval)/total_period)); % p = 1 - pc
mu = n*p;
k=[0:kmax];
prob = 1-sum(exp(-mu)*(mu.^k)./(factorial(k)));% calculating the required
                                               % probability by using the
                                               % Poisson heuristic
prob_final = vpa(prob,8);   % expressing the answer up to 8 decimal places
%%%%%%%%%%%%%%%%%%%%%%% END of CODE %%%%%%%%%%%%%%%%%%%%%%%%
```

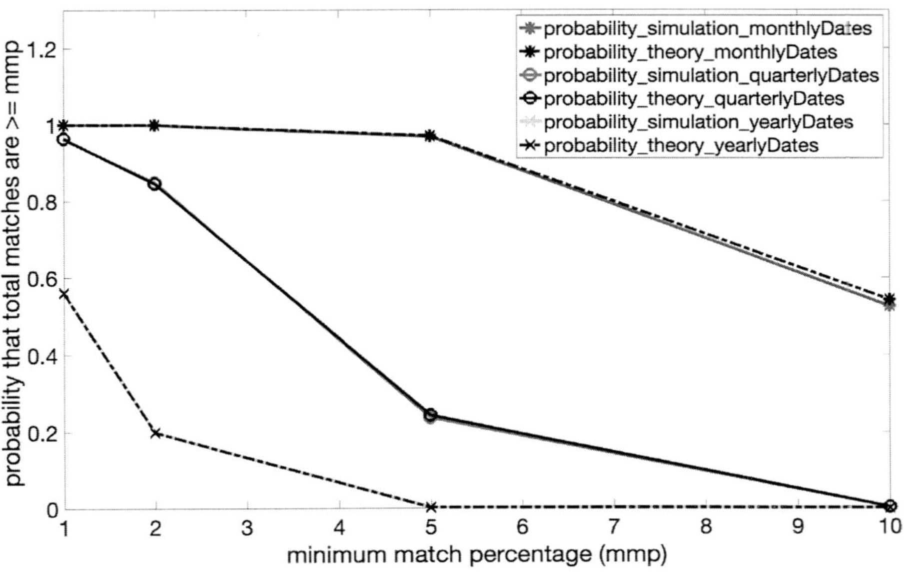

Figure 2.24 Comparison between simulated experiment of dating match-ups and theoretical predictions of the same clearly shows that the Poisson heuristic is a very good approximation when n is large, p is small, and $m = n_p$ is moderate in magnitude. See color plate on page 295.

The results of the above computer experiment are elucidated graphically in Figure 2.24. First, the theoretical predictions of the Poisson heuristic are strikingly similar to the actual simulated probability values. This establishes the veracity of the Poisson heuristic when n is large, p is small, and $\mu = np$ is of moderate magnitude. Second, both the simulated and the Poisson models demonstrate that the probability, $P(X \geq$ "minimum match percentage"), decreases with increasing values of the minimum match percentage (mmp). Third, both the simulated and the Poisson models show that the corresponding probabilities fall significantly with increasing duration of the experiments for the same mmp. This entails that in order to keep clients hooked to the RATATOON application, it may be economically prudent (from a profitable business perspective) to keep the duration (period) of the subscriptions shorter (monthly as opposed to yearly). In the experiments and analyses considered here, we have chosen a fairly large sample size of 100 clients per server to ensure statistical validity of reported findings.

There are several interesting examples illustrating the utility of the Poisson heuristic. Some of the more familiar ones in the literature are those of the *birthday problem* and the *matching problem*. Interested readers are referred to the texts mentioned in the chapter bibliography. These problems have many important applications in cryptography and information security. We will return to the Poisson distribution once again during our discussion of the chapter project. Poisson processes will also be discussed in chapter sections pertaining to Markov chains and queuing models.

v. **Uniform distribution (discrete):** Let $X \sim Unif\left(\left[1,m\right]\right)$ be a random variable on m successive integers starting with 1. Each outcome has an associated identical probability (uniform) prescribed as follows:

$$X \sim Unif\left(\left[1,m\right]\right)$$

$$p_i = P\left(X=i\right) = \frac{1}{m}, \ \forall i \in \left[1,m\right]$$

$$E\left(X\right) = \sum_{x=1}^{m} x\frac{1}{m} = \frac{\left(m+1\right)}{2}$$

$$Var\left(X\right) = E\left(X^2\right) - \left(E\left(X\right)\right)^2 = \frac{m^2-1}{12}$$

A simple example is the outcome of rolling a six-sided die. Each of the $m = 6$ sides has a probability of occurrence equal to $\frac{1}{m} = \frac{1}{6}$. We will again return to the uniform distribution when we discuss continuous random variables.

vi. **Negative-binomial and Pascal distributions:** In many practical situations we may be interested in knowing the chance of the rth success of Bernoulli trials in $v = r + k$ trials, where $k = 0, 1, 2, \ldots.$ This situation is equivalent to the occurrence of exactly k failures prior to r successes. So, if X is the random variable that denotes $v = k + r$ trials for the rth success to occur, we may also rephrase this as counting $v - 1 = r + k - 1$ trials with exactly k failures and a

Figure 2.25 Rolling a die yields an outcome that has equal probability of happening as that of any other possible outcomes.

success in the very next trial. The number of possible ways this may happen is $\binom{r+k-1}{k} = \binom{v-1}{k}$. This beckons the definition of the p.m.f. as follows:

$$X \sim Pa(k; r, p) \quad \text{or} \quad X \sim NB(k; r, p).$$

$$P(X=k) \;=\; \binom{r+k-1}{k}(1-p)^k\, p^r \equiv \binom{-r}{k}\big(-(1-p)\big)^k\, p^r \;\text{ for } k=0,\,1,\,2,\,3,\,\dots$$

Here $\binom{-r}{k}$ is defined as $(-r)(-r-1)\dots(-r-y+1)/y!$. Indeed, it may be verified that $\sum_{k=0}^{\infty} P(X=k) = 1$ is true as demanded by the unitarity axiom of probability. This entails that an infinite sequence of Bernoulli trials is bound to yield r successes. Thus, we may infer that the Pascal or negative binomial distribution is a model for *the waiting time to the rth success.*

Example 2.16   Did Blaise Pascal play badminton?

A game of badminton involves two players Li Pen and Brendon Hart with skill levels $p = 0.6$ and $h = 0.4$, respectively. Here skill levels may be interpreted as probabilities of winning a rally by the respective players. In order to win a game of badminton, 21 individual victories are required by either player. Further, consider that a game can last at most 41 rallies (i.e., a winning scoreline of 21–20 is permissible by the rules of the game). Answer the following questions:

Figure 2.26 A game of badminton between Li Pen and Brendon Hart.

i.   What is the probability that Li Pen will win the game of badminton against Brendon Hart in 26 rallies?

ii. What is the probability that Li Pen will win the game of badminton against Brendon Hart?

iii. What is the probability that a game of badminton between Pen and Hart will end in 26 rallies?

We proceed with the calculations inspired by our formulation of the Pascal distribution above.

i.   A game lasts at least $2v + 1 = 21$ rallies and at most $4v + 1 = 41$ rallies. Thus, $v = 10$. So, if Pen wins at rally number $4v + 1 - r = 26$, we have $r = 15$. Let us define $P_r = P_{15}$ as the probability that Pen wins the game $4v + 1 - r = 26$ rallies. Then, Pen must have won $2v = 20$ out of $4v - r = 25$ rallies and lost 5 out of 25 rallies. Thus,

$$P(\text{Pen wins in 26 rallies}) = P_{15} = \binom{25}{20}(0.6)^{21}(0.4)^5 = 0.0119\,.$$

ii. $P(\text{Pen wins}) = P(\text{Pen wins in 1 rally}) + P(\text{Pen wins in 2 rallies}) + \dots + P(\text{Pen wins in 41 rallies}) = P_{20} + P_{19} + \dots + P_1 + P_0 = 0.9035.$

iii. $P(\text{game ends in 26 rallies}) = P_{15} + H_{15} = 0.0120$, where

$$H_{15} = \binom{25}{20}(0.4)^{21}(0.6)^5\,.$$

Continuous probability models

i. **Exponential distribution:** X is a positive continuous random variable with *rate* parameter $\mu > 0$. X is exponentially distributed with the p.d.f. prescribed below.

$$X \sim \exp(\mu).$$

$$f_X(x) = \begin{cases} \mu e^{-\mu x} & \text{for } x > 0, \\ 0, & \text{otherwise.} \end{cases}$$

$$E(X) = \frac{1}{\mu}.$$

$$Var(X) = \frac{1}{\mu^2}.$$

If we consider the case of Poisson-distributed arrivals in a fixed interval of time, then the inter-arrival times are exponentially distributed random variables. The c.d.f. is $F_X(x) = P(X \leq x) = \int_{-\infty}^{x} f(\zeta)d\zeta$, where f is the p.d.f. mentioned above. It follows that

$$F_X(x) = \begin{cases} 1 - e^{-\mu x} & \text{for } x \geq 0, \\ 0 & \text{for } x < 0. \end{cases} \tag{2.35}$$

Example 2.17   Memorylessness of the exponentially distributed random variable

A continuous random variable X has a memoryless property if

$$P(X > t + s | X > s) = P(X > t) \quad \text{for all} \quad t > 0,$$

regardless of the value of $s > 0$. This is easy to show because if we consider the definition of the conditional probability, then

$$P(X > t + s | X > s) = \frac{P(X > t + s)}{P(X > s)} = \frac{e^{-\mu(t+s)}}{e^{-\mu s}} = e^{-\mu t} = P(X > t).$$

The exponential distribution is the only known continuous distribution with the memoryless property.

Perhaps in light of the above example and the one discussed earlier using the geometric distribution about the memoryless property, it is not difficult to intuit that there must be a connection between the exponential distribution and the geometric distribution. This is further unravelled by the following discussion. Exponentially distributed random variables are used to model time until the occurrence of a rare event.

Example 2.18 Equivalence of geometric and exponential distributions

Recall from equation (2.21) that for $X \sim geom_1(p)$, we have $P(X > k) = (1 - p)^k$. This means $F_X(k) = 1 - (1 - p)^k$. Further, let $T \sim exp(\mu)$, where $F_T(t) = 1 - e^{-\mu t}$. If we suppose $\mu = -\log(1-p)$ and $t = \lfloor n\tau \rfloor^*$ for all $\tau > 0$ and $n = 0, 1, 2, 3, \ldots$, then clearly $F_X\left(\lfloor n\tau \rfloor\right) = 1 - e^{(\log(1-p))\lfloor n\tau \rfloor} = 1 - (1-p)^{\lfloor n\tau \rfloor} = F_T\left(\lfloor n\tau \rfloor\right)$, i.e., the probability distributions of X (geometrically distributed random variable) and T (exponentially distributed random variable) are the same.

The connection may also be established by considering the case when the exponentially distributed random variable $T \in (n-1, n]$ for $n = 1, 2, 3, \ldots$ and $p = 1 - e^{-\mu}$ (equivalently, $\mu = -\log(1-p)$).

$$
\begin{aligned}
P(n-1 < T \leq n) &= F_T(n) - F_T(n-1) \\
&= \left(1 - e^{-\mu n}\right) - \left(1 - e^{-\mu(n-1)}\right) \\
&= p(1-p)^{n-1}. \qquad\qquad (2.36)
\end{aligned}
$$

The RHS of equation (2.36) is identical to the p.m.f. of a random variable that follows the geometrical distribution of type-1 (i.e., $P(X=n)$).

---

* Here $\lfloor\ \rfloor$ is the greatest integer function akin to the *floor operation*. For instance, $\lfloor 2.63 \rfloor = 2$.

There is a more rigorous relationship between the geometric distribution and the exponential distribution that we will simply state here without providing a proof. *Let $Z_n = \dfrac{Y_n}{n}$, where Y_n is a geometric random variable with parameter $p_n = \dfrac{\lambda}{n}$ and $n > \lambda > 0$. Then Z_n converges in distribution[13] to an exponential random variable with parameter λ.*

[13] Let $X_1, X_2, \ldots$ be a sequence of real-valued random variables that converges in distribution to X, $X_n \xrightarrow{\mathcal{D}} X$, if $\lim_{n\to\infty} F_n(x) = F(x)$ for all $x \in \mathbb{R}$ at which $F(x)$ is continuous. $F_i(x)$ is the c.d.f. of X_i, $\forall i = 1, 2, \ldots$, and $F(x)$ is the c.d.f. of X.

Example 2.19 Chance of total power failure in a single-engine jet aircraft

Most single-engine jet aircraft have an auxiliary power unit (APU) as backup power supply in case of an engine failure. Typically, the APU is activated when the main engine fails and while the pilot initiates an *engine restart* procedure. If the APU also fails in flight in addition to the main engine, then we have a total system (power) failure. The lifetime of an operating power unit is exponentially distributed with expected value $\dfrac{1}{\mu}$. Let X denote the time until the first total system failure. It can be shown that $E(X) = \dfrac{2 - e^{-\mu\tau}}{\mu\left(1 - e^{-\mu\tau}\right)} = 500$ flying hours, where $\tau > 0$ is the fixed time to restart the main engine. What is the probability that the aircraft will encounter a total system failure after 100 flying hours?

$$
P(X > 100) \approx e^{-\frac{100}{E(X)}} = 0.8187,
$$
i.e., there is an 82% chance of a total system failure after 100 flying hours.

Figure 2.27 This single engine jet has an APU as a backup during engine failure. How likely will this built-in redundancy prove to be helpful after 100 flying hours?

Below, we have provided a summary of a few other continuous probability distributions.

ii. **Gamma distribution:** Here the parameter $\alpha > 0$ is a shape parameter and $\beta > 0$ is a rate parameter. $\Gamma(\alpha)$ is the well-known gamma function.

$$X \sim Gamma(\alpha, \beta).$$

$$f_X(x) = \begin{cases} \dfrac{\beta^\alpha}{\Gamma(\alpha)} x^{\alpha-1} e^{-\beta x} & \text{for } x > 0, \\ 0, & \text{otherwise.} \end{cases}$$

$$E(X) = \frac{\alpha}{\beta}.$$

$$Var(X) = \frac{\alpha}{\beta^2}.$$

The Gamma distribution is used to model aggregate insurance claims and the amount of rainfall accumulated in a reservoir. It is used for modeling attenuation of signal strength in wireless communication. It finds application in oncology for modeling age distribution of cancer incidences. It is also used in Bayesian statistical models.

Example 2.20   Statistical moments of the gamma function

The p.d.f. of a continuous random variable X with parameter $\lambda > 0$ is given by

$$f(x) = \begin{cases} kxe^{-\lambda x} & x \geq 0, \\ 0 & \text{otherwise.} \end{cases}$$

Find out the value of k, the mean, and the variance.

Hint: Properties of gamma function $\Gamma(p) = \int_0^\infty e^{-x} x^{p-1} \, dx$.

- $\dfrac{\Gamma(p)}{\alpha^p} = \int_0^\infty e^{-\alpha x} x^{p-1} \, dx$ for $\alpha, p > 0$.

- $\Gamma(n) = (n-1)! \quad \forall \, n \in \mathbb{N}$.

Solution: Recall that the p.d.f. of a random variable integrates to unity.

$$\int_0^\infty kxe^{-\lambda x} = 1$$

$$k \times \frac{\Gamma(2)}{\lambda^2} = 1$$

$$k = \frac{\lambda^2}{\Gamma(2)}$$

$$k = \lambda^2.$$

$$f(x) = \lambda^2 x e^{-\lambda x}, \qquad x \geq 0, \lambda > 0.$$

$$E(X) = \int_0^\infty xf(x)\,dx = \lambda^2 \int_0^\infty x^2 e^{-\lambda x}\,dx = \lambda^2 \frac{\Gamma(3)}{\lambda^3} = \frac{2}{\lambda}$$

$$E(X^2) = \int_0^\infty x^2 f(x)\,dx = \lambda^2 \int_0^\infty x^3 e^{-\lambda x}\,dx = \lambda^2 \frac{\Gamma(4)}{\lambda^4} = \frac{6}{\lambda^2}$$

$$Var(x) = E(X^2) - \left(E(X)\right)^2 = \frac{6}{\lambda^2} - \left(\frac{2}{\lambda}\right)^2 = \frac{2}{\lambda^2}.$$

iii. **Beta distribution:** Here both the parameters $\alpha > 0$ and $\beta > 0$ are shape parameters. $B(\alpha,\beta) = \dfrac{\Gamma(\alpha)\Gamma(\beta)}{\Gamma(\alpha+\beta)}$ is the beta function that is defined here in terms of the gamma function.

$$X \sim Beta(\alpha,\beta).$$

$$f_X(x) = \begin{cases} \dfrac{x^{\alpha-1}(1-x)^{\beta-1}}{B(\alpha,\beta)}, & \text{for } x \in [0,1], \\[2mm] 0, & \text{otherwise.} \end{cases}$$

$$E(X) = \frac{\alpha}{\alpha+\beta}.$$

$$Var(X) = \frac{\alpha\beta}{(\alpha+\beta)^2(\alpha+\beta+1)}.$$

The Beta distribution is used in the theory of order statistics, in subjective logic in the form of posteriori probability estimates of binary events, and in wavelet analysis. Beta distribution is also used in project management models.

iv. **Pareto distribution:** Here the parameters $x_m > 0$ and $\alpha > 0$ are scale and shape parameters, respectively. $\Gamma(\alpha)$ is the well-known gamma function.

$$X \sim Pareto(x_m,\alpha).$$

$$f_X(x) = \begin{cases} \dfrac{\alpha x_m^\alpha}{x^{\alpha+1}}, & \text{for } x \in [x_m,\infty), \\[2mm] 0, & \text{otherwise.} \end{cases}$$

$$E(X) = \begin{cases} \infty, & \text{for } \alpha \le 1, \\[2mm] \dfrac{\alpha x_m}{\alpha-1}, & \text{for } \alpha > 1. \end{cases}$$

$$Var(X) = \begin{cases} \infty, & \text{for } \alpha \le 2, \\[2mm] \dfrac{\alpha x_m^2}{(\alpha-1)^2(\alpha-2)}, & \text{for } \alpha > 2. \end{cases}$$

The Pareto distribution was originally used to model the distribution and allocation of wealth among individuals in a society where the greatest fortune is owned by a small fraction of the population.

v.   ***Uniform distribution (continuous):*** Here the parameters $-\infty < a < b < \infty$ define the extent of the uniform distribution.

$$X \sim Unif(a, b).$$

$$f_X(x) = \begin{cases} \dfrac{1}{b-a}, & \text{for } x \in [a, b], \\ 0, & \text{otherwise.} \end{cases}$$

$$E(X) = \frac{a+b}{2}.$$

$$Var(X) = \frac{(b-a)^2}{12}.$$

vi.   ***Normal distribution (a.k.a. Gaussian distribution):*** We will take a more detailed look at this familiar bell-shaped normal distribution in Chapter 5, "Statistical experiments." Here we shall summarize some very essential features. In what follows, μ can be negative or positive but finite, $\sigma^2 > 0$, by definition.

$$X \sim N\left(\mu, \sigma^2\right).$$

$$f_X(x) = \frac{1}{\sqrt{2\pi}\sigma} e^{-\frac{(x-\mu)^2}{\sigma^2}} \text{ for } x \in \mathbb{R}.$$

$$E(X) = \mu.$$

$$Var(X) = \sigma^2.$$

Normal distribution is one of the most widely used probability models in statistics partly because of its relevance in connection to the central limit theorem which we will discuss later.

Example 2.21   *Statistical moments of a continuous random variable*

Let X be a continuous random variable with p.d.f.

$$f(x) = cx(2-x), \ \ 0 \le x \le 2, \text{ where } c \text{ is a constant.}$$

i.   Find $E(X^r)$, where $r \in \mathbb{N}$.

ii.   Find mean and variance of X using the result of part (i).

Solution:

i. $c\int_0^2 \left(2x - x^2\right)\, dx = c\left(4 - \dfrac{8}{3}\right) = \dfrac{4c}{3} = 1 \Rightarrow c = \dfrac{3}{4}.$

$\quad f(x) = \dfrac{3}{4}x(2-x).$

$\quad E\left(X^r\right) = \int_0^2 x^r f(x)\, dx$

$\qquad\qquad = \dfrac{3}{4}\int_0^2 \left(2x^{r+1} - x^{r+2}\right)\, dx$

$\qquad\qquad = \dfrac{3}{4}\left(2\cdot\dfrac{x^{r+2}}{r+2} - \dfrac{x^{r+3}}{r+3}\right)_0^2$

$\qquad\qquad = \dfrac{3}{4}2^{r+3}\left(\dfrac{1}{r+2} - \dfrac{1}{r+3}\right)$

$\qquad\qquad = 3\cdot 2^{r+1}\left(\dfrac{1}{r+2} - \dfrac{1}{r+3}\right)$

$\quad E\left(X^r\right) = \dfrac{3\cdot 2^{r+1}}{(r+2)(r+3)}.$

ii. $\quad \text{Mean} = E(X)$

$\qquad\qquad = \dfrac{3\cdot 2^{1+1}}{(1+2)(1+3)}$

$\qquad\qquad = 1.$

$\quad E\left(X^2\right) = \dfrac{3\cdot 2^{2+1}}{(2+2)(2+3)}$

$\qquad\qquad = \dfrac{6}{5}.$

$\quad Var(X) = E\left(X^2\right) - (E(X))^2$

$\qquad\qquad = \dfrac{6}{5} - 1^2$

$\qquad\qquad = \dfrac{1}{5}.$

Example 2.22   Computing probability of a chance event from the p.d.f.

Let X be a continuous random variable with p.d.f.

$$f(x) = \begin{cases} a\left(1 + x^2\right), & 2 \le x \le 5, \\ 0, & \text{otherwise,} \end{cases} \quad \text{where } a \text{ is a constant.}$$

i. Find the value of a.

ii. Find $P\{X < 4\}$.

Solution:

i.
$$\Rightarrow \int_{2}^{5} a\left(1+x^{2}\right) dx = 1$$

$$\Rightarrow a\left((5-2)+\frac{1}{3}(125-8)\right) = 1$$

$$\Rightarrow a\left(\frac{9+117}{3}\right) = 1$$

$$\Rightarrow a = \frac{3}{126} = \frac{1}{42}.$$

ii.
$$P\{X < 4\} = \int_{-\infty}^{4} f(x)\ dx$$

$$= \frac{1}{42}\int_{2}^{4}\left(1+x^{2}\right)\ dx$$

$$= \frac{62}{126} = \frac{31}{63}.$$

2.4.6 *Definition: Compound probability distribution*

Consider the sum $Y = X_1 + X_2 + ... + X_N$, where (i) N is a random number, (ii) X_is, $i = 1, 2, 3, ..., N$ are independent and identically distributed random variables with c.d.f. F_X,[14] and (iii) each X_i is independent of N.[15] By the law of total probability, the *compound* distribution of Y is prescribed as follows:

$$
\begin{aligned}
f_Y(y) &= P(Y = y) \\
&= \sum_{n=0}^{\infty} P\left(X_1 + X_2 + \cdots + X_N = y \mid N = n\right) P(N = n) \\
&= \sum_{n=0}^{\infty} f_Y^{(n)} P(N = n),
\end{aligned}
$$

(2.37)

where $f_Y^{(n)}$ is the n-fold convolution of f_Y.[16] Next, we will compute the first two moments of a random variable with compound distribution.

$$E(Y) = E_N\left(E_Y(Y \mid N)\right)$$

Law of iterated expectation, equation (1.15)

$$= \sum_{n=0}^{\infty} E(Y \mid N = n) P(N = n)$$

$$= \sum_{n=0}^{\infty} n E(X) P(N = n)$$

[14] X is a random variable with mean μ_X and variance σ_X^2.

[15] N is a random variable with mean μ_N and variance σ_N^2.

[16] We have used the notation $f_X(x)$ and $p(x)$, and $f_X^{(n)}(x)$ and $p^{(n)}(x)$ interchangeably in the case of *discrete* random variables. For the continuous case, we have used $f_X(x)$ and $f_X^{(n)}(x)$ exclusively.

$$= \mu_X \sum_{n=0}^{\infty} n P(N = n)$$

$$= \mu_X \mu_N. \tag{2.38}$$

$$Var(Y) = E_N\left(Var(Y|N)\right) + Var_N\left(E(Y|N)\right)$$

Law of total variance, equation (1.19)

$$= E_N\left(N Var(X)\right) + Var_N\left(N E(X)\right)$$

$$= Var(X) E(N) + \left(E(X)\right)^2 Var(N)$$

$$= \mu_N \sigma_X^2 + \mu_X^2 \sigma_N^2. \tag{2.39}$$

The example presented in this section is very similar to the chapter project and should serve as a building block to solve it.

Example 2.23   Aggregate claims of an insurance policy

Let the number of claims, N, generated by a portfolio of insurance policies over a fixed duration have Poisson distribution with rate parameter $\lambda = 3$ claims per policy period. Individual claim amounts X_i (for all values of $i = 1, 2\ldots, N$) can be 1 or 2 million euros with probabilities $q = 0.6$ and $p = 0.4$, respectively. The aggregate claim $Y = X_1 + X_2 + \ldots + X_N$ is a compound Poisson distributed random variable. Find $P(Y = k)$ for $k = 0, 1, 2, 3, 4$. Also find the expected aggregate claim $E(Y)$.

We begin by noting that Y may take a maximum value equal to 4 as the insurance portfolio is capped at 4 million euros. This means that the number of claims may be $N = 0, 1, 2, 3, 4$. X_i is a Bernoulli random variable with probability of success (claim $= 2$ million euros), $p = 0.4$, and probability of failure (claim $= 1$ million euros), $q = 0.6$. Then, for a certain realization $N = n$ ($n = 1, 2, 3, 4$), the sum $X_1 + \ldots + X_n$ is a binomial distributed random variable $\sum_{i=1}^{n} X_i \sim bin(n, p)$. The n-fold p.m.f.s are listed below.

$$p^{(1)}(1) = 0.6$$

$$p^{(1)}(2) = 0.4$$

$$p^{(2)}(2) \;=\; \binom{2}{0}(0.4)^0(0.6)^2 = 0.36$$

$$p^{(2)}(3) \;=\; \binom{2}{1}(0.4)^1(0.6)^1 = 0.48$$

$$p^{(2)}(4) \;=\; \binom{2}{2}(0.4)^2(0.6)^0 = 0.16$$

$$p^{(3)}(3) \;=\; \binom{3}{0}(0.4)^0(0.6)^3 = 0.216$$

$$p^{(3)}(4) \;=\; \binom{3}{1}(0.4)^1(0.6)^2 = 0.432$$

$$p^{(3)}(5) \;=\; \binom{3}{2}(0.4)^2(0.6)^1 = 0.288$$

$$p^{(3)}(6) \;=\; \binom{3}{3}(0.4)^3(0.6)^0 = 0.064$$

$$p^{(4)}(4) \;=\; \binom{4}{0}(0.4)^0(0.6)^4 = 0.1296$$

$$p^{(4)}(5) \;=\; \binom{4}{1}(0.4)^1(0.6)^3 = 0.3456$$

$$p^{(4)}(6) \;=\; \binom{4}{2}(0.4)^2(0.6)^2 = 0.3456$$

$$p^{(4)}(7) \;=\; \binom{4}{3}(0.4)^3(0.6)^1 = 0.1536$$

$$p^{(4)}(8) \;=\; \binom{4}{4}(0.4)^4(0.6)^0 = 0.0256$$

In order to find the p.m.f., we simply scale the n-fold p.m.f. by the Poisson distributed probability weights with $\lambda = 3$.

i. $$P(Y=0) \;=\; e^{-\lambda} = 0.0498$$
$$P(Y=1) \;=\; 0.6\lambda e^{-\lambda} = 0.0896$$
$$P(Y=2) \;=\; 0.4\lambda e^{-\lambda} + 0.36\frac{\lambda^2 e^{-\lambda}}{2} = 0.1404$$
$$P(Y=3) \;=\; 0.48\frac{\lambda^2 e^{-\lambda}}{2} + 0.216\frac{\lambda^3 e^{-\lambda}}{6} = 0.1559$$
$$P(Y=4) \;=\; 0.16\frac{\lambda^2 e^{-\lambda}}{2} + 0.432\frac{\lambda^3 e^{-\lambda}}{6} + 0.1296\frac{\lambda^4 e^{-\lambda}}{24} = 0.1544$$

ii. Without the capping of 4 million euros on the aggregate claim per policy period, the insurance company realizes that the expected aggregate claim is $E(Y) = \mu_X \mu_N = (2(p) + 1(q))\mu_N = 1.4 \times 3 = 4.2$ million euros. This estimation may have served as a guide to cap the aggregate claim to approximately 4 million euros.

2.4.7   *Definitions: Joint and marginal probability distributions*

Consider a discrete random variable X with distribution F_X and another independent discrete random variable Y with distribution F_Y. Further, let X and Y be defined on the same sample space. The collection of points (x_i, y_j), $i, j = 1, 2, 3, \ldots$, that prescribes the joint event $\{X = x_i, Y = y_j\}$, forms an *event space* with probabilities, known as *joint probability mass function*, written as $P(X = x_i, Y = y_j) = p(x_i, y_j) \equiv f_{XY}(x_i, y_j)$.

The *marginal probability mass functions*, $p(x_i) \equiv f_X(x_i)$ and $p(y_j) \equiv f_Y(y_j)$, can be computed by summing out the complementary dimension.[17] Therefore,

$$p(x_i) \;\equiv\; f_X(x_i) = \sum_{y_j} P\left(X = x_i, Y = y_j\right) \equiv \sum_{y_j} f_{XY}\left(x_i, y_j\right)$$

and

$$p(y_j) \;\equiv\; f_Y(y_j) = \sum_{x_i} P\left(X = x_i, Y = y_j\right) \equiv \sum_{x_i} f_{XY}\left(x_i, y_j\right). \tag{2.40}$$

> [17] X and Y may be regarded as *complementary* to each other in the joint event space.

Obviously, the unitary axiom of probability ensures that $\sum_{y_j}\sum_{x_i} f_{XY}\left(x_i, y_j\right) = 1$. It is important to note here that if $f_{XY}(x, y) = f_X(x) f_Y(y)$, then the random variables X and Y are *independent* of each other (see section 1.4.5 for the definition of *independent* events).[18]

> [18] Further, it can be shown that if X and Y are independent random variables, then $E(XY) = E(X)E(Y)$.

Example 2.24   *Joint distribution of two discrete random variables*

The joint distribution of two random variables X and Y is tabulated below.

	$Y = 1$	$Y = 2$	$Y = 3$	$Y = 4$	$p_X(x) = P\{X = x\}$
$X = 1$	$\dfrac{4}{36}$	$\dfrac{3}{36}$	$\dfrac{2}{36}$	$\dfrac{1}{36}$	
$X = 2$	$\dfrac{1}{36}$	$\dfrac{3}{36}$	$\dfrac{3}{36}$	$\dfrac{2}{36}$	
$X = 3$	$\dfrac{5}{36}$	$\dfrac{1}{36}$	$\dfrac{1}{36}$	$\dfrac{1}{36}$	
$X = 4$	$\dfrac{1}{36}$	$\dfrac{2}{36}$	$\dfrac{1}{36}$	$\dfrac{5}{36}$	
$P\{Y = y\}$					$\sum_x \sum_y P\{X = x, Y = y\} = ?$

Compute the following:

i.   The marginal distributions (marginal p.m.f.) of X and Y.

ii. $P\{X \le 3, 2 \le Y < 4\}$.

Solution:

i.

	$Y = 1$	$Y = 2$	$Y = 3$	$Y = 4$	$p_X(x) = P\{X = x\}$
$X = 1$	$\dfrac{4}{36}$	$\dfrac{3}{36}$	$\dfrac{2}{36}$	$\dfrac{1}{36}$	$\dfrac{10}{36}$
$X = 2$	$\dfrac{1}{36}$	$\dfrac{3}{36}$	$\dfrac{3}{36}$	$\dfrac{2}{36}$	$\dfrac{9}{36}$
$X = 3$	$\dfrac{5}{36}$	$\dfrac{1}{36}$	$\dfrac{1}{36}$	$\dfrac{1}{36}$	$\dfrac{8}{36}$
$X = 4$	$\dfrac{1}{36}$	$\dfrac{2}{36}$	$\dfrac{1}{36}$	$\dfrac{5}{36}$	$\dfrac{9}{36}$
$P\{Y = y\}$	$\dfrac{11}{36}$	$\dfrac{9}{36}$	$\dfrac{7}{36}$	$\dfrac{9}{36}$	$\displaystyle\sum_x\sum_y P\{X = x, Y = y\} = 1$

ii.
$$
\begin{aligned}
P\{X \le 3, 2 \le Y < 4\} &= P\{X \le 3, 2 \le Y \le 3\} \\
&= P\{1 \le X \le 3, 2 \le Y \le 3\} \\
&= \sum_{x=1,y=2}^{x=3,y=3} P\{X = x, Y = y\} \\
&= \sum_{y=2}^{3} P\{X = 1, Y = y\} + \sum_{y=2}^{3} P\{X = 2, Y = y\} + \sum_{y=2}^{3} P\{X = 3, Y = y\} \\
&= \left(\frac{3}{36} + \frac{2}{36}\right) + \left(\frac{3}{36} + \frac{3}{36}\right) + \left(\frac{1}{36} + \frac{1}{36}\right) \\
&= \frac{5}{36} + \frac{6}{36} + \frac{2}{36}.
\end{aligned}
$$

$$
P\{X \le 3, 2 \le Y < 4\} = \frac{13}{36}.
$$

Example 2.25 *Typographical errors in a manuscript*

Consider a bouquet of manuscripts, each of which is three pages long. Based on several editorial proof-checking exercises over many years, it has been observed that typically authors commit three typographical errors per manuscript of this type. In this example, we will investigate the distribution of error-free pages in such manuscripts. As an illustrative example, we have chosen a more tractable case here, of three errors distributed across three printed pages. Let us denote a typographical error by the symbol † and a page as a container in a 3-tuple representation of the aforementioned manuscript family (i.e., (-/†† /†) represents a case when two of the three typographical errors are found on page

2 and one on page 3 while page 1 is observed to be free of any errors). The event space is tabulated below.

(†††/-/-)	(-/†††/-)	(-/-/†††)	(††/†/-)	(††/-/†)
(†/††/-)	(-/††/†)	(†/-/††)	(-/†/††)	(†/†/†)

Let N denote the number of pages that have at least one error. Let X_i denote the number of errors on the ith page. Let us begin by considering the joint distribution of X_1 and N. For each event case, we enlist below the realizations of X_1 and N.

i.   (†††/-/-): $X_1 = 3$, $N = 1$,

ii.   (-/†††/-): $X_1 = 0$, $N = 1$,

iii.   (-/-/†††): $X_1 = 0$, $N = 1$,

iv.   (††/†/-): $X_1 = 2$, $N = 2$,

v.   (††/-/†): $X_1 = 2$, $N = 2$,

vi.   (†/††/-): $X_1 = 1$, $N = 2$,

vii.   (-/††/†): $X_1 = 0$, $N = 2$,

viii.   (†/-/††): $X_1 = 1$, $N = 2$,

ix.   (-/†/††): $X_1 = 0$, $N = 2$, and

x.   (†/†/†): $X_1 = 1$, $N = 3$.

The enlisting of the event space above enables us to write the joint p.m.f. $f_{X_1,N}(x,n)$ by computing the relative frequency of occurrence. The following table captures both the joint p.m.f. and the marginal p.m.f. (a.k.a. the *marginals*). The marginals are calculated from the joint p.m.f. by summing along the rows and the columns, respectively.

		X_1				$f_N(n)$
		0	1	2	3	
	1	$\frac{2}{10}$	0	0	$\frac{1}{10}$	$\frac{3}{10}$
N	2	$\frac{2}{10}$	$\frac{2}{10}$	$\frac{2}{10}$	0	$\frac{6}{10}$
	3	0	$\frac{1}{10}$	0	0	$\frac{1}{10}$
$f_{X_1}(x)$		$\frac{4}{10}$	$\frac{3}{10}$	$\frac{2}{10}$	$\frac{1}{10}$	

Figure 2.28 A day in the life of a copy editor who is busy finding typographical errors in a manuscript. The probability distribution of typographical errors in manuscripts is a useful marker for publishing houses to allocate resources in the copy-editing process.

Likewise, we can calculate the joint p.m.f.s for (X_2,N) and (X_3,N) followed by their respective marginals.

				X_2		$f_N(n)$
		0	1	2	3	
N	1	$\frac{2}{10}$	0	0	$\frac{1}{10}$	$\frac{3}{10}$
	2	$\frac{2}{10}$	$\frac{2}{10}$	$\frac{2}{10}$	0	$\frac{6}{10}$
	3	0	$\frac{1}{10}$	0	0	$\frac{1}{10}$
$f_{X_2}(x)$		$\frac{4}{10}$	$\frac{3}{10}$	$\frac{2}{10}$	$\frac{1}{10}$	

				X_3		$f_N(n)$
		0	1	2	3	
N	1	$\frac{2}{10}$	0	0	$\frac{1}{10}$	$\frac{3}{10}$
	2	$\frac{2}{10}$	$\frac{2}{10}$	$\frac{2}{10}$	0	$\frac{6}{10}$
	3	0	$\frac{1}{10}$	0	0	$\frac{1}{10}$
$f_{X_3}(x)$		$\frac{4}{10}$	$\frac{3}{10}$	$\frac{2}{10}$	$\frac{1}{10}$	

Clearly, all the joint distribution functions for (X_1,N), (X_2,N), and (X_3,N) are identical because it is natural to expect that the occurrences of the typographical errors are page-agnostic. Further, in each case, the marginal f_{X_1}, f_{X_2}, and f_{X_3} are also identical. Additionally, we can easily check by inspection that $f_{X_1N}(0,1) \neq f_{X_1}(0) f_N(1)$. In fact, $f_{X_iN}(x_i,n_j) \neq f_{X_i}(x_i) f_N(n_j)$ for all $i = 1, 2, 3$ and $N = 1, 2, 3$. This means that the random variables X_i and N are *not* independent. Lastly, the marginal p.m.f. $f_N(n)$ implicitly describes the probability distribution of error-free pages in manuscripts under consideration here.

The notion of joint and marginal distribution holds true for continuous probability distributions as well. Two continuous random variables X and Y have a *joint p.d.f.* $f_{XY}(x,y)$ if for any subset A of $\mathbb{R}^2$, we have

$$P\big((X,Y) \in A\big) = \iint_A f_{XY}(x,y)\,dx\,dy, \tag{2.41}$$

where $f_{XY}(x,y)$ is a non-negative function (because it represents a probability measure) and normalizes to unit magnitude upon integration, i.e., $\iint_{\mathfrak{S}} f_{XY}(x,y)\,dx\,dy = 1$, where $\mathfrak{S}$ is an appropriate sample set. The *marginal p.d.f.s* are also defined akin to the discrete counterpart as follows:

$$f_X(x) = \int_y f_{XY}(x,y)\,dy, \qquad f_Y(y) = \int_x f_{XY}(x,y)\,dx. \tag{2.42}$$

Further, the condition for independence of the random variables X and Y is $f_{XY}(x,y) = f_X(x) f_Y(y)$ for all $(x,y) \in \mathfrak{S}$.

Example 2.26   ***Joint distribution of two continuous random variables***

The joint distribution of two continuous random variables X and Y is given by

$$f(x,y) = \begin{cases} e^{-(x+y)}, & 0 < x < \infty,\ 0 < y < \infty, \\ 0, & \text{otherwise.} \end{cases}$$

Compute the following:

i. The marginal distributions (marginal p.d.f.) of X and Y.

ii. $P\{X \le \ln 8,\ \ln 2 \le Y \le \ln 4\}$.

Solution:

i. $f_X(x) = \displaystyle\int_{y=-\infty}^{\infty} f(x,y)$

$\qquad = \displaystyle\int_{y=0}^{\infty} e^{-(x+y)}\, dy$

$\qquad = e^{-x} \displaystyle\int_{y=0}^{\infty} e^{-y}\, dy$

$\qquad = e^{-x} \times \left(\dfrac{e^{-y}}{-1} \right)_{0}^{\infty}$

$\qquad = e^{-x} \dfrac{(0-1)}{-1}.$

$\quad f_X(x) = e^{-x}, \quad 0 < x < \infty.$

Similarly,

$\qquad f_Y(y) = e^{-y}, \quad 0 < y < \infty.$

ii. $P\{0 < X \le \ln 8,\ \ln 2 \le Y \le \ln 4\} = \displaystyle\int_{0}^{\ln 8} \left(\int_{\ln 2}^{\ln 4} f(x,y)\, dy \right) dx$

$\qquad = \displaystyle\int_{0}^{\ln 8} \left(\int_{\ln 2}^{\ln 4} e^{-(x+y)}\, dy \right) dx$

$\qquad = \displaystyle\int_{0}^{\ln 8} e^{-x} \left(\int_{\ln 2}^{\ln 4} e^{-y}\, dy \right) dx$

$\qquad = \left(\displaystyle\int_{\ln 2}^{\ln 4} e^{-y}\, dy \right) \left(\int_{0}^{\ln 8} e^{-x}\, dx \right)$

$\qquad = \left(\dfrac{e^{-y}}{-1} \right)_{\ln 2}^{\ln 4} \times \left(\dfrac{e^{-x}}{-1} \right)_{0}^{\ln 8}$

$\qquad = -\left(e^{-\ln 4} - e^{-\ln 2} \right) \times -\left(e^{-\ln 8} - e^{-0} \right)$

$\qquad = -\left(e^{\ln \frac{1}{4}} - e^{\ln \frac{1}{2}} \right) \times -\left(e^{\ln \frac{1}{8}} - 1 \right)$

$\qquad = \left(-\dfrac{1}{4} + \dfrac{1}{2} \right) \times \left(-\dfrac{1}{8} + 1 \right)$

$\qquad = \left(\dfrac{1}{4} \right) \times \left(\dfrac{7}{8} \right).$

$\qquad\qquad P\{X \le \ln 8,\ \ln 2 \le Y \le \ln 4\} = \dfrac{7}{32}.$

Example 2.27   Revisiting the problem of total system failure of an aircraft's power plant

Let us reconsider the problem we investigated earlier about estimating the risk of a total system failure after 100 flying hours as explained in the example of an exponential distribution (also **see** the example alongside Figure 2.27).* The lifetime of each of the two engines (the main power unit [MPU] and the auxiliary power unit [APU]) is represented by two different random variables X and Z with joint p.d.f. $f_{XZ}(x, z) = \mu^2 e^{-\mu x} e^{-\mu z}$, for $x, z > 0$ and $\mu = \dfrac{1}{E(X)} = \dfrac{1}{E(Z)} = \dfrac{1}{500}$.

Let $Y = X + Z$ be the time until the total system failure. What is the p.d.f. of the time until the first engine failure and the total system failure? Further, what is the risk (in terms of a probability measure) of a total system failure after 100 flying hours have elapsed?

The p.d.f.s of the time until the MPU failure and the time until the APU failure are the respective marginal p.d.f.s $f_X(x)$ and $f_Z(z)$.

$$f_X(x) = \int_0^\infty f_{XZ}(x, z)\, dz = \mu e^{-\mu x}, \quad \text{for } x > 0. \tag{2.43}$$

Likewise, $f_Z(z) = \mu e^{-\mu z}, \quad \text{for } z > 0.$ \tag{2.44}

The risk of a total system failure after 100 flying hours is estimated by calculating the right tail distribution

$$P(Y > 100) = 1 - P(Y \le 100)$$

$$= 1 - P(X + Z \le 100)$$

$$= 1 - \int_0^{100} P(X \le 100 - Z \mid Z = z)\, f_Z(z)\, dz$$

Conditioning and law of total probability

$$= 1 - \int_0^{100} P(X \le 100 - z)\, f_Z(z)\, dz$$

$$= 1 - \int_0^{100} \left(\int_0^{100-z} f_X(x)\, dx \right) f_Z(z)\, dz$$

$$= 0.9825. \tag{2.45}$$

There is a 98.25% risk of a total system failure after 100 flying hours.

---

*In this case, we will not consider an "engine re-start" option. The situation of failure of both the power plants during flight would result in total system failure without an opportunity to salvage the crisis.

The concept of joint and marginal distributions (for both discrete and continuous cases) can be extended in an analogous manner for more than two random variables.

Example 2.28 Application of joint and marginal probability distributions

Let X and Y be independent random variables with distribution $geom_1(p)$. Answer the following questions:

i.   What is the probability distribution of $min(X,Y)$?

ii. Compute $P(Y \geq X)$.

iii. What is the probability distribution of $X + Y$?

iv. Compute $P(Y = y | X + Y = z)$ for $z \geq 2$ and $y = 1, 2, ..., z - 1$.

The calculations to solve the above questions are shown below.

i.   Let $Z = min(X,Y)$. If $min(X,Y) > z$, then $X > z$ and $Y > z$. Therefore,

$$P(Z > z) = P(min(X,Y) > z) = P(X > z, Y > z)$$

$$= P(X > z)P(Y > z)$$

X and Y are independent

$$= (1-p)^z (1-p)^z$$

See equation (2.21)

$$= \left((1-p)^2 \right)^z \tag{2.46}$$

This implies that $Z = min(X,Y) \sim geom_1\left(1 - (1-p)^2\right)$.

ii. Consider the countable events $\{X = 1\} \cap \{Y \geq X\}$, $\{X = 2\} \cap \{Y \geq X\}$, $\{X = 3\} \cap \{Y \geq X\}$, ... that partition the sample space of geometrically distributed random variables X and Y, where $Y \geq X$.

$$P(Y \geq X) = P\left(\bigcup_{x=1}^{\infty} \left(\{X = x\} \cap \{Y \geq X\} \right) \right)$$

$$= \sum_{x=1}^{\infty} P\left(\{X = x\} \cap \{Y \geq x\} \right)$$

Probabilities are additive for disjoint events

$$= \sum_{x=1}^{\infty} P(X = x) P(Y \geq x)$$

Independence of X and Y

$$= \sum_{x=1}^{\infty} p(1-p)^{x-1} (1-p)^{x-1}$$

$$= \frac{1}{2-p}. \tag{2.47}$$

iii. The distribution for $X + Y$ is prescribed by a convolution sum.

$$
\begin{aligned}
P(X+Y=z) &= \sum_{x=1}^{z-1} P(X=x, X+Y=z) \\
&= \sum_{x=1}^{z-1} P(X=x, Y=z-x) \\
&= \sum_{x=1}^{z-1} P(X=x) P(Y=z-x) \\
&= (z-1) p^2 (1-p)^{z-2}.
\end{aligned}
$$

iv. We use the definition of conditional probability in the following calculation:

$$
\begin{aligned}
P(Y=y \mid X+Y=z) &= \frac{P(Y=y, X+Y=z)}{P(X+Y=z)} \\
&= \frac{P(X=z-y) P(Y=y)}{P(X+Y=z)} \\
&= \frac{1}{z-1}.
\end{aligned}
\tag{2.48}
$$

Example 2.29   *Law of total probability for continuous random variables*

Here we will consider an example which has a direct practical application in modeling queues that are discussed in detail later in Chapter 4. Consider two independent random variables X and Y that follow exponential distributions with rate parameters λ and μ. We are interested in computing $P(X < Y)$.

Since $X \sim exp(\lambda)$ and $Y \sim exp(\mu)$, $f_X(x) = \lambda e^{-\lambda x}$ and $f_Y(y) = \mu e^{-\mu y}$. We will use the law of total probability to compute $P(X < Y)$ as follows:

$$
\begin{aligned}
P(X<Y) &= \int_0^\infty P(X<Y \mid Y=y) f_Y(y) \, dy \\
&= \int_0^\infty P(X<y) f_Y(y) \, dy \\
&= \int_0^\infty \int_0^y f_X(x) \, dx \; f_Y(y) \, dy \\
&= \int_0^\infty \int_0^y f_X(x) f_Y(y) \, dx \, dy \\
&= \lambda \mu \int_0^\infty \int_0^y e^{-\lambda x} e^{-\mu y} \, dx \, dy \\
&= 1 - \frac{\mu}{\lambda + \mu} \\
&= \frac{\lambda}{\lambda + \mu}.
\end{aligned}
\tag{2.49}
$$

2.5   Chapter project: Predicting insurance claim aggregates during a policy period

2.5.1   *Interlude: Computing the moments of the compound Poisson distribution and estimating aggregate insurance claims by clients by theoretical analysis*

Consider that $Y_j = X_1 + X_2 + \cdots + X_{N_j}$ is the aggregate of a random number of claims N_j per quarter (policy period), where $N_j \sim Poisson(\lambda_j)$, $j = 1,2,3,4$ (corresponding to each of the four quarters), and $X_i \sim Bernoulli([1,2], p_2)$ are individual claims with probability $p_1 = \dfrac{2}{3}$ and $p_2 = \dfrac{1}{3}$ corresponding to claims denominations of \$100,000 and \$200,000, respectively. Further, $\lambda_1 = 2$, $\lambda_2 = 3$, $\lambda_3 = 1$, $\lambda_4 = 3$. $Z = \sum_{j=1}^{4} Y_j$ is the yearly total of all claims made to the firm. Answer the following questions:

(i)   Identify the distribution of Y_j.

(ii) Compute $E(Z)$ and $Var(Z)$.

(iii) Compute $P(Y_2 > 5)$ and $P(Y_3 > 5)$ analytically (without a computer simulation). Subsequently, comment on the discrepancy between the two results (if any).

Note: A word of caution is in order here. While X_i is indeed a binary random variable, in order for your calculations to work out in accordance with a Bernoulli random process, you may have to carry forth a simple transformation of the observables $[1,2] \to [0,1]$. This will be especially necessary while computing statistical moments such as E(Z)!

Figure 2.29 What are the risks in the insurance business?

2.6   Transformation of a random variable

Consider that we know the distribution of a random variable X. This variable can undergo a transformation $g(X)$ for any realizable value of X and thereby define a new random variable under the law $g(\cdot)$. Let us denote the new random variable Y. What can we say about the probability distribution of Y? Let us attempt to answer this question with an illustrative example.

> ### Example 2.30   *Distribution law of a square-law detector*
>
> Consider a continuous random variable X with c.d.f. $F_X(x)$. X undergoes a transformation under a square-law $Y = g(X) = X^2$. Compute the p.d.f. of Y.
>
> *Solution*: The event $\{Y \le y\}$ is equivalent to $\left\{ -\sqrt{y} \le X \le \sqrt{y} \right\} = \left\{ -\sqrt{y} < X \le \sqrt{y} \right\} \cup \left\{ X = -\sqrt{y} \right\}$ where the union is between disjoint events and $P\left(X = -\sqrt{y} \right) = 0$ because Y is defined by a continuous transformation $g(\cdot)$ of a continuous random variable X. Consequently,
>
> $$F_Y(y) = F_X\left(\sqrt{y}\right) - F_X\left(-\sqrt{y}\right) + P\left(X = -\sqrt{y}\right) = F_X\left(\sqrt{y}\right) - F_X\left(-\sqrt{y}\right) + 0. \quad (2.50)$$
>
> Since the transformation is governed by a square-law, we are only concerned with $y > 0$ and, therefore,

$$
\begin{aligned}
f_Y(y) &= \frac{d}{dy}F_Y(y) \\
&= \frac{1}{2\sqrt{y}}\left(f_X\left(\sqrt{y}\right)+f_X\left(-\sqrt{y}\right)\right).
\end{aligned}
\tag{2.51}
$$

Thus, for instance, if $X \sim N(0,1)$, then the p.d.f. of Y is

$$
f_Y(y) = \frac{1}{\sqrt{2\pi y}}e^{-\frac{y}{2}}\mathbb{1}_{y\geq0}.
\tag{2.52}
$$

We can obtain a general formula for computing the p.d.f. of a random variable under a continuous transformation. Let X be a continuous random variable with p.d.f. $f_X(x)$ and let the *differentiable* function $g(x)$ of the real variable x define the transformation. The p.d.f. of Y is given by

$$
f_Y(y) = \sum_{i=1}^{n}\frac{f_X(x_i)}{\left|g'(x_i)\right|}, \quad g'(x_i)\neq0,
\tag{2.53}
$$

where $g'(x)\equiv\dfrac{dy}{dx}$ and $x_i,\, i = 1, 2, ..., n$, are the n real roots of the equation $y - g(x) = 0$.

In Chapter 7 of this book, on multivariate statistics, in section 7.3, we will obtain a further generalization of this formula for multivariate random variables (e.g., X and Y are jointly transformed by two laws $g(\cdot,\cdot)$ and $h(\cdot,\cdot)$) by inverting the determinant of a certain Jacobian matrix.

2.7   Moment generating functions and their applications

The moment generating function (m.g.f.) of a random variable X is denoted by $m_X(t)$ and is defined for all t, in a neighborhood of zero, as follows:

$$
m_X(t):= E\left(e^{tX}\right) =
\begin{cases}
\displaystyle\sum_x e^{tx}P(X=x), & \text{if } X \text{ is a discrete random variable,} \\
\displaystyle\int_x e^{tx}f_X(x)\,dx, & \text{if } X \text{ is a continuous random variable.}
\end{cases}
\tag{2.54}
$$

The following four results underscore the utility of the m.g.f.s to find the probability distributions of the sum of random variables and also to find the higher moments of random variables.

1. If $X_1, X_2, ..., X_n$ are n independent random variables, and $Y := X_1 + X_2 + \cdots + X_n$, then

$$
m_Y(t) = m_{X_1}(t)m_{X_2}(t)\cdots m_{X_n}(t) = \prod_{i=1}^{n}m_{X_i}(t).
\tag{2.55}
$$

2. If X and Y are two random variables with finite m.g.f.s $m_X(t) = m_Y(t)$ for all t, then X and Y have the same probability distributions.

3. The *joint m.g.f.* of X and Y is

$$
m_{X,Y}(s,t) = E(e^{sX+tY}).
\tag{2.56}
$$

The following always holds:

If X and Y are independent $\Leftrightarrow m_{X,Y}(s,t) = m_X(s)m_Y(t), \forall s, t.$ (2.57)

4. *Computing higher order moments from m.g.f.*: For all $n \geq 0$, $E(X^n) = m_X^{(n)}(0).*$

---

*Here $m_X^{(n)}(0)$ is the nth derivative of the m.g.f. of X evaluated at $t = 0$.

The framework presented above that allows us to compute all the statistical moments $E(X^m)$ is founded on the fact that the exponential function has an infinite radius of convergence. In other words,

$$e^x = 1 + x + \frac{x^2}{2!} + \frac{x^3}{3!} + \cdots$$ (2.58)

converges for all x. A closer inspection allows us to recover $E(X^n)$ for any n by substituting tX for x in equation (2.58)and calculating the appropriate derivative[19]

[19] The derivative is computed with respect to t.

of the expected value $E(e^{tX}) = 1 + tE(X) + \dfrac{t^2 E(X^2)}{2!} + \cdots$ evaluated at $t = 0$.

Let us consider the following example in order to fully appreciate the utilitarian nature of the above results in the context of (i) identifying the distribution of sums of random variables, and (ii) computing all the moments of the random variables.

Example 2.31   *Sum of two Poisson random variables is another Poisson random variable*

Consider that X and Y are independent Poisson random variables with rate parameters λ_1 and λ_2, respectively. What is the probability distribution of $X + Y$? First, let us compute the m.g.f. of X.

$$m_X(t) = E(e^{tX}) = \sum_{k=0}^{\infty} \frac{e^{tk} e^{-\lambda_1} \lambda_1^k}{k!}$$

$$= e^{-\lambda_1} \sum_{k=0}^{\infty} \frac{\left(\lambda_1 e^t\right)^k}{k!}$$

$$= e^{-\lambda_1} e^{\lambda_1 e^t}$$

Taylor series of exponential function

$$= e^{\lambda_1\left(e^t - 1\right)}.$$ (2.59)

Figure 2.30 Dark-colored cars enter the highway lane at rate λ_1 and light-colored cars enter the lane at rate λ_2. The lanes merge after sometime where the rate of flow of cars in the narrow lane is $\lambda_1 + \lambda_2$.

Similarly, $m_Y(t) = e^{\lambda_2\left(e^t - 1\right)}$.

Now, by using the result of equation (2.55), we obtain $m_{X+Y}(t) = m_X(t)m_Y(t) = e^{\lambda_1\left(e^t-1\right)} e^{\lambda_2\left(e^t-1\right)}$, which is the m.g.f. of a Poisson random variable with rate parameter $\lambda_1 + \lambda_2$. Therefore, $X + Y \sim Poisson\left(\lambda_1 + \lambda_2\right)$.

Example 2.32   Computing moments of random variables

Consider a random variable X with p.d.f. $f_X(x) = \dfrac{e^x}{\left(1+e^x\right)^2}$ for $-\infty < x < \infty$. Use the m.g.f. to find $E(X)$ and $Var(X)$.

The m.g.f. $m_X(t) = \displaystyle\int_{-\infty}^{\infty} e^{tx} \dfrac{e^x}{\left(1+e^x\right)^2}\, dx = \int_0^1 \left(\dfrac{1-u}{u}\right)^t du$ for $-1 < t < 1$. Here we have used the transformation $u = \dfrac{1}{1+e^x}$. Now, recall that $\dfrac{d}{dt} a^t = a^t \log a$, where $a^t = e^{t\log a}$. Using this result, we have $m_X'(t) = \displaystyle\int_0^1 \log(\dfrac{1-u}{u})(\dfrac{1-u}{u})^t du$. Thus, $m_X'(0) = \displaystyle\int_0^1 \left(\log(1-u) - \log u\right) du = 0$. Likewise, $m_X''(0) = \dfrac{\pi^2}{3}$. This entails that $E(X) = 0$ and $Var(X) = E(X^2) - \left(E(X)\right)^2 = \dfrac{\pi^2}{3}$.

Example 2.33   Moment generating function to find statistical moments of a geometric random variable

Let $X \sim \text{geom}_1(p)$. Find the m.g.f. of X and use it to find $E(X)$ and $Var(X)$.

Solution:

Let $q = 1 - p$. The p.d.f. of X is

$$p(x) = q^{x-1}p, \quad x = 1, 2, 3, \ldots$$

By definition, m.g.f. is as follows:

$$m_X(t) = E\left(e^{tx}\right)$$

$$= \sum_{x=0}^{\infty} e^{tx} q^{x-1} p$$

$$= pe^t \sum_{x=1}^{\infty} \left(e^t\right)^{x-1} q^{x-1}$$

$$= pe^t \sum_{x=1}^{\infty} \left(qe^t\right)^{x-1}$$

$$= pe^t \left(\frac{1}{1-qe^t}\right).$$

$$m_X(t) = \frac{pe^t}{1-qe^t} = \frac{p}{e^{-t}-q}.$$

We compute the first and second derivatives of the m.g.f.:

$$\Rightarrow m_X^{(1)}(t) = \frac{pe^{-t}}{\left(e^{-t}-q\right)^2}.$$

$$\Rightarrow m_X^{(2)}(t) = pe^{-t}\frac{2e^{-t}}{\left(e^{-t}-q\right)^3} - pe^{-t}\frac{1}{\left(e^{-t}-q\right)^2}$$

$$= \frac{pe^{-t}}{\left(e^{-t}-q\right)^2}\left(\frac{2e^{-t}}{e^{-t}-q}-1\right)$$

$$= \frac{pe^{-t}}{\left(e^{-t}-q\right)^2}\left(\frac{e^{-t}+q}{e^{-t}-q}\right)$$

$$m_X^{(2)}(t) = \frac{pe^{-t}\left(e^{-t}+q\right)}{\left(e^{-t}-q\right)^3}.$$

Recall that

$$E(X) = m_X^{(1)}(0) = \frac{p}{\left(1-q\right)^2} = \frac{p}{p^2} = \frac{1}{p}$$

$$E(X^2) = m_X^{(2)}(0)$$

$$= \frac{p(1+q)}{\left(1-q\right)^3} = \frac{p(1+q)}{p^3}$$

$$E(X^2) = \frac{1+q}{p^2}.$$

$$\mathrm{Var}(X) = E(X^2) - \left(E(X)\right)^2$$

$$= \frac{1+q}{p^2} - \frac{1}{p^2}.$$

$$\mathrm{Var}(X) = \frac{q}{p^2} = \frac{1-p}{p^2}.$$

Example 2.34 *Moment generating function to compute statistical moments of a Poisson random variable*

Let $X \sim \mathrm{Poisson}(\lambda)$. Find the m.g.f. of X and use it to find $E(X)$ and $Var(X)$.

Solution:

The p.d.f. of X is

$$p(x) = \frac{e^{-\lambda}\lambda^x}{x!}, \quad x = 0, 1, 2, \dots$$

By definition, m.g.f. is as follows:

$$m_X(t) = E\left(e^{tx}\right)$$

$$= \sum_{x=0}^{\infty} \frac{e^{tx} e^{-\lambda} \lambda^x}{x!}$$

$$= e^{-\lambda} \sum_{x=0}^{\infty} \frac{\left(\lambda e^t\right)^x}{x!}$$

$$= e^{-\lambda} e^{\lambda e^t}$$

$$m_X(t) = e^{\lambda\left(e^t - 1\right)}.$$

We compute the first and second derivatives of m.g.f.:

$$\Rightarrow m_X^{(1)}(t) = \lambda e^t \left(e^{\lambda\left(e^t - 1\right)}\right)$$

$$\Rightarrow m_X^{(2)}(t) = \left(\lambda e^t\right)^2 \left(e^{\lambda\left(e^t - 1\right)}\right) + \lambda e^t \left(e^{\lambda\left(e^t - 1\right)}\right)$$

$$m_X^{(2)}(t) = \lambda e^t \left(\lambda e^t + 1\right) e^{\lambda\left(e^t - 1\right)}.$$

Again, recall that

$$E(X) = m_X^{(1)}(0) = \lambda$$

$$E\left(X^2\right) = m_X^{(2)}(0)$$

$$= \lambda(\lambda + 1)$$

$$E\left(X^2\right) = \lambda^2 + \lambda.$$

$$\mathrm{Var}(X) = E\left(X^2\right) - \left(E(X)\right)^2$$

$$= \lambda^2 + \lambda - \lambda^2$$

$$\mathrm{Var}(X) = \lambda.$$

2.7.1 *Cumulants and their applications*

Under some circumstances, it may be more appropriate to consider the logarithm of the m.g.f. Consequently, we may define a *cumulant generating function* (c.g.f.) as follows:

$$\kappa_X(t) := \log m_X(t), \tag{2.60}$$

where t lies in the neighborhood of 0.

The c.g.f. is well defined only when the RHS of equation (2.60) has a convergent series expansion. Let us write $\kappa_X(t)$ as a power series in t and equate it with the logarithm of the power series of $m_X(t)$.

$$\kappa_X(t) = \kappa_1 t + \kappa_2 \frac{t^2}{2!} + \cdots + \kappa_r \frac{t^r}{r!} + \cdots = \log m_X(t)$$

$$= \log\left(1 + E(X)t + E\left(X^2\right)\frac{t^2}{2!} + \cdots + E\left(X^r\right)\frac{t^r}{r!} + \cdots\right).$$

$$\tag{2.61}$$

Here $\kappa_r = \kappa^{(r)}(0)$ is the coefficient of $\dfrac{t^r}{r!}$ and known as the rth *cumulant* of X. Now, we may choose t appropriately small in order to write a Taylor series expansion for the logarithm term on the RHS of the above equation.[20] The like powers of t from the LHS and the RHS may then be compared to deduce the following relationships between the moments and the cumulants of X:

$$
\begin{aligned}
E(X) = \mu &= \kappa_1. \\
Var(X) = \mu_2 &= \kappa_2, \\
\mu_3 &= \kappa_3, \text{ where } \mu_3 := E\!\left((X-\mu)^3\right), \\
\mu_4 &= \kappa_4 + 3\kappa_2^2, \text{ where } \mu_4 := E\!\left((X-\mu)^4\right),
\end{aligned}
\tag{2.62}
$$

and so on.

Let $S = X + Y$ be the sum of two independent random variables. The c.g.f. $\kappa_S(t)$ of the sum S is given by

$$
\kappa_S(t) = \kappa_X(t) + \kappa_Y(t). \tag{2.63}
$$

The above result may be extended to the sum of n independent random variables.

Further, in the case of a multivariate random vector variable $X = \begin{pmatrix} X^{(1)} \\ X^{(2)} \\ \cdot \\ \cdot \\ X^{(k)} \end{pmatrix}$, the c.g.f. of X, $\kappa_X(\boldsymbol{\xi})$ is defined as follows:

$$
\kappa_X(\boldsymbol{\xi}) = 1 + \sum_i \xi_i \kappa_i + \sum_{i,j} \xi_i \xi_j \frac{\kappa_{ij}}{2!} + \sum_{i,j,k} \xi_i \xi_j \xi_k \frac{\kappa_{ijk}}{3!} + \cdots, \tag{2.64}
$$

where $\kappa_i := E(X^{(i)})$, $\kappa_{ij} := E(X^{(i)}X^{(j)})$, $\kappa_{ijk} := E(X^{(i)}X^{(j)}X^{(k)})$, and so on, $\boldsymbol{\xi} := (\xi_1\ \xi_2\ \cdots)^T$. It may be worth noting that if $X^{(i)}$ and $X^{(j)}$ are independent, then $\kappa_{\underbrace{i,\ldots,i}_{r},\underbrace{j,\ldots,j}_{s}} = 0,\ \forall r,\ s \geq 1$. The converse is not necessarily true.

We will close this section by simply stating a few applications of cumulants: the *skewness* of a distribution is given by $\dfrac{\kappa_3}{\kappa_2^{3/2}}$ and the *kurtosis* is given by $\dfrac{\kappa_4}{\kappa_2^2}$. Interested readers may refer to advanced topics and applications such as conditional moments and cumulants in the text by Thomas A. Severini (see p. 124).[21]

2.7.2 *Probability generating functions*

The discussion in the preceding section on m.g.f.s underscores the utility of expressing the functions of random variables as power series to compute statistical

---

[20] $\log(1+x) = x - \dfrac{x^2}{2} + \dfrac{x^3}{3} \cdots$ is a Taylor series expansion that converges for all $|x| \leq 1$.

[21] Thomas A. Severini, *Elements of Distribution Theory*, 1st ed. (Cambridge University Press, 2005).

moments. This approach may be generalized to extract probabilities of events of interest to us by computing derivatives of the probability generating function (p.g.f.) of the random variable in question that is written as a power series as shown below.

$$g_X(t) := E(t^X) = \sum_{k=0}^{\infty} \phi_k t^k, \tag{2.65}$$

where $\phi_k := P(X=k)$.

An immediate consequence of the definition above, and which is aligned with a similar strategy elucidated in the previous section on m.g.f., is the following:

$$\phi_k = \frac{1}{k!} g_X^{(k)}(0), \tag{2.66}$$

where $g_X^{(k)}(0)$ is the kth derivative of the p.g.f. $g_X(t)$ evaluated at $t = 0$. It is important to note here that the framework for p.g.f. is relevant to the case of discrete random variables because of the form of the definition of $g_X(t)$ given above.

Example 2.35   *Finding the probability distribution of X using the p.g.f. of X*

Consider a discrete random variable X with the p.g.f. $g_X(t) = \frac{t}{4}(1+3t)$. Find the probability distribution of X.

We will begin by computing $g_X(0)$ and, subsequently, all higher derivatives of the form $g_X^{(k)}(0)$ and thereby enlist the p.m.f. of X.

$$g_X(t) = \frac{t}{4}(1+3t) \quad \Rightarrow \quad \phi_0 = g_X(0) = 0,$$

$$g_X'(t) = \frac{1}{4} + \frac{6}{4}t \quad \Rightarrow \quad \phi_1 = g_X'(0) = \frac{1}{4},$$

$$g_X''(t) = \frac{6}{4} \quad \Rightarrow \quad \phi_2 = \frac{1}{2!} g_X''(0) = \frac{3}{4},$$

$$g_X^{(n)}(t) = 0, \forall n > 2 \quad \Rightarrow \quad \phi_n = 0, \forall n > 2.$$

Therefore, the probability distribution of X can be prescribed as follows:

$$X = \begin{cases} 1 & \text{with probability } \frac{1}{4}, \\ 2 & \text{with probability } \frac{3}{4}. \end{cases} \tag{2.67}$$

In the next example given here, we will consider a one-dimensional random walk of the type discussed in the chapter project of Chapter 1 where we had a squirrel making a jump to the left or right at every instant with a probability p and $q = 1 - p$, respectively. Such a random walk spans the space of integers $\mathbb{I} = \{\cdots, -2, -1, 0, 1, 2, \cdots\}$. We consider a case when the random walk starts at zero, $S_0 = 0$, where S_n is the location (state) of the *random walker* at the end of n time steps. An interesting question in this regard would be to estimate the expected number of steps taken by the random walker before transitioning through location (state) +1 for the first time. This is discussed in the following example, where we make use of the p.g.f.s.

Example 2.36   First passage through a certain state by a one-dimensional random walker – an application of p.g.f.

Recall the chapter project from Chapter 1 where we encountered a random walker (the squirrel) on a one-dimensional island who makes an independent choice to jump either to the left or to the right at every instant of time. Now consider a similar situation wherein a random walker takes a step to the right with probability p and a step to the left with probability q at each time step. Clearly, $p + q = 1$. What is the expected time (in terms of number of steps) before the random walker makes a passage through +1 for the first time? Here we will assume that the random walk begins at the origin (labeled 0).

Let us begin by considering a Bernoulli random variable as follows:

$$X_i = \begin{cases} +1 & \text{when the } i^{th} \text{ step is to the right,} \\ -1 & \text{when the } i^{th} \text{ step is to the left,} \end{cases} \tag{2.68}$$

where $i = 1, 2, \cdots$. We may regard a move to the right as being equivalent to the occurrence of a head (labelled as H) upon tossing a fair coin and a move to the left as being equivalent to the occurrence of a tail (labelled as T). We may now define the states of the random walk as follows:

$$S_n = X_1 + X_2 + X_3 + \cdots + X_n, \tag{2.69}$$

where S_n is the state (location) of the random walker after n time steps. $S_0 = 0$.

First passage: We may now define a random variable N such that $S_N = 1$, $N > 0$ but $S_k \neq 1$ for any $k < N$. This defines the event $\mathfrak{P}_n$ of the first passage through the state 1 in $N = n$ time steps. Further, $\phi_n := P(N = n)$ is the probability of the event $\mathfrak{P}_n$.

Since we are interested in utilizing the p.g.f. of a random variable that is written as a power series in t with terms like t^n, it makes sense to compute $E(t^N)$.

$$\begin{aligned} E\left(t^N\right) &= \sum_{n=0}^{\infty} t^n P(N = n) \\ &= \sum_{n=0}^{\infty} t^n \phi_n \\ &= g_N(t). \end{aligned} \tag{2.70}$$

Subsequently, we may proceed with the computation by conditioning upon the disjoint events corresponding to $X_1 = 1$ and $X_1 = -1$. The law of total expectation prescribes

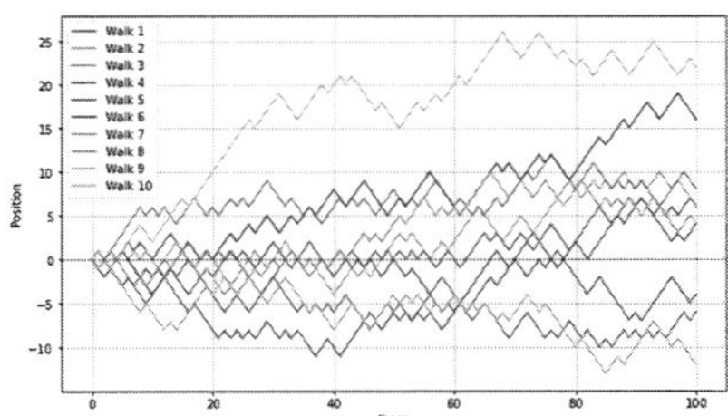

Figure 2.31 Several realizations of a one-dimensional random walk all starting from the state zero. The plots display locations/states on the vertical axis and the time steps on the horizontal axis. A quick inspection will reveal that such random walks may make several passages through a given state during the course of its evolution. A question of interest may be to estimate the average number of steps made by such a random walker before it transitions through a given state for the first time (*first passage* through a state).

$$g_N(t) := E\left(t^N\right) = E\left(t^N \,|\, X_1 = 1\right) P\left(X_1 = 1\right) + E\left(t^N \,|\, X_1 = -1\right) P\left(X_1 = -1\right)$$

$$= tp + E\left(t^N \,|\, X_1 = -1\right) q. \tag{2.71}$$

$$S_{N=1} = 1 \text{ for the first term}$$

In order to evaluate the second term on the RHS of equation (2.71), we may now consider a two-stage process:

i. N_1 time steps to go from state -1 to 0, and

ii. N_2 additional time steps to go from state 0 to the target state 1.

Consequently, $N = 1 + N_1 + N_2$, where 1 accounts for the very first step in transitioning from $S_0 = 0$ to $S_1 = -1$. N_1 and N_2 are independent random variables that have the same distribution as N and are independent of N.

$$E\left(t^N \,|\, X_1 = -1\right) = E\left(t^{1+N_1+N_2} \,|\, X_1 = -1\right)$$

$$= E(tt^{N_1} t^{N_2} \,|\, X_1 = -1)$$

$$= tE\left(t^{N_1} t^{N_2} \,|\, X_1 = -1\right)$$

$$= tE\left(t^{N_1} \,|\, X_1 = -1\right) E\left(t^{N_2} \,|\, X_1 = -1\right)$$

$$N_1 \text{ and } N_2 \text{ are independent}$$

$$= tE\left(t^{N_1}\right) E\left(t^{N_2}\right)$$

$$N_1, N_2 \text{ are independent of } X_1$$

$$= tg_N^2(t). \tag{2.72}$$

$$N_1, N_2 \text{ have the same distribution as } N$$

Using equations (2.71) and (2.72), we obtain the quadratic relation

$$g_N(t) = pt + qtg_N^2(t), \tag{2.73}$$

which has a solution $g_N(t) = \dfrac{1 - \sqrt{1 - 4pqt^2}}{2qt}$ consistent with $g_N(0) = 0$ obtained by evaluating $\lim_{t \to 0} g_N(t) = 0$. Now, since we can choose t to be as close to zero as we desire and $p, q < 1$, it is safe to assume that $4pqt^2 \ll 1$. Then, Taylor's expansion around $x := 4pqt^2 = 0$ gives us the following expression, $\sqrt{1-x} = 1 - \sum_{n=1}^{\infty} \dfrac{(2n-3)!!}{2^n n!} x^n$, where $g_N(t) = \sum_{n=0}^{\infty} t^n \phi_n$, $\phi_{2n} = 0$, $\phi_{2n-1} = \dfrac{(2n-3)!!}{n!} 2^{n-1} p^n q^{n-1}$.*

From this we may deduce that the probability of first passage through $+1$ in $(2n-1)$ time steps or less is $P_{2n-1} = \phi_1 + \phi_3 + \phi_5 + \cdots + \phi_{2n-1}$. Likewise, the probability of first passage (ever) through $+1$ by the random walker is

$$\sum_{n=0}^{\infty} \phi_n \equiv g_N(1) = \left(\frac{1 - \sqrt{1 - 4pqt^2}}{2qt}\right)_{t=1} = \frac{1 - |p - q|}{2q} = \begin{cases} \dfrac{p}{q} & \text{if } p < q, \\ 1 & \text{if } p \geq q. \end{cases}$$

Here we have used the fact that $|x| = \begin{cases} -x & \text{if } x < 0, \\ +x & \text{if } x \geq 0, \end{cases}$ and $\sqrt{1-4xy} = (x-y)$. Consequently, if $p \geq q$, we are certain that the random walker will make a passage through +1, but if $p < q$, then there is a probability $\dfrac{p}{q}$ that the walker will make a passage through +1.

Now, if indeed $p \geq q$, what is the expected waiting time until the first passage through +1, i.e., what is $E(N)$? By definition, we know that

$$E(N) = \sum_{n=1}^{\infty} nP(N=n) = \sum_{n=1}^{\infty} n\phi_n = g'_N(1).$$

$g'_N(1) = \left(g'_N(t)\right)_{t=1} = \dfrac{1}{p-q}$ for $p > q$, but ∞ for $p = q$, Thus, if $p = q = \dfrac{1}{2}$, then $E(N) = \infty$ even though we are certain to make a passage through +1.

---

*The double factorial notation represents the following relation: $(2n + 1)!! = (2n + 1)(2n - 1)\ldots 5 \cdot 3 \cdot 1$ and $(-1)!! = 1$.

2.8 Asymptotic results

We will simply state and present to the readers how to apply the important results of this section. Proofs of some of these theorems are omitted here in this introductory text. Interested readers may refer to some excellent books mentioned in the chapter bibliography to study the proofs.

First, a few clarifications are in order on the issue of integrability. If the random variable X is integrable, i.e., $X \in \mathcal{L}^1$, then $E(|X|) < \infty$. If the random variable X is square-integrable, i.e., $X \in \mathcal{L}^2$, then $E(X^2) < \infty$.

2.8.1   *Markov's inequality*

For a non-negative random variable $X \in \mathcal{L}^1$, and a constant $c > 0$, we have the following upper bound for the probability tail:

$$P(X \geq c) \leq \frac{E(X)}{c}. \tag{2.74}$$

2.8.2   *Chebyshev's inequality*

If $E(X^2) < \infty$, $k \in \mathbb{R}^+$ and $E(X) = \mu$, $Var(X) = \sigma^2$, then the following is true.

$$P(|X - \mu| \geq k) \leq \frac{\sigma^2}{k^2}. \tag{2.75}$$

2.8.3   *Law of large numbers*

There are two important theorems on the law of large numbers which forms the basis of many widely used statistical algorithms like the Monte Carlo methods.

- **Weak Law of Large Numbers (WLLN):** Let X_1, X_2, ..., X_n be independent and identically distributed (i.i.d.) random variables in $\mathcal{L}^2$ with mean μ and variance σ^2.

  $S_n = X_1 + X_2 + \cdots + X_n$. Then for every fixed $\varepsilon > 0$,

  $$P\left(\left|\frac{S_n}{n} - \mu\right| > \varepsilon\right) \to 0, \qquad \text{as } n \to \infty. \tag{2.76}$$

The WLLN is a direct consequence of Chebyshev's inequality. This may be easily checked. $\frac{S_n}{n} = \frac{X_1 + \cdots + X_n}{n}$ is also a random variable whose variance is as follows:

$Var\left(\frac{S_n}{n}\right) = \frac{1}{n^2} Var\left(X_1 + \cdots + X_n\right) = \frac{n\sigma^2}{n^2}$ because all the covariance terms of the form $Cov\left(X_i, X_j\right)$, $i \neq j$, are zero. This is due to the *independence* of the X_is. Now as $n \to \infty$, $Var\left(\frac{S_n}{n}\right) \to 0$ and the WLLN follows.

- **Strong Law of Large Numbers (SLLN):** Let X_1, X_2, ..., X_n be i.i.d. random variables in $\mathcal{L}^1$ with mean μ. $S_n = X_1 + X_2 + \cdots + X_n$. Then,

  $$P\left(\frac{S_n}{n} \to \mu\right) = 1. \tag{2.77}$$

  Here it is assumed that the set $\{\omega\}$ such that $\frac{S_n(\omega)}{n} \to \mu$ is an event.

We now illustrate an application of the law of large numbers which will be our first introduction to the Monte Carlo family of simulations.

Example 2.37   Will the magician collect enough for his evening beer?

A street artist performs a magic show in the main street of Mingletown for three hours. He hopes to earn enough for a bottle of beer which costs ₹350. Throughout the three hours, people give him coins at random. So we will assume that the coins arrive in his bag according to a Poisson distribution. The amount of money each person gives is random with the following distribution:

$$P(INR\ 5) = 0.4$$

$$P(INR\ 10) = 0.4$$

$$P(INR\ 20) = 0.2$$

On average, 5 people per hour give money to the street artist. This implies that the Poisson process has intensity $\lambda = 5$. What is the probability the artist accumulates enough money to get his beer? In other words, what is $\hat{I} = P\left(X_3 \geq 350\right)$, where X_i is the money accumulated after $i = 1, 2, 3$ hours?

We can estimate this easily using the Monte Carlo simulation, which makes use of the strong law of large numbers. The estimate $\hat{I} = \frac{1}{N}\sum_{j=1}^{N} Z_j$, where Z_j is a Bernoulli random variable with output 0 or 1 depending on whether the ith

Figure 2.32 A street magician performing tricks to earn money for a bottle of beer. What are his odds of making enough money?

iteration (out of many thousand iterations) of the Monte Carlo method resulted in the artist bagging enough money for the beer. The law of large number implies that $\hat{l} \to E\left(Z_j\right)$ as $n \to \infty$ almost surely. Succinctly, $\hat{l} = P\left(X_3 \geq 350\right) = E\left(\mathbb{I}_{X_3 \geq 350}\right)$, where we have used the indicator random variable and a useful result from section. 1.8.1. The computational estimate of $E\left(\mathbb{I}_{X_3 \geq 350}\right)$ is basically an ensemble averaging process accomplished over many thousand Monte Carlo iterations as is illustrated in the code below.

```
% Monte Carlo Simulation of Compound Poisson Process
t = 3; lambda = 5; N = 10^6;
beer = zeros(N,1); beer_price = 350;

for i = 1:N
    n = poissrnd(lambda * t);
    if n~=0
        coins = zeros(n,1);
        for j = 1:n
            U = rand;
            coins(j) = (U <= 2/5)*5 + ...
            (U > 2/5 && U <= 4/5)*10 + (U > 4/5)*20;
        end
    end
beer(i) = (sum(coins) >= beer_price);
end
l_hat = mean(beer)   % l_hat = P(X3 >= beer_price)
relErr_hat = std(beer) / (l_hat * sqrt(N))
% relative error of l_hat by crude Monte Carlo simulation
```

The estimate `l_hat = 8.1000e-05` entails that there is very little chance that the artist will accumulate enough money to buy a bottle of beer. His chance of making enough money for beer could be enhanced if the chance of receiving the largest denomination of money, i.e., ₹**20,** is increased at the expense of the lower denominations and/or if the frequency at which the people offered money (i.e., the value of λ) increased and/or if he performed his show for a longer duration.

2.8.4   *Central limit theorem (CLT)*

Consider a random sample of size n denoted by $X_1, X_2, ..., X_n$ that is drawn from a population of mean μ (i.e., $E\left(X_i\right) = \mu, i = 1, 2, ..., n$) and variance σ^2 (i.e., $Var\left(X_i\right) = \sigma^2$). Then the asymptotic distribution of $Z = \dfrac{\overline{X} - \mu}{\sigma / \sqrt{n}}$ as $n \to \infty$ is the standard normal distribution $N(0, 1)$.

We will revisit the central limit theorem in Chapter 5, "Statistical experiments." In the following sections, we will simply use this theorem to arrive at some asymptotic results.

2.8.5   *Normal approximation to the binomial distribution (DeMoivre–Laplace limit theorem)*

Let $X \sim Bin\left(n, p\right)$. Then $\dfrac{X - np}{\sqrt{np\left(1-p\right)}} \to N\left(0,1\right)$ as $n \to \infty$.

A special case of the central limit theorem when applied to investigate the distribution of the sum of Bernoulli random variables is known as the DeMoivre–Laplace limit theorem after Abraham de Moivre.[22]

[22] Abraham de Moivre, *The Doctrine of Chances*, Illustrated ed. (Cambridge University Press, 2013).

Let $S_n := X_1 + X_2 + \cdots + X_n$, where X_i is a Bernoulli random variable with success probability p for all $i = 1, 2, \dots, n$. Consequently, $S_n \sim Bin(n, p)$. Then

$$\lim_{n \to \infty} P\left(\frac{S_n - np}{\sqrt{np(1-p)}} < z \right) = \Phi(z), \tag{2.78}$$

where $\Phi(z) := P(Z < z)$ and $Z \sim N(0,1)$. Equivalently,

$$\lim_{n \to \infty} P\left(a < \frac{S_n - np}{\sqrt{np(1-p)}} < b \right) = \Phi(b) - \Phi(a) = P(a \le Z \le b). \tag{2.79}$$

Figure 2.33 How well did I learn from Prof. Bean?

Example 2.38 *Approval rating of a teacher*

A survey is conducted among 100 students in a class to verify if a teacher is performing well. Since the teacher under review is an average teacher, it turns out that students are equally likely to approve or disapprove of his teaching style. What is the probability that exactly (i) 50 students and (ii) 75 students approve the teacher's style?

Let us first use the binomial distribution to compute this probability. Here $n = 100$ and $p = \dfrac{1}{2}$. Let Y be the number of students who approve of the teacher's style. $Y \sim Bin\left(100, \dfrac{1}{2}\right)$. Therefore,

$$\begin{aligned} P(Y = 50) &= P(Y \le 50) - P(Y \le 49) \\ &= 0.0796. \end{aligned} \tag{2.80}$$

The above calculation is accomplished by using the following Matlab command:

```
>> binocdf(50,100,0.5) - binocdf(49,100,0.5)
```

Likewise,

$$\begin{aligned} P(Y = 75) &= P(Y \le 75) - P(Y \le 74) \\ &= 1.9131 \times 10^{-7}. \end{aligned} \tag{2.81}$$

Now, let us attempt to estimate the same quantity by using the normal approximation $Y \approx N(\mu, \sigma^2)$, where $\mu = np = 100 \times 0.5 = 50$ and $\sigma^2 = np(1 - p) = 50 \times 0.5 = 25$. Applying what is known as the *continuity correction*, $P(Y = 50) = P(49.5 < Y < 50.5)$, we proceed as follows:

$$\begin{aligned} P(49.5 < Y < 50.5) &= P\left(\frac{49.5 - 50}{5} < \frac{Y - \mu}{\sigma} < \frac{50.5 - 50}{5} \right) \\ &= P(-0.1 < Z < 0.1), \text{ where } Z \sim N(0,1) \text{ as a result of the CLT} \\ &= 0.0797. \end{aligned} \tag{2.82}$$

The above calculation of the probability for the standard normal distribution is accomplished using Matlab commands given below.

```
mu = 0;
sigma = 1; % for Z ~ N(0,1)
pd = makedist('Normal','mu',mu,'sigma',sigma);
probability = cdf(pd,0.1) - cdf(pd,-0.1)
```

Likewise,

$$
\begin{aligned}
P(Y=75) &= P(74.5 < Y < 75.5) \\
&= P\left(\frac{24.5}{5} < Z < \frac{25.5}{5}\right) \\
&= 3.09 \times 10^{-7}.
\end{aligned}
\tag{2.83}
$$

A close inspection of the above calculation reveals that the normal approximation is a reasonable estimate of the binomial distributed random process. The approximation is exceptionally good near the mean of the distribution where most of the probability mass is concentrated.

2.8.6 *Normal approximation to the Poisson distribution*

Let $X \sim Poisson(\lambda)$. Then $\dfrac{X-\lambda}{\sqrt{\lambda}} \to N(0,1)$ as $\lambda \to \infty$.

Example 2.39 Predicting earthquakes

The annual number of earthquakes registering at least 3.5 on the Richter scale and having an epicenter within 50 kilometers of downtown Colombo follows a Poisson distribution with mean 6.5. What is the probability that at least 10 such earthquakes will strike next year?*

Let $X \sim Poisson(\lambda = 6.5)$ be the random number of earthquakes that strike Colombo on a given year. $P(X \geq 10) = 1 - P(X \leq 9) = 0.1226$. The Matlab calculations for the above are done as follows:

```
lambda = 6.5;
pd = makedist('Poisson','lambda',lambda);
prob = 1 - cdf(pd,9);
```

Now, even though λ is not very large here, the Poisson estimate can be reasonably approximated by the normal distribution as shown below.

$$
P(X \geq 10) = P(X > 9.5)
$$

Continuity correction

$$
= P\left(\frac{X-\lambda}{\sqrt{\lambda}} > \frac{9.5-6.5}{\sqrt{6.5}}\right)
$$

$$
= P(Z > 1.1767), \text{ where } Z \sim N(0, 1),
$$

$$
= 0.119.
\tag{2.84}
$$

The above is computed in Matlab as follows:

```
mu=0;
sigma = 1;
pd = makedist('Normal','mu',mu,'sigma',sigma);
probability = 1 - cdf(pd,(9.5-6.5)/(sqrt(6.5)));
```

*This example is adapted from the textbook by Richard J. Larsen and Morris L. Marx, *An Introduction to Mathematical Statistics and Its Applications*, 6th ed. (Pearson, 2018).

2.8.7 *Poisson approximation to the binomial distribution*

Consider $S_n \sim Bin\left(n, p_n\right)$ such that $p_n \to 0$, $np_n \to \lambda$ as $n \to \infty$, then $S_n \sim poisson(\lambda)$ in the asymptotic limit.

Example 2.40 *Application of Poisson approximation to a binomial distributed random process*

Suppose we roll two dice 24 times and X be the number of times a pair of aces appear. Compute the following: $P(X = 0)$, $P(X = 2)$. Here $n = 24$ and $p = \frac{1}{36}$, which means $np = \frac{2}{3} = \lambda$. Estimation of $P(X = k)$ for $k = 0, 2$ is shown below using the binomial distribution and the corresponding Poisson approximation is shown alongside.

	Binomial distribution	*Poisson approximation*
	$P\left(X = k\right) = \binom{n}{k} p^k \left(1 - p\right)^{n-k}$	$P\left(X = k\right) = e^{-\lambda} \dfrac{\lambda^k}{k!}$
$P(X = 0)$	$(1 - \dfrac{1}{36})^{24} = 0.5086$	$e^{-2/3} = 0.5134$
$P(X = 2)$	$\binom{24}{2}\left(\dfrac{1}{36}\right)^2 (1 - \dfrac{1}{36})^{22} = 0.1146$	$e^{-2/3}\left(\dfrac{2}{3}\right)^2 \dfrac{1}{2!} = 0.1141$

From the probability estimates tabulated above, we may infer that the Poisson approximation is very close to the predictions of the binomial distribution model.

2.9   Chapter project: Predicting insurance claim aggregates during a policy period

2.9.1 *Prologue: Predicting risk of monetary loss associated with the insurance scheme for the company using a Monte Carlo simulation*

In this section, use the *crude* Monte Carlo simulation (and thereby the law of large numbers) to predict the following:

1. Estimate $P(Y_2 > 5)$ and $P(Y_3 > 5)$ using the crude Monte Carlo simulation. Compare your simulation results here with the analytical results you obtained in section 2.5.1. Comment on your comparisons.

2. Let the total annual income on the sale of insurance premiums be $1,000,000. What is the risk of yearly loss for the company in terms of $P(Z > 1000000)$? You may provide your analysis of the risk by using an appropriate Monte Carlo simulation.

2.10   Selected bibliography

Durrett, Rick. *Elementary Probability for Applications* (1st edition). Cambridge University Press, 2009.

Feller, William. *An Introduction to Probability Theory and Its Applications*, Vol. 1 (3rd edition). John Wiley & Sons, Inc., 1968.

Feller, William. *An Introduction to Probability Theory and Its Applications*, Vol. 2 (2nd edition). John Wiley & Sons, Inc., 1971.

de Finetti, Bruno. *Theory of Probability* (Special edition). John Wiley & Sons, Inc., 2017.

Gupta, S. C. and V. K. Gupta. *Fundamentals of Mathematical Statistics* (11th edition). Sultan Chand and Sons, 2017.

Richards, Ian and Heekyung Youn. *Theory of Distributions: A Non-technical Introduction* (1st edition). Cambridge University Press, 2007.

Ross, Sheldon M. *Introduction to Probability Models* (12th edition). Academic Press, Elsevier, 2019.

Severini, Thomas A. *Elements of Distribution Theory* (1st edition). Cambridge University Press, 2005.

Tijms, Henk. Probability: *A Lively Introduction* (12th edition). Cambridge University Press, 2019.

Williams, David. *Weighing the Odds* (1st edition). Cambridge University Press, 2010.

2.11   Exercise problems

1. (***Expectation of positive integer valued random variables***) Let X be a random variable with values on the positive integer set. Prove that $E\left(X\right) = \sum_{k=1}^{\infty} P\left(X \geq k\right) \leq \infty.$

2. Consider a random variable X with p.d.f. $f_X(x) = 6x(1-x)$ for $x \in \left(0,1\right)$ and zero otherwise. Find $E(X)$ and $Var(X)$.

3. (***Needle and*** π) A famous attempt at estimating the value of π was made by repeatedly dropping a needle of length $L \leq 1$ on the floor made of long continuous tiles of unit width. This was done by estimating the probability that the dropped needle would touch a tile edge and then comparing (equating) this answer with computer-simulated experiments of multiple needles dropped on such a tiled floor. What is the probability that a dropped needle would touch a tile edge?

4. (***Stuck in traffic!***) Cars start successively at the origin and travel at different but constant speeds along an infinite narrow road on which no passing is possible. When a car reaches a slower car it is compelled to follow it at the same speed. In this way platoons will be formed whose ultimate size depends on the speeds of the cars but not on the times between successive departures. Consider the speeds of the cars to be independent random variables with a common continuous distribution. Choose a car at random. Prove the following:

   i.   The probability that the chosen car does not trail any other car tends to half.

   ii.   The probability that it leads a platoon of total size n (with exactly $n-1$ cars following it) tends to $\dfrac{1}{(n+1)(n+2)}$.

   iii.   The probability that the given car is the last in a platoon of size n tends to the same limit as in (ii).

5. (***Why do two bass guitars sound twice as loud as one?***) Consider that the sound wave emanating from a guitar is given by a random unit vector. Waves coming from two independent guitars would superimpose to give a resultant vector.

   i.   Use the *law of the cosines* to write the square of the length of the resulting vector as $2 + 2\cos\theta$, where θ is the angle between the two random vectors.

   ii.   Predict an appropriate distribution for θ.

   iii.   Calculate the expected value of *loudness*[23] of the resulting vector.

6. (***How long will I bemoan my bad luck at the toll plaza?***) While on a road trip through the Grand Himalayan Highway, I come across a toll plaza with multiple parallel counters. I am in a dilemma—*which of the several counter options do I exercise so that I can get off the queue as fast as I can?* Often in such situations, the car right behind (car-x) becomes a marker of my decision — *assuming that the car right behind me chooses a different counter, my assessment of my luck is determined by who among the two of us stays ahead in the queue?* To make the problem simpler to solve, we will consider the following simplifications:

(a) all cars in the toll plaza are of equal size,

(b) all queues are stochastically independent,

(c) time interval between successive moves in any queue is an independent continuous random variable with a common probability distribution,

(d) successive moves constitute Bernoulli trials, where "success" means I move ahead by a unit distance and "failure" means car-x moves ahead on a certain trial, and

(e) the probability of success is $p = 0.5$.

Answer the following questions:

   i.   What are my odds of getting ahead of car-x in the queue while I am at the toll plaza?

   ii.   If I do end up being ahead of car-x at some point, what is my expected waiting time before I am ahead?[24]

   iii.   In what manner would the answers to the above two questions differ if $p \neq 0.5$?

Figure 2.34 Two bass guitarists in full song!

[23] Loudness is proportional to square of the amplitude of vibration.

Figure 2.35 The Grand Himalayan toll plaza (*courtesy*: *The Times of India*).

[24] The longer I have to wait to get ahead of car-x, the longer I will end up bemoaning my bad luck at the toll plaza!

7. (***Twin-engine failure***) The $\mathfrak{M}$-09 is a twin engine jet. Let X be the random time to failure of engine-1 and Y be the random time to failure of engine-2. X and Y are independent random variables with distribution $exp(\mu_1)$ and $exp(\mu_2)$ with mean times to failure $\dfrac{1}{\mu_1} = \dfrac{1}{\mu_2} = 100$ flying hours. What is the probability that there is a dual engine flame out in more than 75 flying hours?

Figure 2.36 What are the odds of a dual engine flame out?

8. Let X and Y be independent random variables, each with an exponential distribution $exp(\lambda)$. What is the p.d.f. of the following random variables?

    i. $Z = X + Y$.

    ii. $W = Y - X^2$

9. (***Application of moment generating function***) Consider a random variable X with p.d.f. $f_X(x) = e^{-2x} + \dfrac{1}{2}e^{-x}$, $x > 0$. Find the m.g.f. of X and use it to find the $Var(X)$.

10. (***Generating functions for branching process***) Branching processes have wide applications. For example, in the area of biology, a cell may die or split into two in each generation leading to the survival or extinction of the species. We may be interested in knowing the chances of survival of the species after n generations. The extinction probability $u_n := P(X_n = 0)$, where X_n is the size of the nth generation. The extinction probability u_n can be computed iteratively by using the *probability generating function* (p.g.f.): $P(z) := E(z^X) = \sum_{j=0}^{\infty} p_j z^j$, where $0 \le z \le 1$ and p_j is the probability $P(X = j)$ that in the lifetime of an individual cell, it produces $j = 0, 1, 2, \dots$ new offspring.

Then,

$$u_n = P(u_{n-1}) \text{ for } n = 2, 3, 4, \dots, \tag{2.85}$$

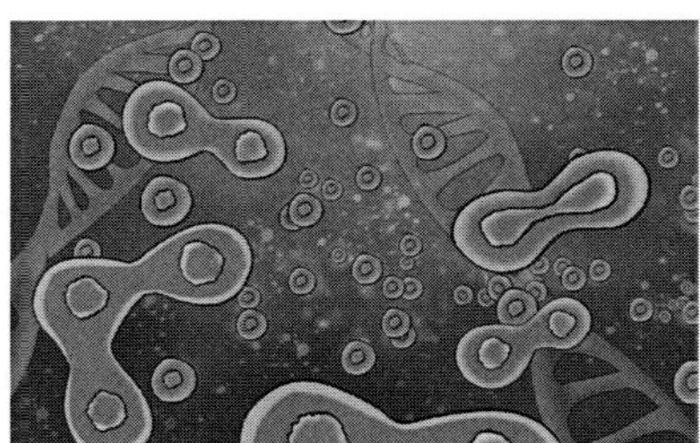

Figure 2.37 Cell division.

with $u_0 = 0$ and $u_1 = p_0$. It follows that in the asymptotic limit $n \to \infty$, u_∞ satisfies the fixed point iteration $u = P(u)$ and is its smallest positive root. Answer the following questions about a branching process:

    i. Deduce equation (2.86) by using the definition of the p.g.f.[25]

    ii. A carcinoma begins with a single cell. In each generation, a carcinogenic cell dies with a probability $\dfrac{1}{3}$ or doubles with probability $\dfrac{2}{3}$. What is the probability that the cancer will die out in the third generation? What is the probability that the cancer will proliferate forever? What are the above probabilities if the cancer initially began with two cells?

[25] You may have to deduce an intermediary step $u_n = \sum_{k=0}^{\infty} u_{n-1}^k p_k$, for $n = 2, 3, \dots$, and use the law of total probability.

3

Discrete Time Markov Chains

Markov chains were first formulated as a stochastic model[1] by Russian mathematician Andrei Andreevich Markov. Markov spent most of his professional career at St. Petersburg University and the Imperial Academy of Science. During this time, he specialized in the theory of numbers, mathematical analysis, and probability theory. His work on Markov chains utilized finite square matrices (stochastic matrices) to show that the two classical results of probability theory, namely, the *weak law of large numbers* and the *central limit theorem*, can be extended to the case of sums of *dependent* random variables.

Markov chains have wide scientific and engineering applications in statistical mechanics, financial engineering, weather modeling, artificial intelligence, and so on. In this chapter, we will look at a few applications as we build the concepts of Markov chains. Additionally, we will also implement a technique (using Markov chains) to solve a simple and practical engineering problem related to aircraft control and automation.

3.1 Chapter objectives

The chapter objectives are listed as follows.

1. Students will learn the definition and applications of Markov processes.

2. Students will learn the definition of the *stochastic matrix* (also known as the *probability transition matrix*) and perform simple matrix calculations to compute conditional probabilities.

3. Students will learn to solve engineering and scientific problems based on discrete time Markov chains (DTMCs) using multi-step transition probabilities.

4. Students will learn to compute return times and hitting times to Markov states.

5. Students will learn to classify different Markov states.

6. Students will learn to use the techniques of DTMCs introduced in this chapter to solve a complex engineering problem related to flight control operations.

Figure 3.1 Russian mathematician Andrei Andreevich Markov (*courtesy*: St. Petersburg Department of V. A. Steklov Institute of Mathematics of the Russian Academy of Sciences, PDMI RAS).

[1] A. A. Markov, Extension of the law of large numbers to dependent quantities (in Russian), *Izvestiia Fiz.-Matem. Obsch. Kazan Univ.*, 2nd ser. **15** (1906): 135–156.

3.2 Chapter project: Automatic prediction of control laws of an aircraft using the Viterbi algorithm

3.2.1 *Prologue: Tracking aircraft control laws*

In this project, we will learn to implement the *Viterbi algorithm* to automatically predict the operational flight control law by analyzing the *pitch* data of the aircraft. The Viterbi algorithm is a practical application of discrete time Markov chains (DTMCs). This project illustrates, in a simplified manner, the framework within which an aircraft's operational performance is monitored in real time by the ground crew. It also demonstrates, as an example, how mathematical technology and engineering redundancies work together toward developing newer and safer flight experiences for passengers. Modern aircraft use fly-by-wire control systems. This means that control surfaces like ailerons, rudders, etc., are maneuvered under the command of electronic signals originating from the flight control computers instead of the pilot's manual inputs. The scope of the fly-by-wire system is determined by the active flight control law. There are primarily three different control laws in an Airbus A-330 aircraft: (i) normal law, (ii) alternate law, and (iii) direct law.

Normal law offers a variety of automated protection to the flight envelope, e.g., automatic stall protection, bank angle protection, and high speed protection. The aircraft can be commanded by the autopilot in this mode. Normal law may be applicable in *ground, flight,* or *flare (landing)* modes. The aircraft's performance is considered to be optimal if it is operational under normal law during its flight. On the other hand, when one or more of the sensors or control surfaces are impaired, then the flight may be forced to operate under alternate law. Some automatic protections like stall protection, bank angle protection, and high speed protection may be lost depending on the exact nature of the failure(s). This may happen as a consequence of faults in the horizontal stabilizer, a single elevator fault, loss of a yaw-damper actuator, loss of slat or flap position sensors, etc. (ALT-1 law), or due to engine flame outs, faults in two inertial or two air-data reference units, damage to all spoilers, aileron faults, etc. (ALT-2 law). Aircraft operation under this alternate control law will often require some direct intervention by the pilot(s). Finally, further degradation of flight control affairs results in the activation of the direct law. Under this law, autopilot function is always lost and most of the automatic flight envelope protections are lost. Pilots will have to manually recover and fly the airplane. This type of degradation of flight controls may result from dual-engine flame outs, dual-elevator failures, etc.

During the course of a flight, tracking and recording the operational control law of an aircraft can reflect the performance profile of the aircraft.[2] One of the ways this real-time health of the aircraft may be captured by ground control and maintenance crew is by analyzing the binary pitch up/down data transmitted to it via the satellite-based *Aircraft Communications Addressing and Reporting System* (ACARS). For example, frequent changes in the pitch of the aircraft (nose up–nose down motions) may be a result of faulty elevators (part of the horizontal stabilizer), thereby activating the alternate or direct control laws. This example illustrates that the pitch data can serve as an important metric for predicting the active control law. In this project, we will use the Viterbi algorithm to predict

Figure 3.2 An Airbus A-330 is shown here steadily climbing towards its cruising altitude. The aircraft is under the command of the autopilot and operating under *normal* flight control law.

Figure 3.3 An erratic pitch profile of an aircraft reveals a likely faulty elevator and a flight operation governed by alternate or direct law.

[2] An aircraft whose flight signature is predominantly governed by normal law can be assumed to be in better operational health than an aircraft whose flight signature is interspersed with several stints of alternate and direct control modes.

the profile of the active control law during flight by analyzing the real-time pitch data profile.

We will return to this case study in a subsequent section of this chapter after we have developed some conceptual background on Markov chains.

3.3   Definition: Markov chain

A Markov chain (or equivalently, a Markov process) is a stochastic process describing a series of possible events whereby the probability of an event at a given instant depends solely on the outcome of the previous event.

In this chapter, we will consider only those Markov processes whose outcomes are discrete events. These are known as *discrete time Markov chains (DTMCs)*. For example, we may think of a game of badminton. The outcome of a given shot in a rally is a discrete event such as a *smash*, a *drop*, or a *lift*, Any given shot by a player likely depends on the previous shot by the opponent, and most likely not so much on any previous shot played by either player. To elucidate this further, consider a *smash* played by player A (player at the far end in Figure 3.4). It is nearly impossible that a return shot by player B (player at the near end in Figure 3.4) will also be a *smash*; it can perhaps be a *lift* or a *drop*. So essentially, a game of badminton can be modeled as a Markov process. We will analyze this particular example in greater detail later in this chapter. Mathematically, we may denote a sequence of events as $\{X_n\}_{n \in I, n \geq 0}$, where the subscript n is a non-negative integer that indexes time. X_n is a random variable which may take a value from the set of possible outcomes (events), $S = \{x_n, x_{n-1}, ..., x_1, x_0\}$. Then, $\{X_n\}$ describes a Markov process when

Figure 3.4 A *smash* by the far-end player cannot be returned by a *smash* by the near-end player in a game of badminton, which can be modeled as a Markov process.

$$P(X_n = x_n | X_{n-1} = x_{n-1}, X_{n-2} = x_{n-2}, ..., X_0 = x_0) = P(X_n = x_n | X_{n-1} = x_{n-1}). \qquad (3.1)$$

Consequentially, a Markov process is characterized by the property of *memorylessness*.

3.3.1   *Stochastic matrix or probability transition matrix, $\mathbb{P}$*

In the previous paragraph, the set S forms the event space comprising all possible outcomes (events) of the Markov process. The elements of this set constitute the *states* of the Markov chain. The probability (or likelihood) of transition between two states is compiled in a matrix form. This matrix is known as the *stochastic matrix* or the *probability transition matrix* and is denoted by p_{ij} (equivalently, $p_{i,j}$) or simply $\mathbb{P}$. Here, the subscripts i, j refer to the fact that we are considering a transition from state i to state j. i, j can take values from $\{1, 2, 3, ..., n\}$ for an n-state system. It may be noted here that the sum of the entries of any row of $\mathbb{P}$ must be identical to one due to the unitarity axiom of probability. The following example will illustrate how we may construct this matrix.

Figure 3.5 A game of roulette in a casino. A gambler's fortune may be modeled as a Markov process.

Example 3.1   Gambler's ruin

Consider a game of gambling in which, on any turn, we win ₹1 with probability $p = 0.4$ or lose ₹1 with probability $p' = (1 - p) = 0.6$. Suppose we adopt a strategy that we quit playing upon making a fortune of ₹10 while the casino throws us out if we have lost all our money. Construct a suitable stochastic matrix. Let $X_n :=$ amount of money we have after "n plays". $S = \{0,1,2,...,10\}$. Then clearly, for all $i = \{1,2,...,9\}$, the following is true based on the rules of the gambling game:

$$P(X_{n+1} = i+1 | X_n = i, X_{n-1} = i_{n-1}, ..., X_0 = i_0) = P(X_{n+1} = i+1 | X_n = i) = p_{i,i+1} = 0.4. \quad (3.2)$$

Likewise, $p_{i,i-1} = 0.6$. Additionally, $p_{0,0} = 1$ because there is no chance of making money when we have lost everything since the casino will throw us out. Similarly, $p_{10,10} = 1$ because ₹10 is the maximum allowable fortune we can accumulate. States 0 and 10 are known as *absorbing* states as there is no *escape* from these two states in this gambling model. Finally, we may now write the stochastic matrix as follows:

$$\mathbb{P} = \begin{array}{c} \\ i=0 \\ i=1 \\ i=2 \\ \\ \\ \\ i=9 \\ i=10 \end{array} \begin{pmatrix} 1 & 0 & 0 & \cdot & \cdot & \cdot & 0 \\ 0.6 & 0 & 0.4 & \cdot & \cdot & \cdot & 0 \\ 0 & 0.6 & 0 & 0.4 & \cdot & \cdot & 0 \\ \cdot & \cdot & \cdot & \cdot & \cdot & \cdot & \cdot \\ \cdot & \cdot & \cdot & \cdot & \cdot & \cdot & \cdot \\ \cdot & \cdot & \cdot & \cdot & \cdot & \cdot & \cdot \\ 0 & 0 & \cdot & \cdot & 0.6 & 0 & 0.4 \\ 0 & 0 & \cdot & \cdot & 0 & 0 & 1 \end{pmatrix} \quad (3.3)$$

3.4   Multi-step transition probabilities

The stochastic matrix $\mathbb{P}$ pertains to a single step in the stochastic transition of the underlying process. But in many practical cases, we may be interested in knowing the likelihood of transitions between states in more than one step. For example, mathematically, the probability of transitioning from state i to state j in $m(>1)$ steps may be expressed as follows.

$$p_{i,j}^m \equiv p^m(i,j) = P(X_{n+m} = j | X_n = i); \quad m > 1. \quad (3.4)$$

Example 3.2   Social motility

Consider that X_n represents the social class of a family in the nth generation. Assume that there are broadly three social groups based on income, namely, lower = 1, middle = 2, upper = 3. Based on a certain demographic and income analysis, the motility within this society was captured succinctly by the following stochastic matrix.

$$\mathbb{P} = \begin{array}{c} \\ 1 \\ 2 \\ 3 \end{array} \begin{pmatrix} 0.7 & 0.2 & 0.1 \\ 0.3 & 0.5 & 0.2 \\ 0.2 & 0.4 & 0.4 \end{pmatrix}. \quad (3.5)$$

Figure 3.6 Class motility in society can also be analyzed by a Markov model.

If Ginny's parents were of middle income class, what is the probability that Ginny belongs to the upper income class and her children belong to the lower income group? Here we are asked to find $P(X_2 = 1, X_1 = 3 | X_0 = 2)$, where X_0 refers to Ginny's parents' generation, X_1 refers to her generation, while X_2 refers to her children's generation. We will derive the answer from first principles and subsequently, claim that the same may be deduced by formal inspection of the appropriate $p_{i,j}$ entries.

$$P(X_2 = 1, X_1 = 3 \mid X_0 = 2) = \frac{P(X_2 = 1, X_1 = 3, X_0 = 2)}{P(X_0 = 2)}$$

$$\text{Def. } P(A \mid B) = \frac{P(A, B)}{P(B)}$$

$$= \frac{P(X_2 = 1, X_1 = 3, X_0 = 2)}{P(X_1 = 3, X_0 = 2)} \times \frac{P(X_1 = 3, X_0 = 2)}{P(X_0 = 2)}$$

Multiply and divide by the same term

$$= P(X_2 = 1 | X_1 = 3, X_0 = 2,) \times P(X_1 = 3 | X_0 = 2)$$

Conditional probability

$$= P(X_2 = 1 | X_1 = 3) \times P(X_1 = 3 | X_0 = 2)$$

Markov property

$$= p_{3,1} p_{2,3} = p_{2,3} p_{3,1} = 0.2 \times 0.2 = 0.04.$$

Entries of $\mathbb{P}$

Terms rearranged

Now that we have seen all the intermediary steps starting from the expression that we originally set out to evaluate, it may be convenient to notice that the answer could have been formally read out by multiplying the probability entries for the successive transitions 2 (middle) $\rightarrow 3$ (upper) and 3 (upper) $\rightarrow 1$ (lower). Let us now suppose that Sheila, from the same society, belongs to the lower income group; what is the probability that her grandchildren will belong to the upper class? This probability can be easily evaluated to be $p^2(1,3)$, i.e., $(1,3)$ entry of the $\mathbb{P}^2$ matrix.

Example 3.3 Constructing a stochastic matrix

Consider the Markov chain $\{X_n\}_{n \geq 0}$ with state space $S = \{1, 2, 3, 4\}$. Find the stochastic matrix P, such that:

i. $p_{1,j} = 0.2$ for $j = 1, 2, 3$

ii. $p_{2,j} = 2p_{2,j+1}$ for $j = 1, 2, 3$

iii. $p_{3,1} = 0.1, p_{3,2} = p_{3,3} = 0.2$

iv. $p_{4,j} = \frac{1}{2} p_{4,j-1}$ for $j = 2, 3, 4$

Solution:

$$p_{14} = 1 - (0.2+0.2+0.2) = 0.4$$

$$p_{24} + p_{23} + p_{22} + p_{21} = p_{24}(1 + 2 + 4 + 8) = 15p_{24} = 1$$

$$\Rightarrow p_{24} = \frac{1}{15}, p_{23} = \frac{2}{15}, p_{22} = \frac{4}{15}, p_{21} = \frac{8}{15}$$

$$p_{34} = 1 - (0.1 + 0.2 + 0.2) = 0.5$$

$$p_{41} + p_{42} + p_{43} + p_{44} = p_{41} + \frac{1}{2}p_{41} + \frac{1}{4}p_{41} + \frac{1}{8}p_{41} = 1$$

$$\Rightarrow p_{41}\left(1 + \frac{1}{2} + \frac{1}{4} + \frac{1}{8}\right) = 1$$

$$\Rightarrow p_{41} = \frac{8}{15}, p_{42} = \frac{4}{15}, p_{43} = \frac{2}{15}, p_{44} = \frac{1}{15}$$

$$P = \begin{bmatrix} 0.2 & 0.2 & 0.2 & 0.4 \\ \dfrac{8}{15} & \dfrac{4}{15} & \dfrac{2}{15} & \dfrac{1}{15} \\ 0.1 & 0.2 & 0.2 & 0.5 \\ \dfrac{8}{15} & \dfrac{4}{15} & \dfrac{2}{15} & \dfrac{1}{15} \end{bmatrix}.$$

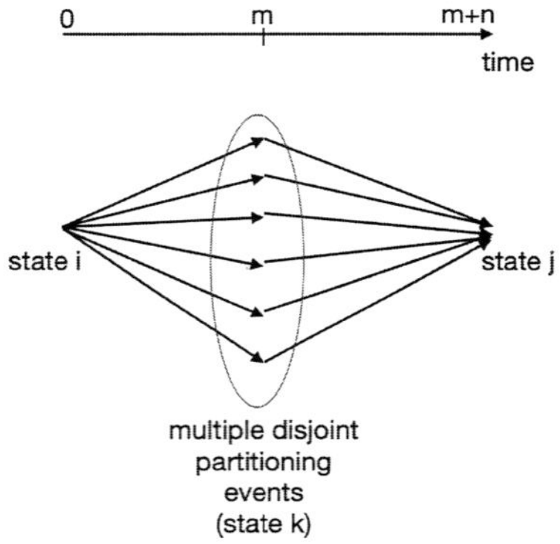

3.4.1 Chapman–Kolmogorov equation

Multi-step transition probabilities for Markov models may also be computed by considering the Chapman–Kolmogorov equation given below.

$$p_{i,j}^{m+n} = \sum_{k \in S} p_{i,k}^{m} p_{k,j}^{n}. \tag{3.6}$$

Let us find out why the above may be true. Consider that while transitioning from state i to state j, we pass through an intermediary state k. This k may be any of the states from the set S (state space). We will consider the intermediary event $X_m = k$ as a partitioning event.[3]

$$p_{i,j}^{m+n} = P(X_{n+m} = j \mid X_0 = i)$$

$$= \sum_{k \in S} P\left(X_{m+n} = j, X_m = k \mid X_0 = i\right)$$

$$= \sum_{k \in S} P\left(X_{m+n} = j \mid X_m = k, X_0 = i\right) P\left(X_m = k \mid X_0 = i\right)$$

$$= \sum_{k \in S} p_{k,j}^{n} p_{i,k}^{m} = \sum_{k \in S} p_{i,k}^{m} p_{k,j}^{n}.$$

Markov property

Figure 3.7 Schematic illustration of the Chapman–Kolmogorov equation to compute multi-step transition probabilities through intermediary partitioning events $X_m = k$, $k \in S$.

[3] See law of total probability:
$$P(A) = \sum_{B_i \in S} P(A \mid B_i) P(B_i),$$
where the B_is are the partitioning events.

Example 3.4   Conditional expectation from a stochastic matrix

Consider the Markov chain $\{X_n\}_{n\geq 0}$ with state space $S = \{1, 2, 3\}$ and the probability transition matrix P.

$$P = \begin{array}{c} \\ 1 \\ 2 \\ 3 \end{array} \begin{array}{ccc} 1 & 2 & 3 \\ \begin{bmatrix} 0.1 & 0.4 & 0.5 \\ 0.3 & 0.3 & 0.4 \\ 0.2 & 0.7 & 0.1 \end{bmatrix} \end{array}.$$

i.   Compute $P\left(X_7 = 3, X_5 = 2 \mid X_4 = 1, X_3 = 2\right)$.

ii. Compute $E\left(X_3 \mid X_2 = 2\right)$.

Solution:

i.   Using the definition of conditional probability, we get

$$P\left(X_7 = 3, X_5 = 2 \mid X_4 = 1, X_3 = 2\right) = \frac{P\left(X_7 = 3, X_5 = 2, X_4 = 1, X_3 = 2\right)}{P\left(X_4 = 1, X_3 = 2\right)}$$

$$= \frac{P\left(X_7 = 3, X_5 = 2, X_4 = 1, X_3 = 2\right)}{P\left(X_5 = 2, X_4 = 1, X_3 = 2\right)}$$

$$\times \frac{P\left(X_5 = 2, X_4 = 1, X_3 = 2\right)}{P\left(X_4 = 1, X_3 = 2\right)}$$

$$= P\left(X_7 = 3 \mid X_5 = 2, X_4 = 1, X_3 = 2\right)$$

$$\times P\left(X_5 = 2 \mid X_4 = 1, X_3 = 2\right).$$

Using the Markov property, we get

$$P\left(X_7 = 3, X_5 = 2 \mid X_4 = 1, X_3 = 2\right) = P\left(X_7 = 3 \mid X_5 = 2\right) \times P\left(X_5 = 2 \mid X_4 = 1\right)$$

$$= p_{23}^2 \times p_{12}.$$

Now using the Chapman–Kolmogorov equation, we get

$$p_{23}^2 = \sum_{k=1}^{3} p_{2,k} p_{k,3} = 0.3 \times 0.5 + 0.3 \times 0.4 + 0.4 \times 0.1 = 0.15 + 0.12 + 0.04 = 0.31.$$

Therefore,

$$P\left(X_7 = 3, X_5 = 2 \mid X_4 = 1, X_3 = 2\right) = p_{23}^2 \times p_{12} = 0.31 \times 0.4 = 0.124.$$

ii.
$$E\left(X_3 \mid X_2 = 2\right) = \sum_{x \in S} x P\left(X_3 = x \mid X_2 = 2\right)$$

$$= 1 \times P\left(X_3 = 1 \mid X_2 = 2\right)$$

$$+ 2 \times P\left(X_3 = 2 \mid X_2 = 2\right)$$

$$+ 3 \times P\left(X_3 = 3 \mid X_2 = 2\right)$$

$$= 1 \times p_{21} + 2 \times p_{22} + 3 \times p_{23}$$

$$= 1 \times 0.3 + 2 \times 0.3 + 3 \times 0.4$$

$$= 0.3 + 0.6 + 1.2.$$

$$E\left(X_3 \mid X_2 = 2\right) = 2.1.$$

The Chapman–Kolmogorov equation will be revisited when we study continuous time Markov chains (CTMCs). This equation will be used to derive the *detailed balance* condition and finds many applications as a tool to investigate behavior of equilibrium systems.

3.5   Distribution of states

Consider that for a given Markov model, the stochastic matrix $\mathbb{P}$ is known. Further, let us suppose that an initial probability distribution of states is also known. We may be interested in knowing the probability distribution of states at a later time (or in the long run).

Mathematically, we may denote the initial distribution of k-states, $\{s_1\, s_2\, \dots\, s_k\}$, by

$$\vec{\mu}^{(0)} = \left(\mu_1^{(0)}\ \mu_2^{(0)}\ \dots\ \mu_k^{(0)}\right) = \left(P\left(X_{01} = s_1\right)\ P\left(X_{02} = s_2\right)\ \dots\ P\left(X_{0k} = s_k\right)\right),$$

where $\sum_{i=1}^{k} \mu_i^{(0)} = 1$ due to the unitarity axiom of probability. Here, the 0s in the superscripts and subscripts represent the *initial* time instant and the indices $i = \{1, 2, \dots, k\}$ refer to the states. Then, the probability distribution of states after n-steps may be written as

$$\vec{\mu}^{(n)} = \vec{\mu}^{(0)}\mathbb{P}^n. \tag{3.7}$$

To find the long-run distribution of states, we take the limit $n \to \infty$ in equation (3.7).

Figure 3.8 Daily weather forecasts can be modeled based on a Markov model.

Example 3.5   Weather model

Consider a simple weather model that predicts weather on a given day as follows: (i) the weather stays the same on any given day as the previous day 75% of the time and (ii) the weather changes from day to day 25% of the time. For simplicity, we may only consider two weather patterns in this model, namely, sunny and rainy. What is the long time weather forecast given that on a certain day it is observed to be sunny?

We must first begin by constructing the stochastic matrix $\mathbb{P}$ based on the transitions between the weather patterns (states): s for sunny and r for rainy.

$$\mathbb{P} = \begin{matrix} & s & r \\ s & \begin{pmatrix} 0.75 & 0.25 \\ r & 0.25 & 0.75 \end{pmatrix} \end{matrix}. \tag{3.8}$$

$$\begin{aligned}
\vec{\mu}^{(1)} &= \vec{\mu}^{(0)}\mathbb{P} = \begin{pmatrix} 1 & 0 \end{pmatrix}\begin{pmatrix} 0.75 & 0.25 \\ 0.25 & 0.75 \end{pmatrix} = \begin{pmatrix} 0.75 & 0.25 \end{pmatrix} \\
\vec{\mu}^{(2)} &= \vec{\mu}^{(0)}\mathbb{P}^2 = \vec{\mu}^{(1)}\mathbb{P} = \begin{pmatrix} 0.625 & 0.375 \end{pmatrix} \\
&\quad\vdots \\
\vec{\mu}^{(\infty)} &= \vec{\mu}^{(0)}\mathbb{P}^\infty = \begin{pmatrix} 0.5 & 0.5 \end{pmatrix}.
\end{aligned}$$

This is the equilibrium distribution of states.* Thus, the long-time weather forecast is that it is equally likely to be sunny or rainy according to this model.

---

*What is $\vec{\mu}^{(\infty)}\mathbb{P}$?

From the aforementioned discussions on the applications of a Markov model, we may glean that the stochastic matrix encodes almost all the information about the underlying Markov process.

Example 3.6   Badminton game, what is a winning strategy?

Consider a Markov model of a game of badminton. For simplicity, let us consider that a player chooses to play one of three shots, namely, *smash (S)*, *drop (D)*, and *lift (L)*. The objective is to devise a winning strategy given a match situation. Based on the data generated over several games, the following table lists the probability of a return shot played by a player given a certain type of shot played by their opponent.

Shot	Return shot	Probability
D	D	1/3
D	L	1/3
D	S	0
L	D	1/5
L	L	1/5
L	S	2/5
S	L	2/5
S	D	1/5
S	S	0

i.   Identify an appropriate state space for the Markov model.

ii. Construct the stochastic matrix $\mathbb{P}$.

iii. Given a *lifted* serve, what are the chances that there is a winner in three shots?

iv. In a rally, if a player receives a lift from their opponent, which shot option maximizes their chance of winning the rally in the return shot?

Solution: The Markov analysis is as follows.

i.   State space, $\mathcal{S} = \{D, L, S, W\}$, where W refers to a *winning* shot.

ii. The stochastic matrix may be computed based on the data table.*

Figure 3.9 Should I *smash* or feign and *drop* to win the rally now?

$$\mathbb{P} = \begin{array}{c} \\ D \\ L \\ S \\ W \end{array} \overset{\begin{array}{cccc} D & L & S & W \end{array}}{\begin{bmatrix} 1/3 & 1/3 & 0 & 1/3 \\ 1/5 & 1/5 & 2/5 & 1/5 \\ 1/5 & 2/5 & 0 & 2/5 \\ 0 & 0 & 0 & 1 \end{bmatrix}} \tag{3.9}$$

iii. $\vec{\mu}^{(3)} = \vec{\mu}^{(0)}\mathbb{P}^3 = \begin{pmatrix} 0.1316 & 0.1476 & 0.1067 & 0.6142 \end{pmatrix}$. $\mu_W^{(3)} = p_{LW}^3 = 0.6142$. This is the (L,W) entry of the $\mathbb{P}^3$ matrix.**

iv. Essentially, we must compute $P(X_2 = W, X_1 = ?|X_0 = L)$, where ? may be any one of L, D, or S. So we will compute each of $(p_{LD} \times p_{DW})$, $(p_{LS} \times p_{SW})$, $(p_{LL} \times p_{LW})$ and pick the strategy that maximizes the resulting probability. It turns out that the respective probabilities are 0.06667, 0.16, 0.04. Hence, a return *smash* will most likely win us the rally in that shot.***

---

*It is a good practice to check that each of the rows of $\mathbb{P}$ adds up to unity!

**In an actual game, an overall optimal strategy must account for maximizing the chances of winning the rally in a certain number of shots as well as possibly continuing the rally long enough to tire out the opponent!

***An AI assisted training session based on such a Markov model will enable a player to devise optimal shot selections during an actual game!

3.6 Recurring events in Markov chains

In a finite state system with stochastic transitions between states, a Markov chain may visit certain states (events) multiple times. The time interval (steps) between such successive visits to a given state may itself be a random number. Repeated visits and the inter-arrival time between such visits may be of interest depending on the application. For example, a kiosk of a bank teller may have a long queue of customers waiting to be served. The number of persons in the queue may represent the states of a Markov chain. The arrival process can be modeled in terms of a Poisson process while the departures may depend on exponentially distributed service times. The bank operations policy may rely on analyzing the time interval between successive visits to state 0 and/or by the sojourn time (total time a customer spends in the system), which may involve analyzing recurring events in the Markov model of the bank teller system.[4]

Figure 3.10 A bank teller queue may be a Markov model with recurring events such as a state when the queue is empty.

[4] It turns out that the Markov model of a teller system may be a CTMC depending on how we model the time elapse process. We will study CTMCs in greater detail in a subsequent chapter.

3.6.1 *Definition: Hitting probability*

Let $\{X_n\}_{n \in \{0, \mathbb{I}^+\}}$ represent a Markov chain with state space $\mathcal{S}$. Let $A \subset \mathcal{S}$. Further, let us define the following:

$T_A :=$ first time the chain hits A starting from outside (or inside) A

$$:= \min\{n \geq 0 | X_n \in A\}, \tag{3.10}$$

with $T_A = 0$ if $X_0 \in A$ and $T_A = \infty$ if $\{n \geq 0 \,|\, X_n \in A\} = \{\}$. Then, we may define the *probability of hitting state A* through a state $l \in A$ starting from a state $k \in \mathcal{S}$ as below:

$$g_k(l) = P\left(X_{T_A} = l \mid X_0 = k\right).$$ (3.11)

Initial state

Final state

The above definition helps us to calculate the chance of hitting a certain state A beginning from a given state. Why is this useful to know? In the context of the bank teller example above, if we find that $g_s(0) = 0$ for any $s > 0$,[5] then this will likely invite the attention of the bank manager to change the operational policy of the bank (e.g., by introducing additional tellers). This may be required to ensure that the bank teller receives a much-needed break from serving customers during a long shift.

[5] Number of customers s in an operational queue is a positive number.

3.6.2 *Iterative formula for hitting probability*

Let $k \in S \setminus A$.[6] We have $T_A \geq 1$ given $X_0 = k$.

[6] $S \setminus A \equiv S - \{A\}$, i.e., the set S minus the contents of the set A.

$$g_k(l) = P\left(X_{T_A} = l \mid X_0 = k\right) = \sum_{m \in S} P\left(X_{T_A} = l, X_1 = m \mid X_0 = k\right)$$

Sum over partitioning events

$$= \sum_{m \in S} P\left(X_{T_A} = l \mid X_1 = m, X_0 = k\right) P\left(X_1 = m \mid X_0 = k\right)$$

See the law of total probability

$$= \sum_{m \in S} \underbrace{P\left(X_{T_A} = l \mid X_1 = m\right)}_{g_m(l)} \underbrace{P\left(X_1 = m \mid X_0 = k\right)}_{p_{km}}$$

Markov property

$$= \sum_{m \in S} g_m(l) p_{km}.$$ (3.12)

Therefore, the iterative formula for $g_k(l)$ is given as follows.

$$\boxed{g_k(l) = \sum_{m \in S} p_{km} g_m(l)}, \text{ where } k \in S \setminus A, \; l \in A.$$ (3.13)

3.6.3 *Definition: Absorbing state*

We have seen absorbing states in earlier examples. Formally, consider the case $p_{kl} = \mathbb{I}_{k=l}$[7] for all $k, l \in A$. Here, the state k is an absorbing state, i.e., $\{X_n\}$ is *trapped* (absorbed) in $A \subset S$.[8]

[7] Here $\mathbb{I}_{k=l} = 1$ only when $k = l$. $\mathbb{I}$ is an *indicator* function.

[8] It may be interesting to note the following:
$$\boxed{\sum_{l \in A} g_k(l) + P(T_A = \infty \mid X_0 = k) = 1}$$

3.6.4 *Iterative formula for mean hitting times and mean absorption times*

We have stated earlier that the time duration between successive visits to a certain state of a Markov chain is a random variable. In many applications, we may be interested to know the expected value of this random variable. Let us define this expected value as

$$h_k(A) := E(T_A | X_0 = k).$$
(3.14)

Clearly, $h_k(A) = 0 \ \ \forall k \in A \subset S$. Further, $\forall k \in S \setminus A$,

$$h_k(A) = E\left(T_A | X_0 = k\right)$$

$$= \sum_{m \in S} E\left(T_A | X_1 = m, X_0 = k\right) P\left(X_1 = m | X_0 = k\right)$$

Law of total expectation

$$= \sum_{m \in A} \underbrace{E\left(T_A | X_1 = m\right)}_{1} p_{km} + \sum_{m \in S \setminus A} \underbrace{E\left(T_A | X_1 = m\right)}_{1 + h_m(A)} p_{km}$$

Markov property

$$= \sum_{m \in A} p_{km} + \sum_{m \in S \setminus A} \left(1 + h_m(A)\right) p_{km}$$

$$= \sum_{m \in S} p_{km} + \sum_{m \in S \setminus A} h_m(A) p_{km}$$

$$= 1 + \sum_{m \in S \setminus A} p_{km} h_m(A) + \sum_{m \in A} p_{km} \overset{0}{\cancel{h_m(A)}}$$

Axiom: $\sum_{i \in S} p_{ki} = 1$

$$= 1 + \sum_{m \in S} p_{km} h_m(A).$$

Adding a term whose value is 0

Summarizing, we have

$$\boxed{h_k(A) = 1 + \sum_{m \in S} p_{km} h_m(A)}, \text{ for all } k \in S \setminus A.$$
(3.16)

In the above derivation, $E(T_A | X_1 = m) = 1$ in the first summation because the Markov chain has already moved one step forward ($X_0 \rightarrow X_1$) and has then hit the desired state $m \in A$. In the second summation, $E(T_A | X_1 = m) = 1 + h_m(A)$ because the chain has moved one step forward and the counting process must be reset again until the chain hits the desired state A.

3.6.5 *Definition: First return time and its mean*

The time (in number of steps) taken to make a *first return* to a certain state $y \in S$ is defined as follows:

$$T_y^r := \min\{n \geq 1 | X_n = y\}; \quad y \in S,$$
(3.17)

with $T_y^r = \infty$ if $X_n \neq y \ \ \forall n \geq 1$. Note $T_y^r = T_y$ if $X_0 \neq y$.

T_y^r is a random variable. Let us define $\mu_x(y) = E(T_y^r | X_0 = x) \geq 1$. Clearly, when $x = y$, we have the definition of mean return time. Following the spirit of the derivation in section 3.6.4, it is possible to derive an iterative formula for the mean return time. Here, we simply state the final result.

$$\boxed{\mu_x(y) = 1 + \sum_{m \in S, m \neq y} p_{xm} \mu_m(y)} \tag{3.18}$$

In the following section, we will work out an illustrative example using the aforementioned iterative formulae.

He was too young to have been blighted
by the cold world's corrupt finesse;
his soul still blossomed out, and lighted
at a friend's word, a girl's caress.
In heart's affairs, a sweet beginner,
he fed on hope's deceptive dinner;
the world's éclat, its thunder-roll,
still captivated his young soul.
He sweetened up with fancy's icing
the uncertainties within his heart;
for him, the objective on life's chart
was still mysterious and enticing—
something to rack his brains about,
suspecting wonders would come out.

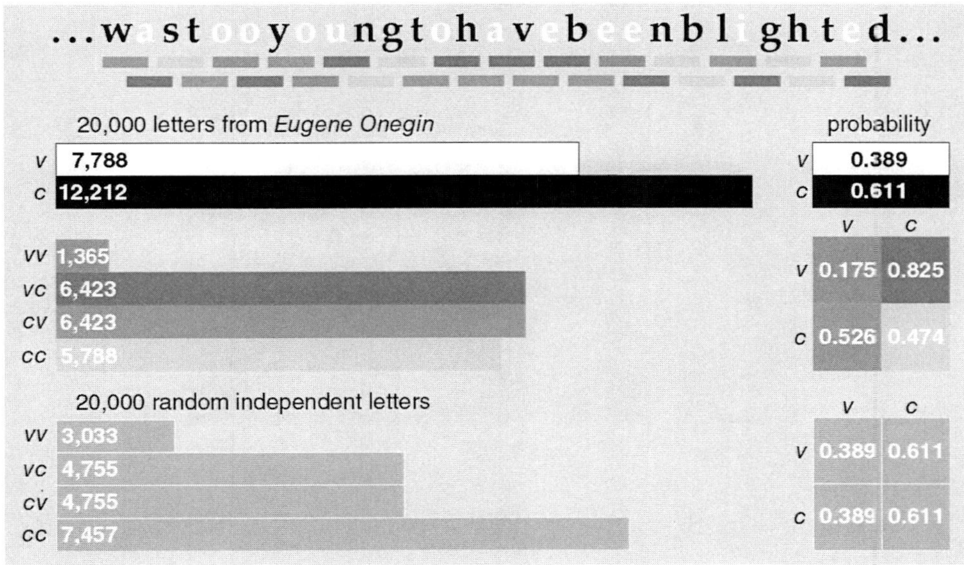

Figure 3.11 Markov conducted a statistical experiment to understand the structure of language by analyzing Alexander Pushkin's poem *Eugene Onegin*. Shown here is a single stanza from the poem (English translated version) with the characteristic rhyming of the words. A Markov model analysis of the language used in the poem (colored illustrations) shows the distinctively different *alternation* noticed in the poem compared to a random sequence of letters (gray illustrations) of the same length. Here *v* refers to vowels and *c* refers to consonants. The stochastic matrices $\mathbb{P}$ are shown on the right. Can you compute the mean hitting times and mean return times to consonants and vowels? (*courtesy:* Brian Hayes, First links in the Markov chain, *American Scientist* **101**, no. 2 (March–April 2013): 92, DOI: 10.1511/2013.101.92). See color plate on page 295.

Example 3.7 Return times in a game of badminton

Let us divide each side of a badminton court into two quadrants (*play zones*) labeled ① and ② as shown in the picture below (see Figure 3.12). These quadrants may be regarded as states of a simple Markov model. Data collected over several games of badminton may enable us to populate a simple stochastic matrix as follows.

$$\mathbb{P} = \begin{array}{c} \\ ① \\ ② \end{array} \begin{array}{cc} ① & ② \\ \begin{bmatrix} 2/3 & 1/3 \\ 1/3 & 2/3 \end{bmatrix} \end{array} \tag{3.19}$$

i.   During a rally, given a shot by a player from play zone ①, after how many shots (on an average) does the shuttle return to quadrant ① on either side of the net?

Now, let us redefine the Markov chain by constructing new states corresponding to the direction of shots played. For example, the states of the new model are ⑪ (corresponding to a shot $1 \rightarrow 1$) and so on.

ii. Construct the stochastic matrix $\mathbb{P}_{\text{new}}$ for this new model.

iii. Given a valid service $1 \rightarrow 1$, what is the average number of shots before either player can expect a shot $2 \rightarrow 1$ from their opponent?

Figure 3.12 A badminton court with designated *play zones* (states). In a rally, the shuttle may be considered to make random visits to these states. A player's preparedness to receive the shuttle, effectively, may be guided by mean return and hitting time estimates, which may be included during their practice sessions before an official game.

Solution:

i.   We begin with the simple model $\mathbb{P}$ and use equation (3.18) to compute $\mu_1(1)$.

$$\mu_1(1) = 1 + p_{12}\mu_2(1) = 1 + \frac{1}{3}\mu_2(1)$$

$$\mu_2(1) = 1 + p_{22}\mu_2(1) = 1 + \frac{2}{3}\mu_2(1)$$

whence $\mu_2(1)=3$ and $\mu_1(1)=2$, i.e., on an average, every second shot in a rally returns to play zone ① given a serve from quadrant ①.

ii. Let us now construct $\mathbb{P}_{new}$. The relevant states are ⑪, ⑫, ㉑, and ㉒. Now, for a state transition ⑪→⑪, a sequence of following shots must be played: $1 \rightarrow 1$ and $1 \rightarrow 1$. Note that the naming convention of the states is set up in such a way that the last numeral of the current state must be the same as the first numeral of the next state. This means transitions like ⑪ → ㉑ and ㉑ → ⑫ are not possible. Consequently, the transition ⑪ → ⑪ can happen with a probability 2/3 because this transition is wholly concurrent with the event whereby the most recent shot is $1 \rightarrow 1$, which has a probability 2/3 associated with it. Similar arguments enable us to construct $\mathbb{P}_{new}$ as follows.

$$\mathbb{P}_{new} = \begin{array}{c} \\ \text{⑪} \\ \text{⑫} \\ \text{㉑} \\ \text{㉒} \end{array} \begin{array}{cccc} \text{⑪} & \text{⑫} & \text{㉑} & \text{㉒} \\ \begin{bmatrix} 2/3 & 1/3 & 0 & 0 \\ 0 & 0 & 1/3 & 2/3 \\ 2/3 & 1/3 & 0 & 0 \\ 0 & 0 & 1/3 & 2/3 \end{bmatrix} \end{array}. \tag{3.20}$$

iii. Now using equation (3.18), we have the following system of equations:

$$\mu_{11}(21)=1+\frac{2}{3}\mu_{11}(21)+\frac{1}{3}\mu_{12}(21),$$

$$\mu_{12}(21)=1+\frac{2}{3}\mu_{22}(21), \text{ and}$$

$$\mu_{22}(21)=1+\frac{2}{3}\mu_{22}(21),$$

whose solution is $\mu_{22}(21)=3$, $\mu_{12}(21)=3$, and $\mu_{11}(21)=6$. Therefore, given a valid service $1 \rightarrow 1$, both players will have to wait 6 shots on an average before they may expect a shot $2 \rightarrow 1$.

3.7 Chapter project: Automatic prediction of aerodynamic control laws of an aircraft using the Viterbi algorithm

3.7.1 *Interlude: The Viterbi algorithm*

Now that we have learnt the fundamental ideas of Markov chains, we are in a position to deduce the Viterbi algorithm. We will use the Viterbi algorithm to predict the control laws prevailing in the aircraft cockpit.

I Stochastic and emission matrices: We will consider a certain stochastic process with the following state space of dimension K, $S = \{s_1, s_2, ..., s_K\}$. Associated with this process is a T dimensional observation set $Y = \{y_1, y_2, ..., y_T\}$ from amongst a possible N dimensional observation space $O = \{o_1, o_2, ..., o_N\}$. Note: $y_n \in O$.

[9] As a trial example, we may think of a state space S = {rainy, sunny, cloudy}, an observational space O = {walking, shopping, cleaning} and a sequence of observations of activity patterns of Billoo, the handyman as Y = {walking, walking, shopping, walking, cleaning} over the past five days. The objective here is to find the most likely sequence of (hidden) states X corresponding to the sequence of observables Y. E.g., one possible likely outcome may be X = {sunny, sunny, cloudy, sunny, rainy}. But instead of guesswork, the readers are encouraged to use the predictions of the Viterbi algorithm to list the weather pattern for the corresponding days.

Figure 3.13 American electrical engineer Andrew Viterbi invented the *Viterbi algorithm, which is a dynamic programming algorithm* originally used for convolutional codes over noisy digital communication systems. It has since found multiple applications in natural language processing, computational linguistics, bioinformatics, speech recognition, etc. (*courtesy*: Tim Rue © Getty Images).

Further, consider an initial probability distribution given by $\Pi = \{\pi_1, \pi_2, ..., \pi_K\}$.[9] **The probability transition matrix** $\mathbb{P}$ **is a** $K \times K$ **matrix with entries**

$$p_{ij}(t) := \text{probability of transitioning from state } s_i \text{ to state } s_j = \text{Prob}(x_t = s_j \mid x_{t-1} = s_i),$$

and the emission matrix $\mathbb{E}$ **is a** $K \times N$ **matrix with entries**

$$e_{ij}(t) := \text{probability of observing } o_j \text{ from state } s_i = \text{Prob}(y_t = o_j \mid x_t = s_i).$$

Succinctly, we will often write $s_i \equiv i$ **and** $o_j \equiv j$, **where it must be understood that** $x_t = i$ **refers to the random variable** x_t **that takes the state** s_i **and** $y_t = j$ **refers to the random variable** y_t **that is assigned the observable** o_j. **The goal of the prediction algorithm is to forecast the most likely sequence of states (events)** $X = \{x_1, x_2, ..., x_T\}$, $x_n \in S$, **given a prescribed sequence of observables** Y, **i.e., we need to compute**

$$\text{argmax}_X \, Prob(X|Y) = \text{argmax}_X \, Prob(Y|X)\, Prob(X) = \text{argmax}_X \, Prob(Y, X).$$

Here $\text{argmax}(f(x))$ **returns the value of** x **at which the function** $f(x)$ **attains its maximum. In this project, we will implement the Viterbi algorithm to predict the most likely sequence of states that corresponds to a sequence of associated observables assuming a Markovian stochastic model (also known as the** *Hidden Markov Model* **[HMM]).**

II Construction and essential calculations of the Viterbi algorithm: In what follows, we will fix the notation $Prob(X_1 = x_1) \equiv Prob(x_1) \equiv \pi_1$. Note that if $T = 2$, then

$$Prob(Y, X) \equiv Prob(y_1, y_2, x_1, x_2)$$

$$= Prob(y_1, y_2, x_2 \mid x_1)Prob(x_1)$$

$$= Prob(y_1, y_2 \mid x_2, x_1)Prob(x_2 \mid x_1)Prob(x_1)$$

$$= Prob(y_1 \mid y_2, x_2, x_1)Prob(y_2 \mid x_2, x_1)p_{12}\pi_1$$

$$= Prob(y_1 \mid x_1, x_2, y_2)Prob(y_2 \mid x_2)p_{12}\pi_1$$

$$= Prob(y_1 \mid x_1)Prob(y_2 \mid x_2)p_{12}\pi_1. \tag{3.21}$$

In general, we have

$$Prob(Y, X) \equiv Prob(Y = y_1, ..., y_T, X = x_1, ..., x_T)$$

$$= \underbrace{Prob(x_1)}_{\pi_1}Prob(y_1 \mid x_1)\underbrace{Prob(x_2 \mid x_1)}_{p_{12}}Prob(y_2 \mid x_2)\cdots Prob(y_T \mid x_T). \tag{3.22}$$

The Viterbi algorithm involves *recursively* **computing the Viterbi entries** $V_{k,t}$

$$V_{k,t} := \max Prob\big((y_1, ..., y_t), (x_1, ..., x_t = k)\big)$$

$$= \text{probability of the best (most likely) sequence of states (ending with state } k, \text{ i.e., } x_t = k) \text{ corresponding to the sequence of observables } (y_1, ..., y_t).$$

II.1 *Recursive computation of $V_{k,t}$*: By comparing the terms on the RHS of equation (3.22) and the definition of the Viterbi entries above, we see that $V_{k,t}$ can be obtained recursively, and consequently using the argmax function, we can find the most likely sequence of events. The algorithm includes calculation of the following three important terms.

- $$V_{k,t} = \max_{\alpha \in S}\left(Prob\left(y_t = j \mid x_t = k\right)p_{\alpha k}V_{\alpha,t-1}\right) = \max_{\alpha \in S}\left(e_{kj}p_{\alpha k}V_{\alpha,t-1}\right)$$

  with $V_{k,1} \overset{set}{=} Prob(y_1 = o_m \mid x_1 = k)\pi_k = e_{km}\pi_k$.

- $$x_T = \underset{\alpha \in S}{\mathrm{argmax}}(V_{\alpha,T}).$$

- $$x_{t-1} = \text{back\_pointer}(x_t, t) = \text{value of } x \text{ used to compute } V_{k,t}\ \forall t > 1.$$

II.2 *Aesthetics of dynamic programming algorithms*: The Viterbi algorithm belongs to a class of algorithms known as *dynamic programming*. This class of algorithms was developed by American applied mathematician Richard Bellman in 1953. The classic problem solved by this family of algorithms is the *traveling salesman problem.*[10] Another classic puzzle that can be solved by dynamic programming methodology is the tower of Hanoi game. These are just a few examples of many diverse applications of the dynamic programming method in general, and the Viterbi algorithm in particular.

In the subsequent sections, we will present you with a strategy to implement the Viterbi algorithm in your computer, test your code using the weather model example presented in the margin in the previous page, and, finally, use the tested version of the algorithm to predict the outcomes of the prevailing control laws in the aircraft as introduced in the prologue.

Figure 3.14 The *tower of Hanoi* game (*courtesy*: based on Science Buddies and Sabine De Brabandere, The Tower of Hanoi, *Scientific American* [2017]).

[10] David L. Applegate et al., *The Traveling Salesman Problem: A Computational Study* (Princeton University Press, 2007). Robert Bosch of Oberlin College and his collaborators have used the traveling salesman problem to generate artwork. This is yet another example of the proximal interrelationship between mathematics, computational algorithms, and aesthetics (*courtesy*: http://www.math.uwaterloo.ca/tsp/data/art/, accessed July 1, 2023). Further reading: Robert Bosch and Adrianne Herman, Continuous line drawings via the traveling salesman problem, *Operations Research Letters* **32** (2004): 302–303.

Software Implementation

Pseudocode of the Viterbi algorithm

INPUT : S, Π, $\mathbb{E}$, $\mathbb{P}$, $\mathbf{Y} = \{\, y_1, y_2, \ldots, y_T \}$.

Part I: Initialization.

```
for each i of K states
    viterbi_prob(i,1)  = π_i * e_{iy_1}
    viterbi_path(i,1)  = 0
end for
```

Part II: Compute Viterbi probabilities and Viterbi path.

```
for each j of T-1 observations starting with T=2
    for each i of K states
```
$$\text{viterbi\_prob}(i,j) = \max_{\alpha \in S}\left(e_{iy_j} * p_{\alpha i} * \text{viterbi\_prob}(\alpha, j-1)\right)$$
$$\text{viterbi\_path}(i,j) = \underset{\alpha \in S}{\mathrm{argmax}}\left(e_{iy_j} * p_{\alpha i} * \text{viterbi\_prob}(\alpha, j-1)\right)$$
```
    end for
end for
```
$$x_T = s_{z_T} \quad \text{where} \quad z_T := \underset{\alpha \in S}{\mathrm{argmax}}\left(\text{viterbi\_prob}(\alpha, T)\right)$$

The appearance of e_{ij} in the computation of `viterbi_path(i,j)` is unnecessary because it is non-negative and independent of α (so you may choose to skip it).

Part III: *Retracking the most likely path* **X**.

```
for each j of T-1 observations from T to 2
    x_{j-1} = s_{z_{j-1}} where z_{j-1} = viterbi_path(z_j, j)
end for


OUTPUT:  X = {x_1, x_2, . . . , x_T}
```

The student may test the veracity of the Viterbi implementation by first attempting the trial example introduced earlier.[11]

[11] Use the Markovian model explained above to predict the weather for the last five days. Assume the initial weather distribution $\Pi = \{0.43, 0.57\}$, the probability transition matrix $\mathbb{P} = \begin{pmatrix} 0.2 & 0.8 \\ 0.4 & 0.6 \end{pmatrix}$, where state 1 is *rainy* and state 2 is *sunny*, and the probability emission matrix $\mathbb{E} = \begin{pmatrix} 0.2 & 0.4 & 0.4 \\ 0.3 & 0.25 & 0.45 \end{pmatrix}$, where the columns (observations) are labeled in order of *walking*, *shopping*, and *cleaning*, respectively.

[12] $(i) \rightarrow (i)$ even if $p_{ii} = 0$.

3.8 Classification of Markov states and advanced topics

The behavior of a Markov chain is characterized by the properties of the stochastic matrix $\mathbb{P}$ and its states. The states of a Markov chain can be classified based on the entries of $\mathbb{P}$.

3.8.1   *Definition: Communicating states*

A state $j \in \mathcal{S}$ is *accessible* from a state $i \in \mathcal{S}$, i.e., $(i) \rightarrow (j)$, if there exists a finite integer $n \geq 0$ such that $p_{ij}^n := P(X_n = j | X_0 = i) > 0$.[12]

Further, if $(i) \rightarrow (j)$ and $(j) \rightarrow (i)$, then $(i) \leftrightarrow (j)$, i.e., i and j *communicate*. When two states communicate with each other, they are said to belong to the same *class*.

Example 3.8   Communicating states

Consider below a stochastic matrix of a Markov chain with states labeled as 1, 2, 3, and 4.

$$\mathbb{P} = \begin{array}{c} \\ 1 \\ 2 \\ 3 \\ 4 \end{array} \begin{array}{cccc} 1 & 2 & 3 & 4 \\ \left[\begin{array}{cccc} 1/3 & 1/3 & 1/3 & 0 \\ 1/2 & 0 & 0 & 1/2 \\ 2/5 & 1/5 & 0 & 2/5 \\ 1/4 & 1/4 & 1/2 & 0 \end{array}\right] \end{array}$$

In this model:

i. $(3) \leftrightarrow (3)$ even though $p_{33} = 0$.

ii. $(2) \rightarrow (3)$ even though $p_{23} = 0$ because $p_{24} > 0$ and $p_{43} > 0$; hence, there exists N such that $p_{23}^N > 0$. In fact, $p_{23}^2 = 0.4167 > 0$.

     Further, $p_{32} = 1/5 > 0$. Therefore, $(2) \leftrightarrow (3)$.

Since the binary relation $\leftrightarrow$ satisfies *reflexivity, symmetry,* and *transitivity* properties, $\leftrightarrow$ is an *equivalence relation.*[13]

3.8.2 *Definition: Irreducible and reducible Markov chains*

A Markov chain is *irreducible* if all states belong to one class, i.e., if all states communicate with each other.[14]

Example 3.9 Consider a Markov chain with a stochastic matrix

$$
\mathbb{P} = \begin{array}{c}
 \\ 1 \\ 2 \\ 3 \\ 4 \\ 5
\end{array}
\begin{array}{ccccc}
1 & 2 & 3 & 4 & 5 \\
\left[\begin{array}{ccccc}
0 & 1 & 0 & 0 & 0 \\
1/2 & 0 & 1/2 & 0 & 0 \\
0 & 2/3 & 0 & 0 & 1/3 \\
1/3 & 0 & 0 & 0 & 2/3 \\
0 & 0 & 0 & 1 & 0
\end{array}\right]
\end{array} .
$$

In this example, all states communicate with each other.

A Markov chain that is not irreducible is said to be *reducible*, i.e., there is at least one state (or a group of states) from which the chain cannot revisit other states which are not in that group.

In Figure 3.15, we have used a graphical representation of a Markov chain showing the states within circles and the probabilities of transition between states are represented by the numbers along with the arrows.

3.8.3 *Mean number of returns to a state*

Let $q_{ij} \equiv p_{ij}^{n} = P(X_n = j | X_0 = i)$ for some $n \geq 1$ represent the probability of return to state j in a finite time starting from state i. Now let us define the number of visits to state j by the chain $\{X_n\}_{n \in I, n \geq 0}$ as follows.

$$
\begin{aligned}
E(R_j | X_0 = i) &= \sum_{m=0}^{\infty} m P(R_j = m | X_0 = i) \\
&= \sum_{m=1}^{\infty} m q_{ij} q_{jj}^{m-1} \left(1 - q_{jj}\right) \\
&= \left(1 - q_{jj}\right) q_{ij} \sum_{m=1}^{\infty} m q_{jj}^{m-1} \\
&= \left(1 - q_{jj}\right) q_{ij} \frac{1}{\left(1 - q_{jj}\right)^2} \\
&= \frac{q_{ij}}{1 - q_{jj}}.
\end{aligned}
\tag{3.23}
$$

Here we have used the identity $\sum_{m=1}^{\infty} m r^{m-1} = \dfrac{1}{\left(1-r\right)^2}$, where $|r| \leq 1$. The terms to the RHS of the second equality may require further explanation. To begin with, the

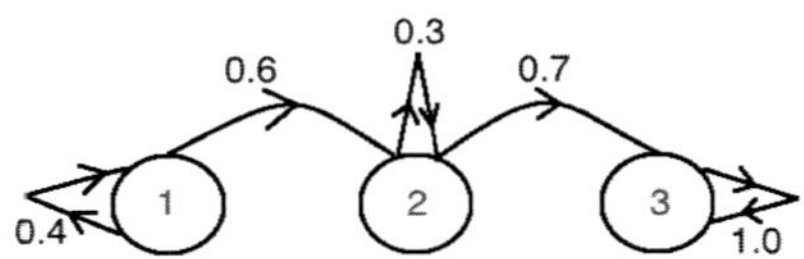

Figure 3.15 In this Markov model, state 3 is an absorbing state and does not communicate with states 1 and 2.

first visit to state j from state i must happen in n steps with probability q_{ij}. This must be followed by m-1 revisits to state j starting from state j with probability q_{jj}^{m-1}. This is true because the count for the revisits to state j happens beginning with state j as the chain is reset as $X_0 = j$ after the first visit to state j. Since the summand of interest pertains to m visits to state j (and no more), we must account for the probability $(1 - q_{jj})$ of no additional visits to state j after the mth visit.

3.8.4   *Definition: Recurrent states*

[15] Let $\{X_n\}_{n \geq 0,\ n \in I}$ be a Markov chain with finite state space S; then $\{X_n\}_{n \geq 0,\ n \in I}$ has at least one recurrent state.

A recurrent state $i \in S$ is said to be *positive recurrent* if $\mu_i(i) < \infty$ and is *null recurrent* if $\mu_i(i) = \infty$.

$i \in S$ is *recurrent* if $q_{ii} = p_{ii}^n = 1$.[15] Additionally,

1.   state i is recurrent if and only if $E(R_i | X_0 = i) = \infty$.

2.   state i is recurrent if and only if $P(R_i = \infty | X_0 = i) = 1$.

3.8.5   *Definition: Transient states*

A state $i \in S$ is *transient* when it is *not* recurrent, i.e.,

$$P(R_i = \infty | X_0 = i) < 1.$$

Further,

1. $i \in S$ is transient if and only if

$$E(R_i | X_0 = i) < \infty.$$

2. $i \in S$ is transient if and only if

$$\sum_{n=1}^{\infty} p_{ii}^n < \infty.$$

3.8.6   *Periodicity of a Markov chain*

[16] Periodicity is a class property. For example, if states i and j belong to the same class, then they have the same period.

The *period* of a state is the greatest common divisor (denominator) of all integers $n > 0$ for which $p_{ii}^n > 0$.[16]

A Markov chain is *aperiodic* if it has period one.

Example 3.10

Consider a Markov chain with $\mathbb{P} = \begin{array}{c} \\ 1 \\ 2 \\ 3 \end{array} \begin{array}{ccc} 1 & 2 & 3 \\ \left[\begin{array}{ccc} 0 & 1 & 0 \\ 0 & 0 & 1 \\ 1 & 0 & 0 \end{array}\right] \end{array}$. Here $p_{ii} = 0$, and $p_{ii}^2 = 0$ but $p_{ii}^3 = 1 > 0$ and so on. Thus, the chain has period three.

3.9   Chapter project: Automatic prediction of aerodynamic control laws of an aircraft using the Viterbi algorithm

3.9.1   *Epilogue: Results of the Viterbi code for predicting the aircraft control laws*

Consider the following probability transition matrix $\mathbb{P}$ and probability emission matrix $\mathbb{E}$ that is available from the Airbus database.

$$\mathbb{P} = \begin{pmatrix} 0.7 & 0.1 & 0.2 \\ 0.4 & 0.5 & 0.1 \\ 0.2 & 0.3 & 0.5 \end{pmatrix}, \quad \mathbb{E} = \begin{pmatrix} 0.6 & 0.4 \\ 0.3 & 0.7 \\ 0.2 & 0.8 \end{pmatrix},$$

and $\Pi = \{0.8, 0.1, 0.1\}$. At a certain time, the company receives the following sequence of pitch measurements at five minute intervals. Devise a model using the Viterbi algorithm to predict the corresponding sequence of control laws that will likely be activated during the same time instant.

Pitch data: 'up', 'down', 'down', 'down', 'down', 'up', 'up', 'down', 'down', 'down', 'down'.

3.10 Selected bibliography

Billingsley, P. *Probability and Measure* (2nd edition). John Wiley and Sons, 1990.

Durrett, Richard. *Essentials of Stochastic Processes* (2nd edition). Springer, 2012. DOI: 10.1007/978-1-4614-3615-7.

Norris, J. R. *Markov Chains*. Cambridge University Press, 2017. DOI: https://doi.org/10.1017/CBO9780511810633.

Privault, Nicolas. *Understanding Markov Chains: Examples and Applications* (2nd edition). Springer, 2018. DOI: https://doi.org/10.1007/978-981-13-0659-4.

3.11 Exercise problems

1. Consider the Markov chain $\{X_n\}_{n \geq 0}$ with state space $S = \{1, 2\}$ and transition matrix

$$\mathbb{P} = \begin{array}{c} 1 \\ 2 \end{array}\begin{array}{cc} 1 & 2 \\ \begin{bmatrix} 0.3 & 0.7 \\ 0.5 & 0.5 \end{bmatrix} \end{array}.$$

   i. Compute $P(X_7 = 2, X_5 = 1 | X_4 = 1, X_3 = 2)$.

   ii. Compute $E(X_2 | X_1 = 2)$.

2. (*Mean hitting times*) Our bunny, whose name is Honey, hops around on a triangle. At each step he moves to one of the other two vertices at random (his decision is based on the flip of a fair two-sided coin). What is the expected time taken by Honey to get from vertex 1 to vertex 2?

3. Consider a Markov chain $\{X_n\}_{n \geq 0}$ on the state space $\{0, 1, 2, 3, 4\}$ with stochastic matrix

$$
\mathbb{P} = \begin{array}{c} \\ 0 \\ 1 \\ 2 \\ 3 \\ 4 \end{array}
\begin{array}{c}
\begin{array}{ccccc} 0 & 1 & 2 & 3 & 4 \end{array} \\
\begin{bmatrix}
0 & 1/4 & 1/4 & 1/4 & 1/4 \\
1 & 0 & 0 & 0 & 0 \\
0 & 1 & 0 & 0 & 0 \\
0 & 0 & 1 & 0 & 0 \\
0 & 0 & 0 & 0 & 1
\end{bmatrix}
\end{array}.
$$

    i.   Draw the graph of this chain.

    ii.   Find the periods of states 0, 1, 2, and 3.

    iii. Which states are absorbing, recurrent, and transient?

    iv. Is the Markov chain reducible? Why?

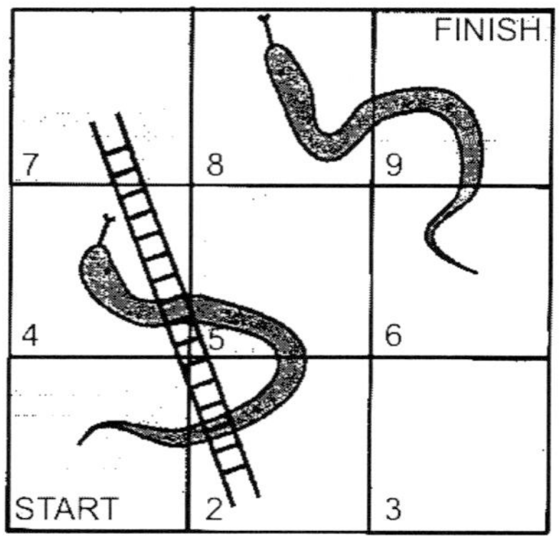

4.   (**Snakes and ladders**) Consider a nine-square snakes and ladders board as shown in Figure 3.16. At each turn, a player tosses a fair coin and advances one or two steps forward depending on whether the outcome is a tail or a head, respectively. Upon landing at the base of a ladder, the player can climb to the top of the ladder, whereas falling at the mouth of a snake brings them down to the tail of the snake.

    i.   Construct an appropriate Markov model and $\mathbb{P}$.

    ii.   In how many turns on an average can the game be completed by a player?

    iii. What is the probability that a player who has made it to square 6 will complete the game before falling to the "START"?

5.   (**Population genetics**)[17] In a certain genetics model, we consider an *n-by-n* array of cells. Each cell is initially colored any one of k different colors. At each step, a cell is chosen at random. This cell then chooses one of its eight neighbors at random and assumes the color of that neighbor. At the boundaries, we may consider a periodic wrapping of left–right and top–bottom columns and rows. The missing diagonal neighbor of any corner cell may be replaced by the cell that is in the diametrically opposite corner. With these boundary adjustments, each cell in the array is adjacent to exactly eight other cells. A state in this Markov chain is a description of the color of each cell. The number of states is k^{n^2}. Even for small n, the size of the state space is enormous. We will analyze this model with the help of a computer simulation. Begin with a random initial configuration of $k = 2$ colors with $n = 20$ and comment on the long-run behavior of the model in terms of the prevalent colors. Repeat the simulation by taking $k = 5$ and $n = 50$. State your observations with possible reasons.

6.   (**Hops to freedom**) Our squirrel from an earlier chapter is stuck in a maze with four cells, labeled as 1, 2, 3, and 4 as shown in Figure 3.17. The outside world is the squirrel's pathway to freedom (consider freedom as state 0). The route to freedom can only be accessed through cell 4. The squirrel starts initially in cell 2. From each cell, the squirrel can move to either of the neighboring cells with equal probability. We assume that at each move the squirrel acts independent of the past (our squirrel is not keen to learn from

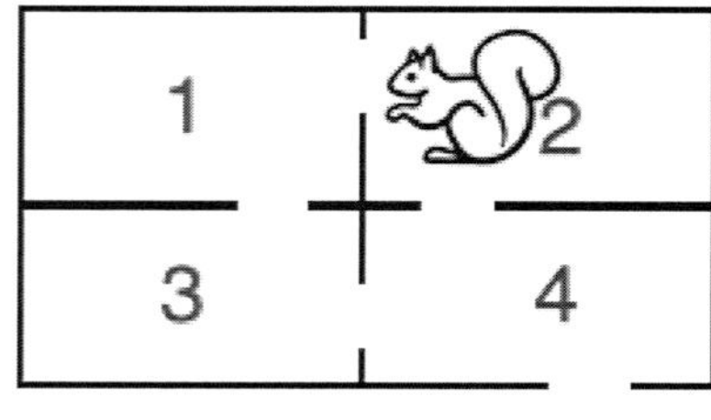

Figure 3.16 How fast will this game end on an average?

[17] S. Sawyer, Results for the stepping stone model for migration in population genetics, *Annals of Probability* **4** (1979): 699–728.

Figure 3.17 How many hops to freedom?

its past mistakes). How many moves will our squirrel make (on an average) before attaining freedom?

7. (**Wandering Daisy**) Our friend Daisy has decided to take a day off from work on a cloudy day. She is having thoughts about visiting the apple orchard just outside her hamlet but is concerned about the impending rain. Let us analyze her prospects and see if she would make it to the orchard given this uncertainty in her mind.

Deterministic mind: Daisy starts walking from her home towards the orchard. Half way through she changes her mind and starts returning home because she thinks that it might rain soon. The clouds begin clearing up soon or so it seems. So half way through her return, she changes her mind and starts walking towards the orchard again. Once again half way through her journey she starts returning home and so on. Construct a mathematical model of her location at every inflection point and comment on her eventual destination. *Hint: Identify Daisy's home as the point "zero" and the orchard as the point "one". Formulate a sequence $\{X_n\}$, where X_n denotes the position at the nth inflection point. What is the limiting value of X_n as $n \to \infty$?*

Stochastic mind version 1:[18] Daisy starts at "zero" (home), goes half way through, and then flips a fair coin. If the coin comes out heads, she continues towards "one" (orchard), and if the coin comes out tails, she turns back toward "zero." Again half way through whatever direction she is headed, she flips a fair coin and either continues in that direction or goes in the opposite way based on the outcome of the toss. Construct a mathematical model of her location at every inflection point and comment on her eventual destination. *Hint: Formulate a sequence X_n, where X_n denotes the position at the nth inflection point. Is X_n a Markov chain? Why? What is the limiting value of X_n as $n \to \infty$?*

Stochastic mind version 2: Now, each time Daisy has to decide on the direction, she uses a biased coin with probability $p(\neq 1/2)$ for heads and then goes a fraction α of the distance in that direction. Construct a mathematical model of her location at every inflection point and comment on her eventual destination.

8. (**Invariant or stationary distributions**) Given a probability measure $\bar{\pi}$, we say that $\bar{\pi}^{(\infty)}$ is *invariant* or *stationary* if

$$\bar{\pi}^{(\infty)}\mathbb{P} = \bar{\pi}^{(\infty)}.$$

Consider a Markov chain $\{X_n\}_{n \geq 0}$ with states arranged on the vertices of a triangle. The transition probabilities between the states are as follows: $P_{12} = 1, P_{23} = 1/2, P_{31}, P_{33} = 1/2, P_{22} = 1/2$.

i. Construct $\mathbb{P}$.

ii. Find the stationary distribution of states $\bar{\pi}^{(\infty)}$.

9. (**Diffusion as a Markov process**) Gas molecules move about randomly in a box which is divided into two halves symmetrically by a partition. A hole is made in the partition. Suppose there are N molecules in the box.

[18] The random sequence X_n generated by Daisy in the stochastic case is a model that is applicable in many real-life practical situations. E.g., let Y_m be a random sequence denoting the level of water in a tank at time instances m, $m \geq 0, m \in \mathbb{I}$. Let $\tau_1, \tau_2, \tau_3, \ldots$ be the times at which the sequence $\{Y_m\}$ has a local minimum or maximum. Let $X_m = Y_{\tau_m}$. Suppose the tank has a global minimum at "zero" and a global maximum normalized to be "one". Then the sequence $\{Y_{\tau_m}\}$ must behave analogous to Daisy's wanderings. The model for the tank can be similarly thought of as a model for stock prices, amount of rainfall, inventory level, or any other randomly varying sequence in a bounded interval.

i.   Show that the number of molecules on one side of the partition just after a molecule has passed through the hole evolves as a Markov chain.

ii.   What are the transition probabilities?

iii.   What is the invariant distribution of this chain?

10. (***Time reversal and entropy***) Past and future are independent of each other in a Markov process. So this entails an inherent time symmetry.[19] However, convergence to stationary distribution of states is asymmetrical in time, i.e., a highly organized state decays to a disorganized one (the invariant distribution) when viewed backward. This is analogous to a situation that embodies an increase in entropy. Therefore, a time symmetry in the absolute sense demands that we begin at equilibrium. Consider a Markov chain $\{X_n\}_{0 \le n \le N}$ with

$$\mathbb{P} = \begin{pmatrix} 0 & 2/3 & 1/3 \\ 1/3 & 0 & 2/3 \\ 2/3 & 1/3 & 0 \end{pmatrix}$$

and $\bar{\pi} = \begin{pmatrix} 1/3 & 1/3 & 1/3 \end{pmatrix}$ is invariant.

i.   Compute the stochastic matrix $\hat{\mathbb{P}}$ of the chain $Y_n = X_{N-n}$.

ii.   Is the chain reversible? (*Hint: Check if* $\hat{\mathbb{P}} = \mathbb{P}$?)

4

Continuous Time Markov Chains and Queues

QUEUES FEATURE IN our daily lives like never before. From the checkout counter in the community grocery store to customer support over the phone, queues are theatres of great social and engineering drama. Entire business operations of many leading companies are geared towards providing hassle-free customer support and experience – timely and effective resolution of client queries about services on a regular basis. Alternatively, it could be effective traffic management and resource optimization for a multiplex cinema operator involved in ticket sales. Sometimes it may not involve humans at all, like in the case of a database query to a computer server for specific information that may be routed through a job queue. How a queue moves in time and how services are offered over epochs determine how businesses will be able to make profit or how efficiently computer servers will execute tasks. All these have a huge technological and economical impact. No wonder we have seen huge investments by concerned stakeholders to upgrade and upscale hardware and software infrastructure to re-engineer queues towards greater system efficiency and profitability. The mathematical technology of queues is crafted out of models that investigate and replicate stochastic behavior of engineering systems. This is the subject of our study in this chapter.

Figure 4.1 The wait is on because queues are everywhere. But the seventeenth-century English poet John Milton lends us hope. He writes: *They also serve who only stand and wait.*

4.1 Chapter objectives

The chapter objectives are listed as follows.

1. Students will study and apply continuous time Markov chains (CTMCs) to solve engineering problems.

2. Students will model and analyze queues using probability distributions.

3. Students will learn the interrelationship between the probability transition matrix and the stochastic generator matrix.

4. Students will learn to derive Kolmogorov's backward and forward equations, use the principle of detailed balance, and find stationary distributions of stochastic processes using a stochastic generator matrix.

5. Students will deduce birth and death stochastic processes and analyze their equilibrium and/or asymptotic behavior.

6. Students will solve an engineering problem using a classical queuing model.

4.2   Chapter project: Queues and crowd management at COVID test centers

4.2.1   *Prologue: Crowd management at COVID test kiosks*

Crowd management at test centers is a major concern for healthcare administrators. This issue has been amplified during the COVID pandemic when transmission rates have been very high at times and huge numbers of people have been infected on a regular basis, especially during peak periods of the disease. Administrators at test centers are faced with a dual challenge: (i) they have to contain the number of people presenting themselves at the test kiosks to a manageable figure to minimize the risk of disease transmission from overcrowding and (ii) manage the exploding cost of clinical care and rapidly rising expenditure of procuring test kits and setting up clean test kiosks that have been sanitized. The latter point, along with the fact that availability of employees to conduct tests during a raging pandemic comes with a premium, demands intelligent engineering and management of hospital and test facilities.

The sick who have symptoms or those who have tested positive using home kits or quick rapid antigen tests (RATs) may present themselves to test kiosks set up at hospital premises in order to either confirm the results of their preliminary screening tests (which may not have been very accurate) or evaluate if their disease state may require them to be admitted in a hospital for more dedicated care and treatment. The test kiosks at hospitals have access to more accurate and fast testing equipment to evaluate the severity and/or progress of the disease and will provide better decision points for doctors if a tested patient must be admitted for clinical care.

In this project, we will source real data from one such hospital facility in New York, located at the premises of the Brooklyn City Hospital. The data used in the case study includes day-wise and week-wise information over the period January 1 through December 31, 2021. This data includes the number of people who presented themselves at the hospital test center after a preliminary positive result of tests conducted at their homes. This data has been sourced from the official website.[1] For convenience we have organized the data in a tabular matrix as provided at the end of this chapter in Table 4.1 (Table-4.1.csv).

[1] https://data.cityofnewyork.us/Health/COVID-19-Daily-Counts-of-Cases-Hospitalizations-an/rc75-m7u3, accessed September 13, 2023.

4.3   Introduction: A gentle initiation to queues

We will discuss a simple n-server queuing system as a motivation to study continuous stochastic processes within the framework of continuous Markov chains. This discussion will depend on Poisson arrivals of clients. Therefore, it is prudent to revisit the definition of a Poisson process.

Poisson process: Consider an arrival process that counts the number of arrivals over time. $N(t)$, $t \geq 0$, is a Poisson process with rate $\lambda > 0$ that counts the number of arrivals if the following statements hold true:

1. $N(0) = 0$.

2. $N(t)$ has independent increments, i.e., the numbers of arrivals in two non-overlapping intervals are stochastically independent.

3. $N(\delta t) \sim Poisson(\lambda \delta t)$, where $N(\delta t)$ refers to the number of arrivals in duration δt.

A simple calculation (see section 5.3.6) shows that the inter-arrival times between Poisson arrivals follow an exponential distribution with the same rate parameter. This will be an important consideration in many of our discussions in this chapter. Let us return to our example of the n-server queue.

Figure 4.2 A queue in a bank with multiple tellers. See color plate on page 296.

Example 4.1 n-server FIFO queue

Consider a scenario where we have n servers that provide service in such a way that the service times of each server is an independent and identically distributed (i.i.d.) random variable that is distributed according to the exponential law with rate parameter μ. Clients enter a queue according to a Poisson process with rate λ. The service principle follows a *first in first out* (FIFO) law applied to the queue. Further, any arrival that finds all servers busy leaves without service. The latter condition (hereafter referred to as cond++) is different from the one shown in Figure 4.2 but we will consider it here in our calculations as an additional constraint. We ask the following question. If an arrival finds that all servers are busy, then what is the expected number of busy servers observed by the next arriving client? For convenience, we will refer to this client as client #13-A.

Solution: Before we proceed with the calculations, let us pause and reflect on the situation a little. When client #13-A arrives, it could find that all servers are busy if no service was completed in the intervening time between its arrival and the exit of the previous customer who departed the queue on finding all servers to be busy. Alternatively, client #13-A could find any of $0 \leq m \leq (n-1)$ servers to be busy depending on the number of services that were completed in the duration between its arrival and the exit of the previous client who left out of frustration. And since there is no guarantee when client #13-A actually enters the queue and the fact that the service times of any of the servers are not uniform, this is truly a stochastic process.

To proceed with our calculations, let us consider T_k to be the expected number of busy servers found by client #13-A if there are currently k busy servers. Here the word "currently" refers to the epoch of the exit of the frustrated client (in case there are no service completions during the intervening time) or the epoch of the latest service completion in the intervening time before the arrival of client #13-A. We want to estimate T_n. It turns out that the definition of T_k above is equivalent to the following: T_k is the expected number of busy servers found by client #13-A *in a k-server system* when there are currently k busy servers. The above statement is true because the fact that there will be at least $(n - k)$ idle servers can be ignored because of the *memoryless (Markovian)* property of exponential service times and exponential inter-arrival times. The boundary condition $T_0 = 0$ is self-evident and needs no further explanation. Our next objective will be to find T_1 again for a 1-server system. In such a case, either 1 or 0 server can be found to be busy by client #13-A. Therefore,

$$T_1 = (1) \times \text{Prob(client \#13-A finds one server busy)}$$

$$+(0) \times \text{Prob(client \#13-A finds zero servers busy)}$$

$$= (1)\frac{\lambda}{\lambda+\mu} + (0)\frac{\mu}{\lambda+\mu}$$

See equation (2.49)

$$= \frac{\lambda}{\lambda+\mu}. \tag{4.1}$$

For the general case, with k busy servers, we will obtain T_k by conditioning upon what happens next. If we label an event of a new arrival or a completion of a service to an alarm clock going off, then with k busy servers, we will need k numbers of $exp(\mu)$ alarm clocks and 1 number of $exp(\lambda)$ alarm clock. We will use the following two partitioning events to condition our calculation of T_k using the law of total expectation:

1. Event $\mathfrak{E}_1$: A service completion happens first during the intervening time before client #13-A arrives.

2. Event $\mathfrak{E}_2$: An arrival happens first (the arrival of client #13-A) before any server becomes available.

Additionally, if τ_i, $i = 1, 2, ..., k$, denotes time to complete a service by server i, then the time till the next service completion after the exit of the frustrated client is distributed as $min(\tau_1, \tau_2, ..., \tau_k) \sim exp(k\mu)$. Therefore, Prob (event $\mathfrak{E}_1$) $= \frac{k\mu}{\lambda+k\mu}$ and Prob (event $\mathfrak{E}_2$) $= \frac{\lambda}{\lambda+k\mu}$. Now we will use the law of total expectation.

$$T_k = E(\text{\#busy servers} \mid \mathfrak{E}_1 \text{ \& currently } k \text{ busy servers}) \, \text{Prob}(\mathfrak{E}_1 \mid \text{currently } k \text{ busy servers})$$

$$+ E(\text{\#busy servers} \mid \mathfrak{E}_2 \text{ \& currently } k \text{ busy servers}) \text{Prob}(\mathfrak{E}_2 \mid \text{currently } k \text{ busy servers})$$

$$= T_{k-1}\frac{k\mu}{\lambda+k\mu} + k\frac{\lambda}{\lambda+k\mu}. \tag{4.2}$$

Due to condition cond++

Here we have used the phrase "#busy servers" as a short-hand for *number of busy servers*. Now, we may use the above recurrence relation to find T_2, T_3, ..., and we list a few of them below.

$$T_2 = T_1 \frac{2\mu}{2\mu + \lambda} + \frac{2\lambda}{2\mu + \lambda}$$

$$= \frac{\lambda}{\lambda + \mu} \frac{2\mu}{2\mu + \lambda} + \frac{2\lambda}{2\mu + \lambda}.$$

$$T_3 = T_2 \frac{3\mu}{3\mu + \lambda} + \frac{3\lambda}{3\mu + \lambda}$$

$$= \frac{\lambda}{\lambda + \mu} \frac{2\mu}{2\mu + \lambda} \frac{3\mu}{3\mu + \lambda}$$

$$+ \frac{2\lambda}{2\mu + \lambda} \frac{3\mu}{3\mu + \lambda} + \frac{3\lambda}{3\mu + \lambda}.$$

And in general,

$$T_n = \frac{n\lambda}{n\mu + \lambda} + \sum_{i=1}^{n-1} \frac{i\lambda}{i\mu + \lambda} \prod_{j=i+1}^{n} \frac{j\mu}{j\mu + \lambda}. \tag{4.3}$$

The above example is in fact an illustration of a continuous time stochastic process and is associated with a continuous time Markov chain with rate parameters $q_{i,i+1} = \lambda_i = \lambda$ and $q_{i,i-1} = \mu_i = \mu$ and a state space $\mathcal{S} = \{0, 1, 2, ..., n\}$ that denotes the number of busy servers (or number of people in the system). These notations and the concept of a continuous time Markov chain (CTMC) will be the subject of our study in the following sections.

Example 4.2 *Stochastic arrivals follow a Poisson distribution*

Consider identical and independent arrivals at a fixed rate λ on a linear time axis beginning with the epoch $t = t_0 = 0$ from the state $\mathcal{E}_0$. $\mathcal{E}_k$ denotes the event of the kth arrival at epoch $t = t_k$. Here the subscript refers to the number of arrivals up to that instant. This stochastic process defines a jump transition from $\mathcal{E}_j$ to $\mathcal{E}_{j+1}$ between two successive arrivals. Whatever the state $\mathcal{E}_j$ at a certain epoch $t_j \leq t < t_{j+1}$, the probability of a jump (an arrival) between epochs t and $(t + h)$ (for small $h > 0$) is $\lambda h + o(h)$, whereas the probability of more than one jump (arrival) is $o(h)$. Define a random variable $Z(t)$ that counts the number of arrivals in an arbitrary interval of time of length t. Show that

$$p_n(t) = P\big(Z(t) = n\big) = \frac{(\lambda t)^n}{n!} e^{-\lambda t}. \tag{4.4}$$

Figure 4.3 An artist's rendition of queues at the checkout counters in a shopping mall. At peak hour and during festive season, the mall manager will be interested to know the average number of people in the queue at any given point of time in order to marshal his employee resources judiciously. See color plate on page 296.

Solution: We will work with $n \geq 1$. Consider the event at epoch $t + h$ that is designated by the state $\mathfrak{E}_n$. The probability of this event is $p_n(t + h)$ as per the definition above. This event can happen in three mutually exclusive manners as enumerated below.

1. *Event ε_1*: The system was in state $\mathfrak{E}_n$ at epoch t and no arrival happened between t and $t + h$. The probability of this event $P(\varepsilon_1) = p_n(t)p_0(h) = p_n(t)\{1 - \lambda h\} + o(h)$.

2. *Event ε_2*: The system was in state $\mathfrak{E}_{n-1}$ at epoch t and exactly one arrival happened between t and $t + h$. $P(\mathcal{E}_2) = p_{n-1}(t)p_1(h) = p_{n-1}(t)\{\lambda h\} + o(h)$.

3. *Event ε_3*: The number of arrivals between epochs t and $t + h$ is more than one and the probability of such an event $P(\mathcal{E}_3) = o(h)$ as defined in the problem statement.

Since the events $\varepsilon_1, \varepsilon_2, \varepsilon_3$ are mutually exclusive, the probabilities simply add up and we have the following result:

$$p_n(t+h) = p_n(t)\{1 - \lambda h\} + p_{n-1}(t)\{\lambda h\} + o(h), \tag{4.5}$$

which may be rewritten as

$$\frac{p_n(t+h) - p_n(t)}{h} = -\lambda p_n(t) + \lambda p_{n-1}(t) + \frac{o(h)}{h}. \tag{4.6}$$

Upon evaluating the above in the limit $h \to 0$, we have the following recurrence relation:

$$p_n'(t) = -\lambda p_n(t) + \lambda p_{n-1}(t), \ n \geq 1. \tag{4.7}$$

When $n = 0$, the only possible route at our disposal is through the event $\mathcal{E}_1$, which leads to $p_0'(t) = -\lambda p_0(t)$. The boundary condition for $n = 0$ is $p_0(0) = 1$

whence we get $p_0(t) = e^{-\lambda t}$. Using this result in the recurrence relation (4.7) and solving for $p_1(t)$, we obtain $p_1(t) = \lambda t e^{-\lambda t}$ that is in agreement with equation (4.4). Proceeding in this recursive manner, we can deduce the generic relation (4.4) for $p_n(t)$.

Note: The little o notation used in the above example is defined as follows: $\lim_{h \to 0} \frac{o(h)}{h} = 0.$[*]

---

[*]Equivalently, we say $f(n) = o(g(n))$ if $\lim_{n \to \infty} \frac{f(n)}{g(n)} = 0$.

4.4   Continuous time Markov chains (CTMCs)

We will begin this section by revisiting the Markov property in the context of a continuous time stochastic process. We will then deduce an equation whose solution generates the stationary distribution of the stochastic process. This result will enable us to investigate asymptotic behavior of queuing systems with wide economic implications for businesses to conduct their operations efficiently by allocating optimal resources.

4.4.1   *Markov property for continuous time processes*

Consider a continuous time stochastic process $\{X(t)\}_{t \geq 0}$ that takes on discrete values (states) from the state space $\mathcal{S}$. The Markov property[2] for continuous time stochastic processes can be stated as follows.

$$P(X(t) = j | X(s) = i, X(t_{n-1}) = i_{n-1}, ..., X(t_1) = i_1) = P(X(t) = j | X(s) = i), \quad (4.8)$$

where $0 \leq t_1 \leq t_2 \leq \cdots \leq t_{n-1} \leq s \leq t$ and $i_1, i_2, ..., i_{n-1}, i, j \in \mathcal{S}$ are the $(n+1)$ states in the state space $\mathcal{S}$ for all $n \geq 1$, $n \in \mathbb{I}^+$.

[2] Markov property is also colloquially known as the memoryless property.

4.4.2   *Definition: Continuous time Markov chains*

A continuous time stochastic process $\{X(t)\}_{t \geq 0}$ is called a CTMC if it obeys the Markov property (4.8). Additionally, a CTMC may be time-homogeneous (or stationary) if it satisfies the condition discussed below.

4.4.3   *Time-homogeneity of Markov chains*

We say that a CTMC is *time-homogeneous* if for any $s \leq t$ and any states $i, j \in \mathcal{S}$, the following is true.

$$p_{i,j}(t-s) \equiv P(X(t) = j | X(s) = i) = P(X(t-s) = j | X(0) = i)$$

$$= P(X(t_1) = j | X(s + t_1 - t) = i). \quad (4.9)$$

The important thing to note in the above statement is that the conditional probabilities do not depend on any particular epoch but only on the interval $(t - s)$. Such a time-homogeneous process is also a *stationary* process.[3] It is essential to emphasize that not all CTMCs are time-homogeneous (or stationary) but in this chapter we will be mostly concerned with time-homogeneous stochastic processes.

[3] Intuitively, this means that *whenever* the process enters state i, the way it evolves stochastically from that epoch is the same as if the process started in state i at time 0.

4.4.4 *Holding time of a Markov chain*

When the stochastic process enters state i, the time it spends in that state before it leaves state i is called the *holding time* T_i. If the arrival rate of this process is λ_i, then $T_i \sim exp(\lambda_i)$ as was the case in the introductory example of this chapter.

Figure 4.4 Probability distributions associated with a Poisson point process.

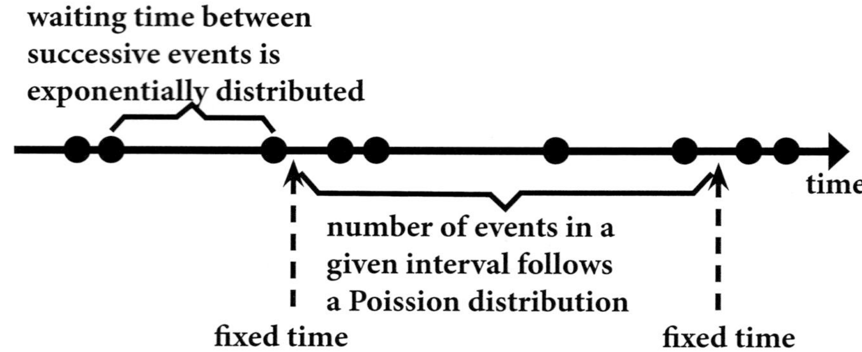

Recall that for a discrete stochastic process such as the discrete time Markov chain, the probability of transition from state i to state j, denoted by $p_{i,j}$, along with the initial set of probability distribution, completely determines the state of the system at all latter times. For a CTMC, the rate at which events occur not only characterizes the epochs of state transitions[4] but also determines the long-run distribution of states of the system. We will devote the next few sections to illustrate these facts in a rigorous manner.

In order to deduce a model of the generator of this CTMC, we will have to begin by defining the rates as follows. Let $q_{i,j}$ be the rate at which the system goes from state i to state j. This is akin to the rates at which the exponential alarm clocks of section 4.3 go off. These rates are functions of time as are the probability transitions, i.e., $p_{i,j} \equiv p_{i,j}(t)$ and $q_{i,j} \equiv q_{i,j}(t)$. Since the $p_{i,j}$s are probabilities, it should make sense to define them as follows.

$$p_{i,j} := \frac{q_{i,j}}{\sum_{j \in \mathcal{S}} q_{i,j}}. \tag{4.10}$$

If we denote $v_i = \sum_{j \in \mathcal{S}} q_{i,j} < \infty$, then the rates may be defined as follows.

$$q_{i,j} := v_i p_{i,j}. \tag{4.11}$$

By definition we have $q_{i,i} = 0$. Further, if $v_i = 0$, then state i is an absorbing state. The entries $p_{i,j}$ constitute the *stochastic matrix* $\mathbb{P} \equiv \mathbb{P}(t)$. The entries $q_{i,j}$ constitute the *generator matrix* for reasons that will become clear during the ensuing discussion. It may be noted that v_i s and $p_{i,j}$ s can be computed from the $q_{i,j}$ s. Further, because $q_{i,j}$ s are essentially rates, they are positive quantities.

4.4.5 *Chapman–Kolmogorov equation for CTMC*

Analogous to the discrete case (see equation (3.6)), the Chapman–Kolmogorov equation for CTMC is presented below.

---

[4] This is true in the sense of a probability distribution.

$$p_{i,j}(t+s) = \sum_{k\in\mathcal{S}} p_{i,k}(t)p_{k,j}(s) = \sum_{k\in\mathcal{S}} p_{i,k}(s)p_{k,j}(t) \qquad (4.12)$$

is the $(i,j)^{th}$ entry of the matrix $\mathbb{P}(t+s) = \mathbb{P}(t)\mathbb{P}(s)$. The Chapman–Kolmogorov equation will be used to derive the most important result of this chapter known as the Kolmogorov backward and forward equations. This is explained in the following paragraphs.

4.4.6   *Kolmogorov equations and the generator matrix*

Let us reconsider some of the ideas we used in solving the last example problem of section 4.3. In that case, since we had an arrival process, and we assumed that we would allow for only one arrival at a time for all practical purposes, the transitions between the states happened in increments of one. However, we will relax this restriction; in the case of a general CTMC (which may not be an arrival process only but may capture something far broader in scope), we will allow for successive transitions between states i and j that need not differ only by one. In the set-up of our previous example, j was necessarily $i+1$. However, the transitioning events are still the same in the infinitesimally small duration h, namely, ε_1, ε_2, and ε_3 with the

same probabilities. The only difference in the present case is that $\lambda = \lambda_i = v_i = \sum_{\substack{j\in\mathcal{S} \\ i\neq j}} q_{i,j}$

as this is the most generic form of the rate of transition emanating from state i. It may be noted that $q_{i,i} = 0$ as stated earlier. One may pose a very pertinent question here – *why is the rate of transition from state i given as a sum over $q_{i,j}s$?* In order to understand this formulation, we invite the reader to think through the motivating example of this chapter on the n-server queuing model. In that case, we had k servers, so the time until one of those servers completed the service was distributed as $exp(k\mu)$. In other words, the servers became available at the rate of sum of $k\mu$s, which is analogous to the rate $\sum_{j\in\mathcal{S}} q_{i,j}$, the sum over all the rates of transitions from state i.

In order to derive a differential equation for the probability transitions, we will begin with the Chapman–Kolmogorov relation (4.12).

$$
\begin{aligned}
p_{i,j}(t+h) &= \sum_{k\in\mathcal{S}} p_{i,k}(h)p_{k,j}(t) \\
&= p_{i,j}(t)\{1 - v_i h + o(h)\} + \sum_{k\neq i} p_{i,k}(h)p_{k,j}(t) \\
&= p_{i,j}(t)\{1 - v_i h + o(h)\} + \sum_{k\neq i} p_{k,j}(t)\{p_{i,k}v_i h + o(h)\}. \qquad (4.13)
\end{aligned}
$$

The first term on the RHS of the above equation stems from an event of the type ε_1 (see last example of section 4.3), whereas the second term arises from events of the types ε_2 and ε_3. The computation of $p_{i,k}(h)$ is performed by applying the law of total probability and upon recalling the fact that

$$
\begin{aligned}
p_{i,k}(h) &= P(X(h)=k|X(0)=i,\Delta\mathcal{E}_2)P(\Delta\mathcal{E}_2) \\
&\quad + P(X(h)=k|X(0)=i,\Delta\mathcal{E}_3)P(\Delta\mathcal{E}_3) \\
&= p_{i,k}(v_i h) + p_{i,k}^2(o(h)) + p_{i,k}^3 o(h) + \dots \\
&= p_{i,k}(v_i h) + o(h). \qquad (4.14)
\end{aligned}
$$

Here $\Delta \mathcal{E}_i$, $i = 2,3$ is the event of one jump transition (for $i = 2$) and more than one jump transitions (for $i = 3$) when the system migrates from state i to k as is explained in the last example of section 4.3. Rearranging the terms in equation (4.13) and dividing all the terms by h followed by evaluating the terms in the limit $h \to 0$, we get

$$p'_{i,j}(t) = \sum_{k \neq i} q_{i,k} p_{k,j}(t) - v_i p_{i,j}(t). \tag{4.15}$$

Here we have used the definition of the transition rates $q_{i,k} = v_i p_{i,k}$. In matrix form, equation (4.15) can be expressed as follows.

$$\mathbb{P}'(t) = \mathbb{G}\mathbb{P}(t), \tag{4.16}$$

or equivalently,

$$(\mathbb{P}'(t))_{i,j} = (\mathbb{G}\mathbb{P}(t))_{i,j}. \tag{4.17}$$

Equations (4.15)–(4.17) are called the *Kolmogorov backward equations*. The infinitesimal generator matrix $\mathbb{G} \equiv (g_{i,j})$ is summarized below.

$$g_{i,j} \;=\; q_{i,j} = v_i p_{i,j}, \;\; \forall i \neq j, \tag{4.18}$$

$$g_{i,i} \;=\; -v_i, \tag{4.19}$$

with boundary conditions $\mathbb{P}(0) = \mathbb{I}$, where $\mathbb{I}$ is the identity matrix.

Likewise, we can derive the *Kolmogorov forward equations*, which we simply state below in matrix form.

$$\mathbb{P}'(t) = \mathbb{P}(t)\mathbb{G}. \tag{4.20}$$

The solutions to the Kolmogorov equations (4.15)–(4.20) with the prescribed boundary conditions can be computed very easily in the form of matrix exponentials.

$$\mathbb{P}(t) = e^{t\mathbb{G}} := \mathbb{I} + t\mathbb{G} + \frac{(t\mathbb{G})^2}{2!} + \cdots \tag{4.21}$$

In the case of a finite state space system, the solution (4.21) can be well approximated by truncating the terms of the infinite sum in (4.21). The solution $\mathbb{P}(t) = e^{t\mathbb{G}}$ underscores the importance of the generator matrix $\mathbb{G}$ vis-à-vis a CTMC because it completely generates the solution $\mathbb{P}(t)$. Consequently, $\mathbb{G}$ plays an important role in obtaining the stationary distribution of a Markov chain. This implies that in the case of CTMC (unlike in the case of DTMC), the rates of state transitions completely determine the solutions of the system. Further, the CTMC with the generator (or rate) matrix $\mathbb{G}$ (or equivalently $(g_{i,j})$) bears with it an *embedded* DTMC with transition probabilities prescribed by the matrix $(p_{i,j})$.

4.4.7 *Stationary distribution of CTMC*

Consider a CTMC $\{X(t)\}_{t \geq 0}$ with finite state space $\mathcal{S}$,[5] generator $\mathbb{G}$, and matrix of probability transition functions $\mathbb{P}(t)$. The $|\mathcal{S}|$-dimensional row vector $\pi \equiv (\pi_i)_{i \in \mathcal{S}}$

---

[5] Here we will assume that the cardinality of the set $\mathcal{S}$ is $|\mathcal{S}|$.

with $\pi_i \geq 0$, $\forall i$ and with the constraint $\sum_{i \in S} \pi_i = 1$ is a *stationary* distribution of the Markov chain if

$$\pi = \pi \mathbb{P}(t), \ \forall t \geq 0. \tag{4.22}$$

Example 4.3 $\ \pi\mathbb{G} = 0$ ***describes the stationary solution of CTMC***

Use the generator matrix $\mathbb{G}$ to deduce a condition for stationarity of a CTMC.

Solution: If π is a stationary distribution of a Markov chain, then the following statements are true.

$$\pi = \pi \mathbb{P}(t), \ \forall t \geq 0,$$

$$= \pi \sum_{n=0}^{\infty} \frac{(t\mathbb{G})^n}{n!}.$$

See equation (4.21) where $\mathbb{P} = e^{t\mathbb{G}} = \cdots$

$$\Leftrightarrow \pi - \pi = \sum_{n=1}^{\infty} \frac{t^n}{n!} \pi \mathbb{G}^n$$

This symbol stands for *if and only if*

$$\Leftrightarrow 0 = \pi \mathbb{G}^n, \ \forall n \geq 1,$$

Sum of positive terms is zero $\Leftrightarrow$ summand is zero.

$$\Leftrightarrow \boxed{\pi \mathbb{G} = 0.} \tag{4.23}$$

Using equation (4.23) to find the stationary distribution of states is computationally more convenient than using equation (4.22). In the former case, one simply has to solve a system of $|S|$ linear equations. Of course, the solution must be compliant with the fundamental axiom of probability $\sum_{i \in S} \pi_i = 1$.

For a CTMC $\{X(t)\}_{t \geq 0}$, the stationary probability distribution is also the limiting probability ($t \to \infty$), i.e.,

$$\lim_{t \to \infty} p_{i,j}(t) = \lim_{t \to \infty} P(X(t) = j | X(0) = i) \equiv \pi_j. \tag{4.24}$$

4.4.8 *Global balance equations* $\pi\mathbb{G} = 0$

Let us investigate the equation (4.23) further. In the component form, equation (4.23) is $\sum_{i \in S} \pi_i g_{i,j} = \pi_j g_{j,j} + \sum_{i \neq j} \pi_i g_{i,j} = 0$. These $|S|$ equations can be written more explicitly by referring to the equations (4.18)–(4.19).

$$\pi_j v_j = \sum_{i \neq j} \pi_i q_{i,j}, \tag{4.25}$$

Figure 4.5 A huge pile of books that can stand firm against a gust of wind. Not only are individual books in balance with their neighbors, but the entire pile is in balance as a whole. This bears the notion of local as well as global balance.

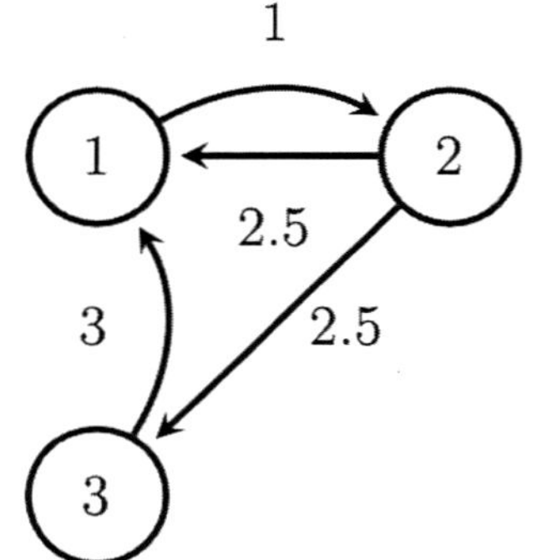

Figure 4.6 Schematic representation of the generator (rate) matrix.

where $v_j = \sum_{i \in S} q_{j,i}$. Each of the terms in the above equation can be interpreted in the following manner.

1. π_j: This term refers to the long run proportion of time the system stays in state j.

2. $v_j = \sum_{i \in S} q_{j,i}$: This term estimates the rate of departure from state j when the system is in state j.

3. $\pi_j v_j$: This term computes the long-run rate of leaving state j.

4. $\pi_i q_{i,j}$: This term calculates the long-run rate of going from state i to state j.

5. $\sum_{i \neq j} \pi_i q_{i,j}$: This term estimates the long-run rate of going to state j starting from a state other than state j.

Therefore, equation (4.25) is a statement of balance between flux out of state j and flux into state j, i.e., it is a statement of *dynamic equilibrium*. This is why equation (4.23) is known as the *global balance equation* or simply the *balance equation*.

Example 4.4   State of a machine prototype

A new prototype washing machine using a revolutionary technology is under development. During the development cycle, it passes through one of three distinct states, namely, normal, test, and repair modes. The rate diagram of the states depicting the entries of the generator matrix is shown in Figure 4.6. Find the stationary distribution of states.

Solution: The generator matrix can be deduced from the rate diagram as follows.

$$\mathbb{G} = \begin{pmatrix} -1 & 1 & 0 \\ 2.5 & -5 & 2.5 \\ 3 & 0 & -3 \end{pmatrix}.$$ In order to find the stationary distribution of states, we must solve $\pi \mathbb{G} = 0$ and $\pi_1 + \pi_2 + \pi_3 = 1$. The former simplifies to $\pi_2 = \dfrac{3\pi_3}{2.5}$ and $\pi_1 = 5\pi_2$. When we use this simplification in the latter equation, we obtain $\pi_3 = \dfrac{5}{41}$.

Substituting this in the previously deduced results, we obtain $\pi_2 = \dfrac{6}{41}$ and $\pi_1 = \dfrac{30}{41}$. Thus, the stationary distribution of states of the prototypical washing machine is $\left(\dfrac{30}{41} \ \dfrac{6}{41} \ \dfrac{5}{41} \right)$ corresponding to the normal, test, and repair modes.

4.4.9   *Detailed (or local) balance*

For a CTMC with a generator matrix $\mathbb{G} \equiv \left(q_{i,j} \right)$, if we can find a distribution of states π such that for every pair of states i and j, the following relation holds

$$\pi_i q_{i,j} = \pi_j q_{j,i}, \tag{4.26}$$

then it can be easily shown, by summing over all the states in $\mathcal{S}$, that π is a stationary distribution and satisfies the global balance equation (4.25). Let us sum over all the states i to obtain $\sum_{i \in \mathcal{S}} \pi_i q_{i,j} = \sum_{i \in \mathcal{S}} \pi_j q_{j,i} = \pi_j \sum_{i \in \mathcal{S}} q_{j,i} = \pi_j v_j$, which is the equation (4.25).[6] The detailed balance may not always hold even when the global balance holds. The converse is always true as shown above.

When does detailed balance not hold for sure? A quick way to check this is to inspect the rate or the generator matrix $\mathbb{G}$ and see if there are any rates $q_{i,j}$ and $q_{j,i}$ such that $q_{i,j} > 0$ and $q_{j,i} = 0$ (or conversely $q_{i,j} = 0$ and $q_{j,i} > 0$). If any of these pathological cases exist for any pair i,j, then the local analysis using the detailed balance condition to estimate the stationary distribution will be futile.

Theorem: A CTMC is *reversible* if and only if the detailed balance condition holds for every pair (i,j) in $\mathcal{S}$.

4.4.10 *Application of detailed balance condition*

For the following two well-known continuous stochastic processes, the detailed balance does hold and is useful to find the stationary probability distribution of states. They are

1. birth–death processes and

2. $M/M/1$ queues (Poisson arrivals and exponential service time of a one-server system).

We discuss these two stochastic processes in detail in the subsequent sections of this chapter.

4.5 Birth and death processes

A birth and death process is a homogeneous stochastic Markov process where the transitions between two successive epochs[7] constitute either a jump by one (i.e., state i goes to state $i + 1$) or a drop by one (i.e., state i goes to $i - 1$). In other words, only *nearest-neighbor* transitions are permissible. Since this stochastic process is a continuous time process, the transitions between the states happen at a prescribed rate.

$$
\begin{aligned}
q_{i,i+1} &= \lambda_i > 0,\ i \ge 0, \quad \text{(birth rates) and} & (4.27)\\
q_{i,i-1} &= \mu_i > 0,\ \ i \ge 1, \quad \text{(death rates).} & (4.28)
\end{aligned}
$$

The state space $\mathcal{S}$ is over the space of all whole numbers.

Deduce the stationary distribution of the birth–death process defined in equations (4.27) and (4.28).

Solution: Since births and deaths are well-defined and non-trivial at every point on the state space, it is easy to validate by inspection that the local analysis of the detailed balance will be useful to find the stationary distribution of states.

[6] The equivalent detailed balance for DTMC is $\pi_i p_{i,j} = \pi_j p_{j,i},\ \forall i, j \in \mathcal{S}$.

Figure 4.7 Everything is in perfect balance.

[7] Here an epoch is a time instant when an event (or a transition of state) occurs, such as an arrival or a death.

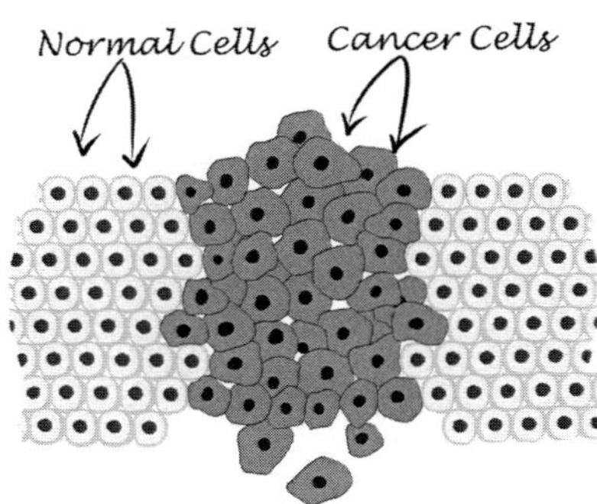

Figure 4.8 Carcinogenic growth of cells and apoptosis can be studied using birth and death models.

Therefore, we begin by writing the detailed balance equations at every point of state transition as follows.

$$\pi_i q_{i,i+1} = \pi_{i+1} q_{i+1,i}$$

$$\Rightarrow \pi_i \lambda_i = \pi_{i+1} \mu_{i+1}, \quad \forall i \geq 0;$$

$$\Rightarrow \pi_{i+1} = \frac{\lambda_i}{\mu_{i+1}} \pi_i$$

$$= \frac{\lambda_i \lambda_{i-1}}{\mu_{i+1} \mu_i} \pi_{i-1}$$

Recursively

$$\vdots$$

$$= \frac{\lambda_i \cdots \lambda_0}{\mu_{i+1} \cdots \mu_1} \pi_0. \tag{4.29}$$

Continuing recursively

Additionally, the π_is must satisfy the axioms of probability; specifically, the relation $\sum_{i \in \mathcal{S}} \pi_i = 1$ must hold. Re-indexing $i = j - 1$ in equation (4.29) and using the aforementioned normalization of the probabilities, we have

$$\pi_0 + \sum_{j=1}^{\infty} \pi_j = \pi_0 + \sum_{j=1}^{\infty} \frac{\lambda_{j-1} \cdots \lambda_0}{\mu_j \cdots \mu_1} \pi_0 = 1$$

$$\Rightarrow \pi_0 \left\{ 1 + \sum_{j=1}^{\infty} \frac{\lambda_{j-1} \cdots \lambda_0}{\mu_j \cdots \mu_1} \right\} = 1$$

$$\Rightarrow \pi_0 = \frac{1}{1 + \sum_{j=1}^{\infty} \dfrac{\lambda_{j-1} \cdots \lambda_0}{\mu_j \cdots \mu_1}}. \tag{4.30}$$

We must ensure that the infinite power series in the denominator of (4.30) is finite.

Figure 4.9 A repairman servicing a machine in a factory line.

Example 4.6   Servicing of machines by a repairman

Let us consider a factory that houses and operates m machines. The machines are also serviced by a repairman in case any of them breaks down and needs repair. The repairman services one machine at a time on a first-come-first-serve basis, so when more than one machine is dysfunctional, then the idle machines go in a service queue. Each of these machines can go from a working state to a service state (due to a breakdown/failure) within a randomly distributed time

$T_f \sim exp(\lambda)$ (measured from any given time). Further, the service time T_s of each machine is distributed according to a rate μ exponential distribution, $T_s \sim exp(\mu)$. Answer the following questions:

1. What is the stationary distribution of states of working machines?

2. What is the expected number of machines in the waiting line?

Solution: Let us define the state of the system S_k when k of the m machines are idle due to a failure. This is a birth and death process because only one of the two transitions is possible: (i) $S_k \rightarrow S_{k+1}$ when a new failure happens, and (ii) $S_k \rightarrow S_{k-1}$ when a repair is successfully completed and a machine is brought back to function. $\pi_k = \text{Prob}(\text{Number of non-functional machines} = k)$. $\tilde{\pi}_k = \pi_{m-k} = \text{Prob}(\text{Number of functional machines} = k)$. We will begin by writing the detailed balance relations:

$$
\begin{aligned}
\pi_i q_{i,j} &= \pi_j q_{j,i}, \\
\pi_i q_{i,i+1} &= \pi_{i+1} q_{i+1,i}, \\
\pi_i (m-i)\lambda &= \pi_{i+1}\mu.
\end{aligned}
\tag{4.31}
$$

The jump rate $q_{i,i+1}$ is $(m-i)\lambda$ can happen owing to any of the $(m-i)$ working machines becoming dysfunctional but the death rate is constant $q_{i+1,i} = \mu$ because the repairman works on only one machine at a time until it is restored. Setting $j = i + 1$, equation (4.31) can be rewritten as

$$
\pi_{j-1} = \frac{\mu}{\lambda}\frac{1}{m-j+1}\pi_j.
\tag{4.32}
$$

In order to deduce a recursive relation for the stationary distribution of working machines, we set $j = m$ in equation (4.32).

$$
\pi_{m-1} = \frac{\mu}{\lambda}\frac{1}{1!}\pi_m,
\tag{4.33}
$$

$$
\pi_{m-2} = \left(\frac{\mu}{\lambda}\right)^2 \frac{1}{1\times 2}\pi_m,
\tag{4.34}
$$

$$
\cdot =
\tag{4.35}
$$

$$
\cdot =
\tag{4.36}
$$

$$
\cdot =
\tag{4.37}
$$

$$
\pi_{m-k} = \left(\frac{\mu}{\lambda}\right)^k \frac{1}{k!}\pi_m.
\tag{4.38}
$$

Next we will use one of the axioms of probability,

$$
\sum_{k=0}^{m}\pi_{m-k} = 1,
$$

to find $\pi_m = \dfrac{1}{1+\sum_{k=1}^{m}\frac{1}{k!}\left(\frac{\mu}{\lambda}\right)^k}$. The stationary distribution of working machines is listed below.

$$\pi_{m-k} = \left(\frac{\mu}{\lambda}\right)^{k}\frac{1}{k!}\frac{1}{1+\sum_{k=1}^{m}\frac{1}{k!}\left(\frac{\mu}{\lambda}\right)^{k}}. \tag{4.39}$$

This is known as the famous *Erlang's loss formula*.

i.   Special cases emerge depending on different values of k; e.g., $k = m$ in equation (4.39) gives us π_0, which can be interpreted as the long-run probability that the repairman is idle (all machines are working).

ii.   The expected number of machines in the service queue (awaiting repair) is given by

$$
\begin{aligned}
M_q &= \sum_{k=0}^{m-1} k\pi_{k+1} \\
&= \sum_{k=1}^{m} k\pi_k - (1-\pi_0).
\end{aligned} \tag{4.40}
$$

Summing equation (4.31) over $i = 0$ through m yields $m\lambda - \lambda\sum_{k=1}^{m} k\pi_k = \mu(1-\pi_0)$, which in conjunction with equation (4.40) gives us the expected number of machines in the service queue.

$$M_q = m - \frac{\lambda+\mu}{\lambda}(1-\pi_0). \tag{4.41}$$

We leave the readers here with an exercise to work out a model and its solution akin to Example 4.6 above but with $r < m$ repairmen who can work concurrently to repair the idle machines.

4.6   Queues (Erlang-T models)

All it provided was hope for people to cling to and a reason to stay in the queue.[8] At some point in our lives, a good majority of us have gone through this emotion while waiting in a queue either at a train ticket counter or a voting booth or perhaps any of the many queues we may have inhabited, albeit for a fleeting moment when compared to the vastness of our lived experience. A common thread of thoughts that we share in such moments is: *how long do I have to stay in the queue? Or how many people on an average are in the queue at a certain time?* These questions and their answers have not only an ontological basis but also a very practical material value. Systems and businesses operate around finding optimal answers to such questions. We will formally study single and multi-server queues in this section and also in the chapter project.

4.6.1   *M/M/n queue and Kendall's notation*

The server–queue models that will be discussed in this chapter have Poisson arrivals.[9] We will use Kendall's notation here: $M/M/n$, where the first M from the left stands for the *memoryless* property of the exponentially distributed inter-arrival time with rate parameter λ, the second M stands for the *memoryless* property of the exponentially distributed service times with rate parameter μ, and n refers to the

[8] Excerpt from *The Queue* by Basma Abdel Aziz.

[9] Refer to Poisson Arrival See Time Averages (PASTA) – here the state of the queue–server system is invariant in distribution to the location of the observer (the observer can be within the system/can be arriving in the queue or the observer can be fully outside the system).

number of servers. Let $X(t)$ be the random variable that denotes the number of customers in the system at time t.

We will begin our analysis when $n = 1$. We summarize below the main attributes of the $M/M/1$ system.

1. It is a single server system.

2. Customers enter the system and arrive in a queue if the server is busy. If the server is available, then they proceed straight to service.

3. Arriving clients are served by the server on a first-come-first-serve basis.

4. Upon completion of service, the clients depart the system.

The $M/M/1$ model is a birth and death system with state space $S = \{0,1,2,..\}$. The birth and death rates are $\lambda_i = \lambda$, $\forall i \geq 0$ and $\mu_i = \mu$, $\forall i \geq 1$. Recall from the birth and death model of the previous section, i.e., equations (4.29) and (4.30), the necessary condition for a stationary distribution is $1 + \sum_{i=1}^{\infty} \left(\dfrac{\lambda}{\mu}\right)^i < \infty$ or $\sum_{i=0}^{\infty} \left(\dfrac{\lambda}{\mu}\right)^i < \infty$. This geometric series converges if $\dfrac{\lambda}{\mu} < 1$. This means that for a stationary distribution of states to exist (steady state), $\lambda < \mu$. Thus, the stability criterion for the $M/M/1$ queue is

$$\boxed{\text{arrival rate } \lambda < \text{service rate } \mu.} \tag{4.42}$$

Figure 4.10 British statistician David George Kendall known for his pioneering work in queuing theory (*author*: Konrad Jacobs; *source*: Archives of the Mathematisches Forschungsinstitut Oberwolfach).

Example 4.7 Stationary distribution of a stable M/M/1 queue

Find the stationary distribution of states for an $M/M/1$ queue with arrival and service rates λ and μ, respectively. Assume that $\lambda < \mu$.

Solution: Based on equation (4.30), we deduce $\pi_0 = \dfrac{1}{\sum_{i=0}^{\infty}\left(\dfrac{\lambda}{\mu}\right)^i} = \left(\dfrac{1}{1-\dfrac{\lambda}{\mu}}\right)^{-1} = 1 - \dfrac{\lambda}{\mu}.$

Here we have used the result of the sum of an infinite geometric series.

Likewise, using equation (4.29), we have

$$
\begin{aligned}
\pi_i &= \frac{\lambda_{i-1} \cdots \lambda_0}{\mu_i \cdots \mu_1}\,\pi_0 \\[2mm]
&= \left(\frac{\lambda}{\mu}\right)^i \pi_0 \\[2mm]
&= \left(\frac{\lambda}{\mu}\right)^i \left(1 - \frac{\lambda}{\mu}\right), \quad \forall i \geq 1
\end{aligned}
$$

Therefore, $\pi_i \sim \text{geom}_0\left(1 - \dfrac{\lambda}{\mu}\right).$ $\tag{4.43}$

Thus, π_i has the distribution of a geometric random variable that counts the number of failures* before first success with probability $p = 1 - \frac{\lambda}{\mu}$.

---

*Here failures must be counted as the number of people ahead of the currently arriving person who will be served before the currently arriving person is served.

4.6.2 *Little's law*

In a general queuing system, let us say the nth arrival spends W_n units of time in the system (time spent in the queue + service time). W_n is known as the *sojourn time* of the nth arrival. The average sojourn time is $E(W) \approx \lim_{n\to\infty} \frac{1}{n}\sum_{i=1}^{n} W_i$.[10] The average rate of arrivals is $\lambda = \lim_{t\to\infty} \frac{N(t)}{t}$, where $N(t)$ denotes the number of arrivals by time t. Further, we define by $L(t)$ the total number of clients in the system (clients in queue + clients being served) at time t and we define by $\overline{L} = \lim_{t\to\infty} \frac{1}{t}\int_0^{\infty} L(\tau)\,d\tau$ the (temporal) average number of clients in the system.[11] Then Little's law states that the following relation holds.

$$\overline{L} = \lambda E(W). \tag{4.44}$$

$\overline{L}$ is thus the average size of the system (in terms of the number of clients in the queue and the number of clients being served).

4.6.3 *Mean sojourn time E(W) and mean queue size $E(\pi_Q)$ of an M/M/1 queue*

Let us once again consider the $M/M/1$ queue as above. We now know that the stationary distribution of such a queue is a geometric distribution with parameter $p = 1 - \frac{\lambda}{\mu}$, $\pi \equiv \pi_i \sim \text{geom}_0(1-\rho)$, where $\rho = \frac{\lambda}{\mu}$. $E(\pi) = \frac{1-p}{p} = \frac{\rho}{1-\rho}$ is the expected number of clients in the stationary $M/M/1$ queue. If W_n is the sojourn time of client n, then by applying Little's law (4.44) and invoking ergodicity, we get

$$E(\pi) = \lambda E(W) = \frac{\rho}{1-\rho}, \tag{4.45}$$

where W is the time spent by a client in the system. $W = T_Q + T_S$, where T_Q is the random time spent by a client in the queue (not including the time spent during service) and T_S is the random time spent by the client while being served. From equation (4.45), we obtain the mean sojourn time

$$E(W) = \frac{1}{\lambda}\frac{\rho}{1-\rho} = \frac{1}{\mu(1-\rho)}. \tag{4.46}$$

Taking the expectation of $W = T_Q + T_S$ and using $E(T_S) = \frac{1}{\mu}$, we obtain the average time spent by a client in the queue (discounting the time of service) as follows.

$$E(T_Q) = E(W) - E(T_S) = \cdots = \frac{\rho}{\mu(1-\rho)}. \tag{4.47}$$

By applying Little's law exclusively to the queue, we get the average number of people in the queue (queue size) in the long run as follows.

$$E(\pi_Q) = \lambda E(T_Q) = \cdots = \frac{\rho^2}{1-\rho}. \tag{4.48}$$

[10] $E(W)$ refers to the average sojourn time among a large number of clients who entered (and exited) the system.

[11] The temporal average is also well approximated by an ensemble average (expected value) because a **stationary** CTMC is always **ergodic**. Conversely, if the ergodicity condition is satisfied by a CTMC, then it surely has a stationary solution.

Further, since $\pi = \pi_Q + \pi_S$, we obtain the average number of clients being served in the long run as follows.

$$E(\pi_S) = E(\pi) - E(\pi_Q) = \cdots = \rho = \frac{\lambda}{\mu}. \qquad (4.49)$$

Similar results can be deduced for a generic case, such as that of an $M/M/n$ queue or an $M/G/n$ queue that may not have an exponential service time. We will see an application of the former in the chapter project and similar examples in the exercise problems of this chapter.

Example 4.8 Delay in a communication channel

A communication channel has a capacity of 1000 bits per second. This channel is used to carry 8 bit characters. The channel can handle a total call volume of 6000 characters per minute. Estimate the average numbers of characters waiting to be transmitted and the mean sojourn time of transmission.

Solution: The arrival rate $\lambda = \frac{6000}{60} = 100$ characters/sec. The service rate $\mu = \frac{1000}{8} = 125$ characters/sec. The utilization factor is $\rho = \frac{\lambda}{\mu} = 0.8$. $E(\pi_Q) = \frac{\rho^2}{1-\rho} = 3.2$ is the average number of characters in the queue. The mean sojourn time is

$$E(W) = \frac{E(\pi)}{\lambda} = \frac{\frac{\rho}{1-\rho}}{\lambda} = \frac{\frac{0.8}{0.2}}{100} = \frac{4}{100} = 0.04 \text{ seconds} = 40 \text{ milliseconds}.$$

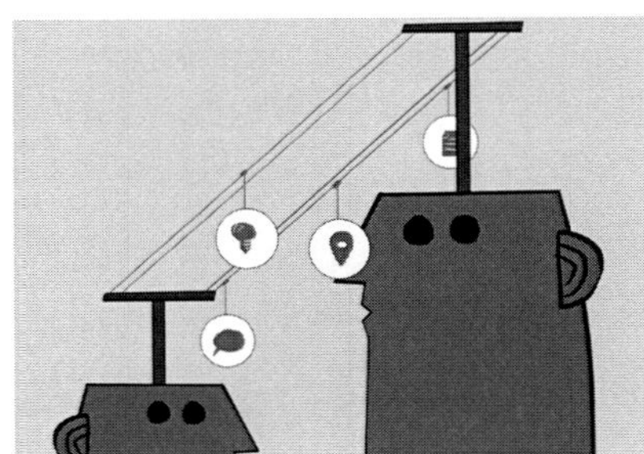

Figure 4.11 Modern technology and business operations rely heavily on efficient communication channels.

4.7 Chapter project: Queues and crowd management at COVID test centers

4.7.1 Interlude: The mathematical technology of analyzing and managing queues

As people start showing up at the test kiosk according to a Poisson process, the counters at the kiosks start testing the patients and the queue starts growing. We will use the $M/M/n$ queuing model to answer the following questions. The arrival rate λ must be estimated from the data provided in Table 4.1 (Table-4.1.csv). The service time at the test counters can be assumed to follow an exponential distribution with rate $\mu = 12$ per hour.

1. Give a condition in terms of the system utilization factor ρ so that the queue is stable (i.e., the queue does not grow in an unbounded manner).

2. Based on the data provided in the table, estimate the arrival rate λ.[12]

3. What is the minimum number of test counters required to satisfy the stability condition of question 1?

4. Deduce the expression for the steady-state probability distribution π_k for the number of patients k in the system upon a random arrival. Consider both the cases when $k \leq n$ and $k > n$.

5. What is the probability π_0 that all test counters are idle upon a random arrival? In other words, what is the probability that no patient is currently being tested?

[12] In this section, consider a *serpentine* queue, i.e., a single queue. In the final section of the chapter project, we will ask the readers to compare this type of queue with a multiple queue version of the system.

6. Compute the probability that all test counters are busy upon a random arrival.

7. What is the average number of patients in the system at any given instant? Use this result and Little's theorem to find the mean sojourn time of a patient.

4.7.2 *M/G/n queues and the Pollaczek–Khinchin formula*

Once again we will restrict our analysis to $n = 1$ in this section. More importantly, we will relax the memoryless constraint of service times and consider a general probability distribution for the service times with a mean service time of $\frac{1}{\mu}$, $\mu > 0$. The arrivals continue to follow a Poisson distribution with rate $\lambda > 0$. The utilization factor $\rho = \frac{\lambda}{\mu} < 1$ ensures that we have a stable queue with a stationary distribution of states (number of clients in the system). We will not derive the formula for the stationary distribution but simply compute the mean sojourn time and the average size of the queue at steady state.

At the outset, let us define the relevant random variables.

1. T_{s_i}: service time experienced by client i,

2. N_q: number of clients in the queue when client $i = (N_q + 1)$ arrives,

3. T_q: the time spent by the ith client in the queue,

4. R_i: residual service time of the client in service at the instant when client i arrives,[13]

5. W_i: sojourn time of a client i in the system,

6. B_i: event when the ith client arrives and finds the server busy and N_q clients ahead of it in the queue, and

7. I_i: event when the ith client arrives and finds the server available for service (idle).

The probability that the ith client finds the server busy upon arrival is $\rho = \lambda E(T_s) = \frac{\lambda}{\mu} < 1$ for a stable queue as stated earlier. Since we are primarily interested in analyzing the stationary state of the system, we will drop the subscript i from the variables. The sojourn time at steady state may be expressed as follows.

$$W \;=\; T_q + T_s, \tag{4.50}$$

$$E(W) \;=\; E(T_q) + \frac{1}{\mu}. \tag{4.51}$$

Note that we have not included R in W because it is already absorbed in the term T_q. We will use the law of total expectation to compute

$$E(T_q) = \rho E(T_q \mid B) + (1 - \rho) E(T_q \mid I) = \rho E(T_q \mid B),$$

which results from the fact that $T_q \equiv 0$ when the event I happens. In order to compute the average time spent in the stationary queue by a client, we use the following result:

---

[13] $R_i \equiv 0$ when client i enters an empty system.

$$E\left(T_q\middle|B\right) = E\left(R + T_{S_1} + T_{S_2} + \cdots + T_{S_{N_q}}\middle|B\right)$$

$$= E\left(R\middle|B\right) + E\left(T_{S_1} + T_{S_2} + \cdots T_{S_{N_q}}\middle|B\right)$$

$$= E\left(R\middle|B\right) + E\left(T_s\middle|B\right)E\left(N_q\middle|B\right)$$

See exercise on *random sums* in Chapter 1

$$= E\left(R\middle|B\right) + \frac{\lambda}{\mu\rho}E\left(T_q\right). \tag{4.52}$$

Here $E(N_q|B)$ is evaluated by the law of total expectation $E(N_q) = E(N_q|B)P(B) + E(N_q|I)P(I)$, where the second term on the RHS is zero and $E\left(N_q\middle|B\right) = \mu E\left(T_q\right)$ using Little's law $E\left(N_q\right) = \lambda E\left(T_q\right)$. Here we have used $\rho = P\left(B\right)$. Thus, we have

$$E\left(T_q\right) = \frac{\rho E\left(R\middle|B\right)}{1-\rho}. \tag{4.53}$$

We will now invoke a useful theorem without providing a proof and use it to find the average time spent by a client in the stationary queue.[14] The mean residual service time $E(R|B)$ is given by $\dfrac{1}{2}\dfrac{E\left(T_s^2\right)}{E\left(T_s\right)} = \dfrac{\mu}{2}E\left(T_s^2\right)$. Using this result, we find

$$\boxed{E\left(T_q\right) = \frac{\lambda E\left(T_s^2\right)}{2\left(1-\rho\right)}.} \tag{4.54}$$

[14] Interested readers may refer to W. C. Chan, T. C. Lu, and R. J. Chen, Pollaczek-Khinchin formula for the M/G/I queue in discrete time with vacations, *IEE Proc.-Comput. Digit Tech.* **144**, no. 4 (1997), pp. 222–226.

The mean sojourn time is given by

$$\boxed{E\left(W\right) = \frac{\lambda E\left(T_s^2\right)}{2\left(1-\rho\right)} + \frac{1}{\mu}.} \tag{4.55}$$

Equations (4.54) and (4.55) are known as the **Pollaczek–Khinchin formula**. The average number of the clients in the system $\overline{N}$ can be obtained by applying Little's law.

$$\boxed{\overline{N} = \lambda E\left(W\right) = \rho + \frac{\lambda^2 E\left(T_s^2\right)}{2\left(1-\rho\right)}.} \tag{4.56}$$

Example 4.9   An M/G/1 queue in action

Consider a single server $M/G/1$ queue at steady state. Arrivals happen according to a Poisson distribution with rate $\lambda = 6$ per hour. The service time is uniformly distributed between 7 and 9 minutes. Answer the following questions:

   i.   What is the average number of clients in the queue?

   ii.   What is the average waiting time in the queue?

   iii.   What is the average number of clients in the system?

   iv.   What proportion of time is the server idle?

Solution: We will fix the time units to minutes for consistency. Therefore, $\lambda = \dfrac{1}{10}$ per minute. $E(T_s) = \dfrac{7+9}{2}$. Therefore, the service rate is $\mu = \dfrac{1}{8}$ per minute.

$$Var(T_s) = \sigma_s^2 = \frac{(9-7)^2}{12} = \frac{1}{3}.$$

   i.   Average number of clients in the queue

$$\bar{N}_q = \lambda E(T_q) = \frac{\lambda^2 E(T_s^2)}{2(1-\rho)} = \frac{\lambda^2 \sigma_s^2 + \lambda^2 \dfrac{1}{\mu^2}}{2(1-\rho)} = \frac{\lambda^2 \sigma_s^2 + \rho^2}{2(1-\rho)} \approx 1.61.$$

   ii.   Average waiting time in the queue $E(T_q) = \dfrac{\bar{N}_q}{\lambda} \approx 16.1$ minutes.

   iii.   Average sojourn time in the system is given by the *Pollaczek–Khinchin* equation (4.55). $E(W) \approx 24.1$ minutes. Now using Little's law, $N = \lambda E(W) = 2.41$ is the average number of clients in the system.

   iv.   Proportion of the time the server stays idle is equal to $1 - \rho = 1 - \dfrac{\lambda}{\mu} = 1 - 0.8 = 0.2$.

There are many different queuing models commensurate with a variety of scenarios. Interested readers may want to check out the textbooks mentioned in the chapter bibliography for a detailed treatment on queuing models.

4.8   Chapter project: Queues and crowd management at COVID test centers

4.8.1   *Epilogue: Simulating the queue using Matlab's Simulink*

We will simulate the behavior of the queue at the test kiosks using the Simulink feature of Matlab. Using these simulations we will be able to compare the behavior of two different types of queues, namely, the *serpentine queue* and the *multi-line queue*.[15] The Simulink environment is available within the SimEvents module of Matlab. In order to access these built-in Simulink queuing models please ensure that the Simulink toolbox is installed in your computer.

Description of the Simulink queuing model

The Simulink model is available with two parallel versions of a simple model of n service counters: (i) one that uses n separate queues and (ii) one with a single *serpentine queue* that provides service to all patients. Patients arrive at random with exponentially distributed inter-arrival times and they are simulated using entities in SimEvents. The exponential service times are also simulated using the relevant SimEvents sub-module. The average inter-arrival time is set to a default value $\dfrac{1}{\lambda} = 2$ hours and the average service time is set to a default value of

[15] For example, the multi-line queue of a 3-server system will have three different queues leading to each of the servers. A randomly arriving client will choose the shortest available queue.

$\dfrac{1}{\mu} = 1$ hour. For the multi-line queue, each patient is cloned after a generation so that the different line configurations can be exercised identically. The Matlab-generated schemata is shown in Figure 4.12.

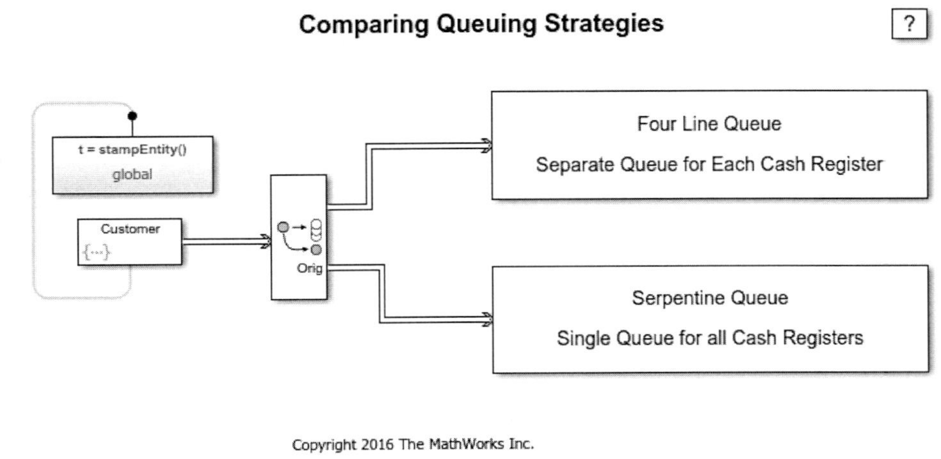

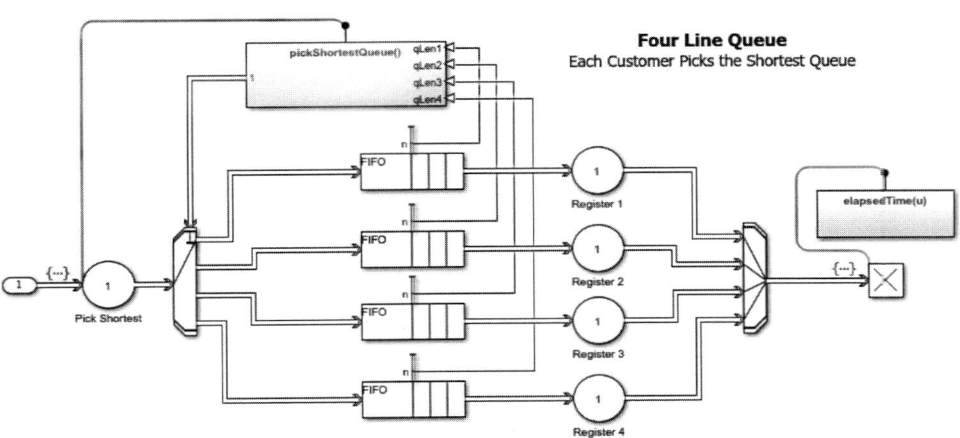

Figure 4.12 Queuing model of the SimEvent module of Simulink.

Multi-line queues

In this model, multi-line queues feed the n service counters. Arriving clients (patients) are routed to the shortest queue. Each queue then feeds the service representing a checkout together. This server holds the patient for the amount of time that was set up during its generation. Figure 4.13 represents the sub-model of a multi-line queue that is embedded into the main simulation model.

Figure 4.13 Multi-line queuing model. The example shown here corresponds to $n = 4$.

Serpentine queue

In this model, a single queue is involved that feeds all n servers via a switch that routes customers to a free service counter when one becomes available. Figure 4.14 represents this model.

Figure 4.14 Serpentine queuing model.

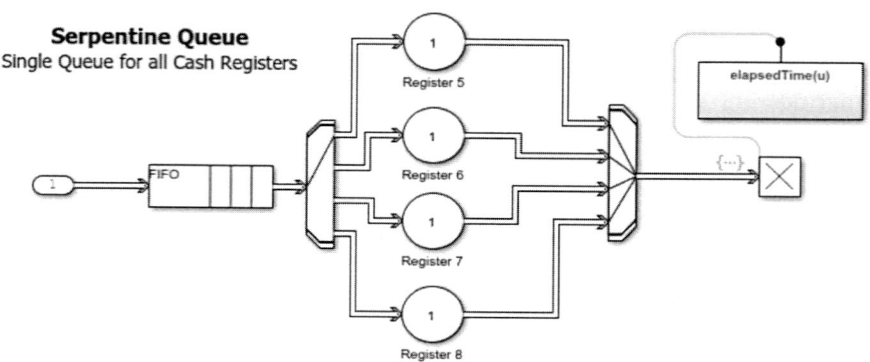

Output of Simulink

The output of this simulation is presented in the form of plots as shown in Figure 4.15, which displays the average sojourn time of a patient for both cases, i.e., the multi-line queuing model and the serpentine queuing model.

Figure 4.15 Average sojourn time for the multi-line queue (left) and the serpentine queue (right).

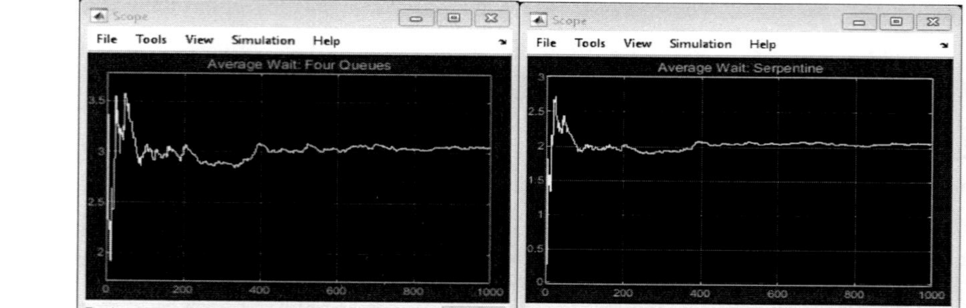

Steps to simulate the queue using Simulink

We now provide the detailed steps to run the simulations using Simulink.

Step 1: Execute the command `openExample('simevents/Queuing StrategiesExample')` in the MATLAB "Command window". This should prompt a new window that contains the model as shown in Figure 4.16:

Figure 4.16 Simulink model.

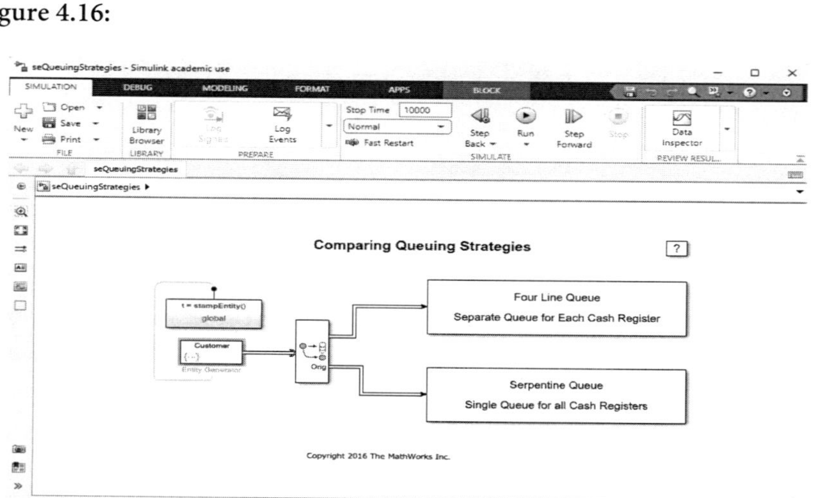

Step 2: Now in order to change the parameters (arrival rate and service rate) as required, double-click on the block model "Customer (Entity Generator)" shown in Figure 4.16. A dialogue box titled "Block Parameter: Entity Generator" (see Figure 4.17) will appear.

Figure 4.17 "Entity generation" section.

Step 3: Change the value of the mean in "Entity generation" (see Figure 4.17) and "Event actions" (see Figure 4.18) as required. Finally, click on "Apply" followed by "OK".

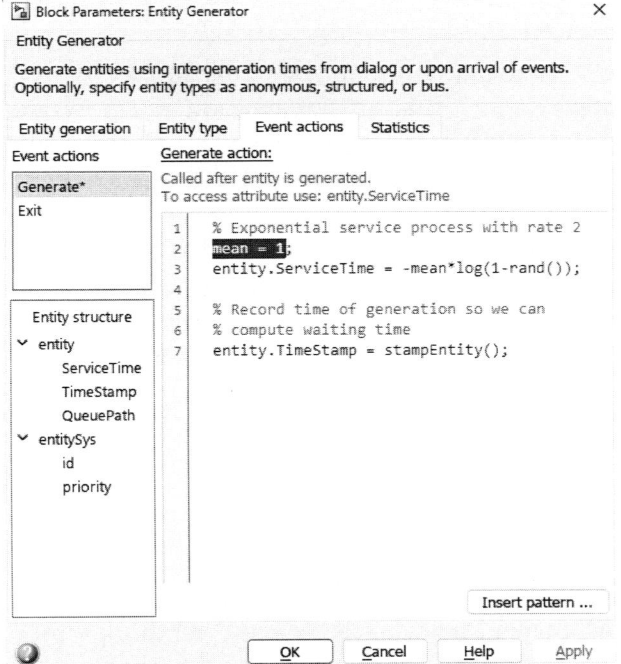

Figure 4.18 "Event actions" section.

Step 4: Now "RUN" the Simulink model (as seen in Figure 4.16). This will produce the results in the form of plots (see Figure 4.19).

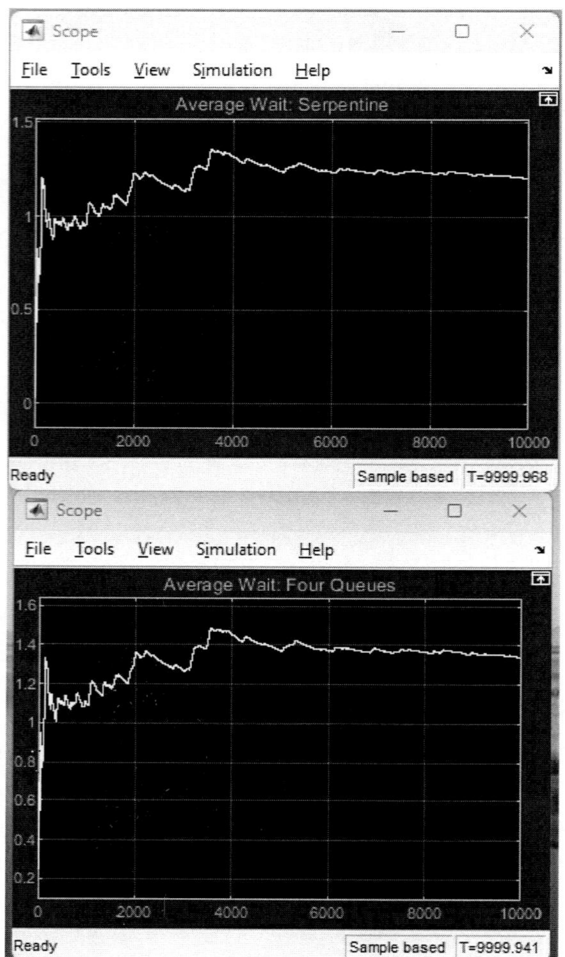

Figure 4.19 "Scope window" displaying the results for sojourn time.

Consider the answer for the minimum number of servers in question 3 to be $\hat{n}$. Now use the Simulink model to simulate the $M/M/\hat{n}$ with the estimated values of λ and μ. Run the simulation for the $M/M/\hat{n}$ model for both cases – multi-line and serpentine – in order to compute the average waiting time in the system. Discard the statistics corresponding to initial transients and then collect the steady-state values. The steady state of the system can be identified by inspection. The documentation of the Simulink model can be accessed through https://in.mathworks.com/help/simevents/ug/comparing-queuing-strategies.html. Follow the steps suggested above to perform the simulation and answer the following questions:

1. Simulate the model using the estimated value of λ in question 2 (section 4.7.1) and the given value of μ.

2. Verify the simulated value of the average sojourn time of the patient in the system with the theoretically calculated value in question 7 (section 4.7.1).

3. What do you observe from the simulation for the multi-line queue and the serpentine queue? Which queue type do you think is more preferable for both patients and health administrators? Explain your answer.

4.9 Selected bibliography

Bhat, B. R. *Stochastic Models: Analysis and Applications* (2nd edition). New Age International Publishers, 2019.

Bose, Sanjay K. *An Introduction to Queuing Systems* (1st edition). Springer Science + Business Media, 2002.

Ching, Wai-Ki and Michael K. Ng. *Markov Chains: Models, Algorithms, and Applications* (1st edition). Springer, 2006.

Durrett, Richard. *Essentials of Stochastic Processes* (3rd edition). Springer, 2016.

Feller, William. *An Introduction to Probability Theory and Its Applications.* Vol. 2 (2nd edition). John Wiley & Sons, Inc., 1971.

Trivedi, Kishor Shridharbhai. *Probability and Statistics with Reliability, Queuing, and Computer Science Applications* (2nd edition). Wiley, 2001.

4.10 Exercise problems

1. (***Traffic in a call center booth***) Assume that on average a call to a customer care number lasts for three minutes. If the customer support number is busy, callers are placed on hold and pushed to a queue. The hold can last for a maximum of three minutes after which the caller is automatically dropped from the queue and is advised to call again later. What is the maximum call rate that can be supported by one customer care number?

2. (*M/M/k* ***queues***) Consider a queuing system with Poisson arrivals of jobs at rate λ and exponential service rate μ at each of $k \geq 1$ counters. Deduce an expression for the average number of jobs in the system. How many servers are busy on an average? Further, derive an expression for the probability that a randomly arriving job enters a queue (and does not directly go to service).

3. (*M/M/1* ***queue – chances of finding a fixed queue size***) Consider an *M/M/1* queue in a steady state. What are the chances that a new arrival will find n clients in the system?

4. (***Parking lots***) Consider a parking lot with N vehicle slots. Incoming traffic follows a Poisson process with rate λ. When the lot is full, traffic is not allowed in the lot and is instead diverted to a different lot. Each vehicle that parks in the lot stays there for a random time that is distributed according to an exponential distribution with rate μ. Construct a model for computing the probability of finding exactly $n \leq N$ spaces occupied.

5. (*M/M/1* ***queues – optimizing service rate to maximize profit***) A single server queuing system provides service at rate μ where the service time is exponentially distributed. The cost of providing service is estimated as 10μ per hour. Additionally, a gross profit of ₹10 is made for every client served by the system. If the system has a capacity $N = 10$ (i.e., no more than 10

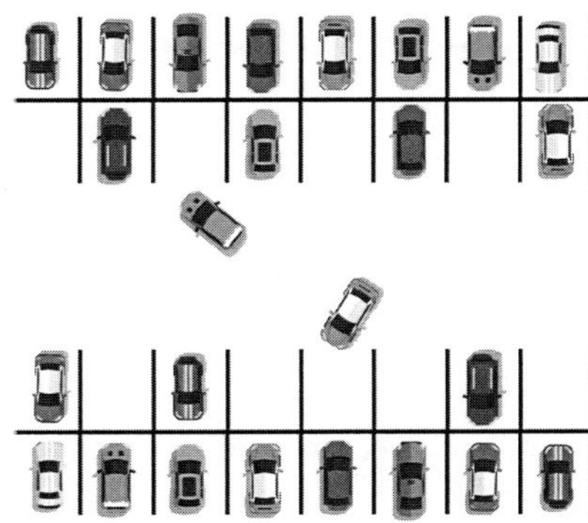

Figure 4.20 Distribution of cars in a parking lot.

clients can be allowed in the system at any given point of time), estimate the rate μ that maximizes the total profit.

6. (**Lazy service**) Consider an $M/M/1$ queue that operates like a classical one except when the queue is empty. In this exceptional case, normal service is resumed only when at least 10 customers are again present in the queue. Answer the following questions:

    i.   What is the probability that the system is empty?

    ii. What is the probability distribution for the number of customers in the system?

    iii. What is the mean number of customers in the system?

    iv. What is the average time to service resumption from the instant the queue is empty?

7. (**Impatient client**) A bank has a single server. Clients arrive at the bank gate according to a rate $\lambda = 2$ Poisson process. The client enters the bank only if the server is free, else he walks away in search of a new bank. Further, the service time of the server is distributed according to a distribution G with mean $\mu_G = 3$. Appropriate units may be assumed. Compute the following:

    i. What is the rate at which clients enter the bank?

    ii. What fraction of arriving clients actually enters the bank?

8. (**Impatient client with a fickle mind**) Answer question 7 with the caveat that the arriving client enters the bank either when the server is free or with a probability $p = \frac{1}{2}$ when the server is busy.

9. (**Renewal process**) $\{N(t)\}_{t \geq 0}$ is a *counting process* marking the arrival of $N(t)$ point-events up to time t and let $\mathcal{T}_1, \mathcal{T}_2, \ldots$ denote the inter-arrival times between successive arrivals and be i.i.d. random variables. Then $N(t)$ is called a *renewal process*. Thus, a renewal process is a counting process such that at the instant of an event occurrence a renewal is said to have happened. This is so because the inter-arrival times have the same distribution irrespective of the epoch of the previous event. The renewal epochs are denoted by R_i, where $R_0 = 0$ and $R_n = \sum_{i=1}^{n} \mathcal{T}_i$, $n \geq 1$. Further, let $\mu = E(\mathcal{T}_n)$, $n \geq 1$. Prove that with probability equal to one, we have the following result:

$$\lim_{t \to \infty} \frac{N(t)}{t} = \frac{1}{\mu}. \tag{4.57}$$

Figure 4.21 How often does Alex have to change his baby's diapers?

10. (**Changing diapers**) Alex must take care of his four-month-old baby, and part of that job entails changing the baby's diapers as soon as it is soiled. If a diaper lasts for a uniformly distributed time between 1 hour and 2 hours before it is soiled, then at what rate does Alex have to change his baby's diapers in the long run? Further, if Alex does not have a back-up supply of diapers handy and must have to fetch a new one from the storage, which takes him anywhere between 15 and 30 minutes (again uniformly distributed), then what is the average rate at which Alex changes the baby's diapers?

WEEK\DAYS	Monday	Tuesday	Wednesday	Thursday	Friday	Saturday	Sunday	Total
WEEK-01	-	-	-	-	415	1107	1065	2587
WEEK-02	1894	1759	1605	1640	1455	1009	1007	10369
WEEK-03	1755	1615	1534	1557	1402	1073	904	9840
WEEK-04	1462	1475	1352	1405	1324	1012	908	8938
WEEK-05	1763	1420	1504	1432	1099	875	883	8976
WEEK-06	344	873	1505	1388	1245	900	567	6822
WEEK-07	1345	1090	1061	931	934	671	585	6617
WEEK-08	860	1110	1006	591	855	681	792	5895
WEEK-09	1117	1073	1023	1027	943	625	734	6542
WEEK-10	1174	936	956	934	888	594	595	6077
WEEK-11	1107	1086	958	964	937	627	644	6323
WEEK-12	1063	986	941	857	902	660	710	6119
WEEK-13	1215	1027	1048	1147	999	632	574	6642
WEEK-14	1043	938	959	861	847	638	497	5783
WEEK-15	1039	933	868	813	803	564	462	5482
WEEK-16	853	775	722	640	595	411	357	4353
WEEK-17	641	518	513	489	438	308	270	3177
WEEK-18	498	391	418	392	322	190	220	2431
WEEK-19	316	297	243	268	225	121	106	1576
WEEK-20	207	216	196	147	145	94	99	1104
WEEK-21	151	138	129	134	115	81	63	811
WEEK-22	105	102	81	62	67	32	34	483
WEEK-23	35	59	70	62	66	44	35	371
WEEK-24	64	55	58	46	44	28	35	330
WEEK-25	54	37	46	52	59	29	33	310
WEEK-26	40	50	48	51	61	37	28	315
WEEK-27	49	53	75	63	60	41	32	373
WEEK-28	62	125	147	110	111	63	79	697
WEEK-29	176	152	185	171	183	109	141	1117
WEEK-30	276	233	309	326	319	188	214	1865
WEEK-31	423	406	384	444	427	263	294	2641
WEEK-32	518	489	522	517	520	373	321	3260
WEEK-33	626	536	546	578	499	348	422	3555
WEEK-34	628	595	555	543	505	338	276	3440
WEEK-35	642	613	570	525	464	296	343	3453
WEEK-36	583	531	487	478	429	255	334	3097
WEEK-37	274	434	466	630	504	335	504	3147
WEEK-38	633	609	550	451	559	357	464	3623
WEEK-39	545	397	509	513	452	309	333	3058
WEEK-40	518	355	369	455	408	280	458	2843
WEEK-41	512	507	487	406	451	250	308	2921
WEEK-42	363	464	409	426	322	194	260	2438
WEEK-43	364	314	294	341	261	162	198	1934
WEEK-44	363	238	288	311	230	158	199	1787
WEEK-45	314	302	300	349	290	198	205	1958
WEEK-46	401	358	363	347	372	241	258	2340
WEEK-47	428	430	432	428	353	279	293	2643
WEEK-48	517	482	460	189	364	302	318	2632
WEEK-49	661	577	700	629	579	382	463	3991
WEEK-50	849	895	982	912	983	619	870	6110
WEEK-51	2390	3421	4123	4514	4275	3243	3437	25403
WEEK-52	8933	9446	9452	9606	5572	2344	7215	52568
WEEK-53	15029	14203	14298	13483	8744	-	-	65757
Grand Total	57222	56124	57106	55635	45421	24970	30446	**326924**

Table 4.1 Number of infected persons who presented themselves at the COVID test kiosks (Table-4.1.csv).

5

Statistical Experiments

STATISTICAL EXPERIMENTS ENABLE us to make *inferences* from data about parameters that characterize a population. Generally speaking, inferences may be of two types, namely, *deductive* inference and *inductive* inference.[1] Deductive inference pertains to conclusions based on a set of premises (propositions) and their synthesis. Deductive reasoning has a definitive character. For example, all men are mortal (first proposition); Socrates is a man (second proposition); hence, Socrates is mortal (deductive conclusion). On the other hand, inductive inference has a probabilistic character. One conducts an experiment and collects data. Based on this data, certain conclusions are drawn that may have a broader applicability beyond the contours of the particular experiment performed by the researcher. This generalization of the conclusions drawn from the particular experiment constitutes the framework of inductive reasoning. For example, measurement of heights of a small group of people belonging to a certain population is conducted. Based on the calculations of this small sample set, and upon finding that for this small group the average height of men is greater than the average height of women, it is inferred that the men of this population are generally taller than the women.

The formal practice of inductive reasoning dates back to the thesis of Gottfried Wilhelm Leibniz (see Figure 5.1). He was the first to propose that probability is a relation between hypothesis and evidence (data). His thesis was founded on three conceptual pillars: *chance* (probability), *possibilities* (realizable random events), and *ideas* (generalization of inferences by induction).[2] We have encountered the first two concepts in earlier chapters of this textbook. In this chapter, we will delve into the third theme whereby we will discuss methods to draw conclusions from data derived from statistical experiments based on the principles of inductive reasoning.

Figure 5.1 Portrait of the German polymath Gottfried Wilhelm Leibniz (*courtesy*: Getty Images).

[1] Alexander M. Mood, Franklin A. Graybill, and Duane Boes, *Introduction to the Theory of Statistics*, 3rd ed. (McGraw Hill Education, 2017).

[2] Ian Hacking, *The Emergence of Probability*, 2nd ed. (Cambridge University Press, 2006).

5.1 Chapter objectives

The chapter objectives are listed as follows.

1. Students will learn the concepts of random samples, population parameters, statistics, sampling distributions, and hypotheses.

2. Students will learn to deduce the probability distributions of test statistics and apply these sampling distributions to perform different tests of hypotheses.

3. Students will learn to analyze the asymptotic behavior of certain sampling distributions.

4. Students will learn to conduct analysis of variance (ANOVA).

5. Students will be trained to provide error estimates of inferences and confidence intervals of model parameters in statistical experiments.

6. Students will learn to organize data sourced from reconstruction projects following natural disasters and perform an ANOVA experiment to prioritize post-disaster reconstruction efforts.

5.2   Chapter project: Prioritizing post-disaster reconstruction measures

5.2.1   *Prologue: What factors severely impede the efficient implementation of post-disaster reconstruction efforts?*

Reconstruction projects following catastrophic disasters like floods, hurricanes, earthquakes, etc., are often negatively impacted by a plethora of factors. These include availability of capital investment with the local government, availability of manpower resources for conducting rehabilitation and reconstruction efforts, existing laws relating to land acquisition for building temporary and permanent housing for displaced persons, etc.

The goal of this project is to highlight the issues and challenges in post-disaster reconstruction (PDR) efforts and to determine the significant differences between the issues and challenges in different locations where PDR projects are carried out. As a chief construction engineer of an international non-governmental organization, you are tasked with devising an emergency strategy for tackling the issues concerning the efficient implementation of PDR projects. Your decision-making process relies on an extensive database across six international cities[3] where project engineers have rated the most pressing issues that are responsible for delay in PDR projects. Your first task (and the objective of this project) is to identify those issues that are common across geographical locations and address them as a priority. In order to accomplish this, you are provided with a database of responses by construction engineers from six different international cities. Construction engineers who worked in these cities in the past have rated the significance of the respective issues on a scale of 1–10, with 1 being *strongly disagree* and 10 being *strongly agree*. The issues that will be investigated are

- *shortage of relief workers and technical staff,*
- *land ownership and related laws,*
- *funding and aid for PDR projects, and*
- *community participation in rebuilding efforts.*

In order to accomplish this task, you will use the F statistic, which is constructed by considering two different estimates for the sample variance. The ratings of the construction engineers from different locations (as mentioned above) constitute the sample data for this statistical experiment. Before analyzing this data, we must develop conceptual knowledge about sampling distributions (like the F distribution which is used in performing an ANOVA experiment).

Figure 5.2 An artist's rendition of the devastation created by an earthquake. Post-disaster reconstruction and rehabilitation measures may include relief operations, land acquisition, re-development costs, and community engagement and coordination.

[3] Port-au-Prince (Haiti), Tacloban City (Philippines), Latur (India), New Orleans (USA), Kathmandu (Nepal), and Bagh City (Pakistan).

5.3 Elements of statistical sampling

The quintessential idea is to conduct our statistical experiments on a smaller group instead of performing them on the entire population. The manner in which this smaller group is selected is crucial to the accuracies of the inferences drawn from the experiments. In this section, we will study some important sampling techniques and some important theoretical results that hold true in the asymptotic limit of large enough sample size.

5.3.1 *Definition: Random sample*

Consider a population $\mathfrak{P}$ from which we will pick our sample of size n randomly. The n random variables X_1, X_2, ..., X_n constitute a *random sample* if they are *independent* and *identically distributed* (i.i.d.), i.e., (i) if the X_is are independent of each other, and (ii) if each of the X_is follows the same probability distribution (identically distributed).

If the sample is *not* chosen in a random manner from $\mathfrak{P}$, then the statistical techniques which we will learn in this chapter will not apply. Further, the inferences drawn from the experiments may be incorrect.

5.3.2 *Definition: Statistic*

A *statistic* is a function of the observations made from a random sample. It is a random variable because it depends on the randomly selected sample.

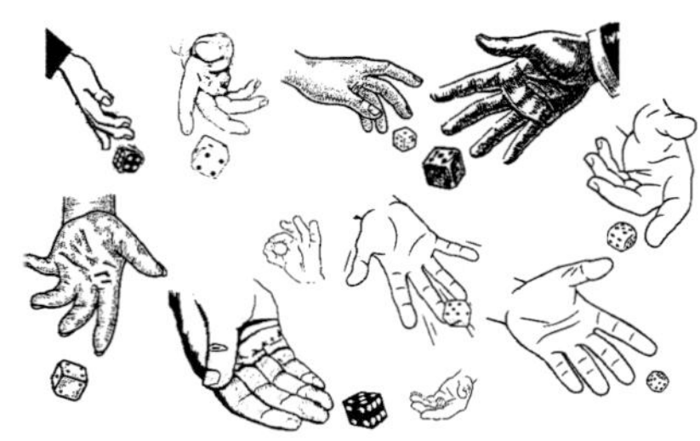

Figure 5.3 The outcomes of the throws of multiple dice are independent and identically distributed. The outcome of each throw is independent of others because the kinematic motions of the hands are not dependent on one another as they belong to different individuals. Yet there is a certain *sameness* about the structure and motion of the hands (and the shape of the dice) due to which the outcome of each throw results in one of six numbers with the same probability $\frac{1}{6}$ (identically uniformly distributed).

> ### *Example 5.1 Random sample and statistic*
>
> Let us consider that we want to know the proportion p of people in a certain organization who are interested to avail themselves of health coupons to obtain access to a nearby gym facility. There are over 10000 employees in the organization, and soliciting an answer from every individual may not be logistically practical. So the organization randomly picks 100 people irrespective of gender, age, departmental affiliations, etc., and records the preferences of each member of this sample. It then computes and finds that 71 of these 100 people intend to avail themselves of the health coupons. Therefore, $\hat{p} = 0.71$ may be regarded as a reasonable estimate of p. It may be noted that had we chosen a different set of 100 people, the value of $\hat{p}$ might have been slightly different. So the statistic $\hat{p}$ is itself random as it is a function of the random sample.

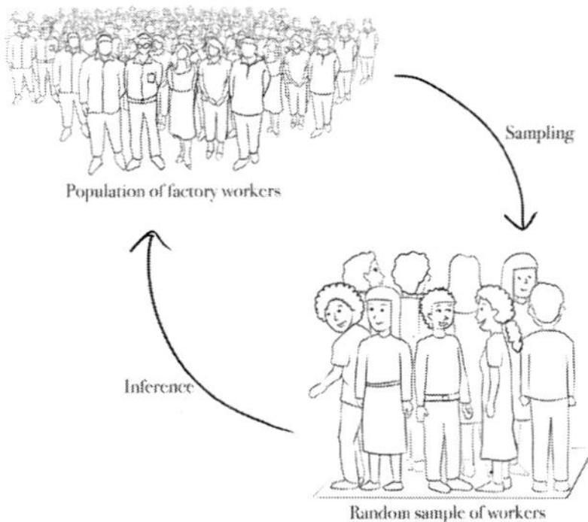

Figure 5.4 Random sampling of the whole population of factory workers enables us to make observations and estimates from a smaller representative group. We can then make inference about the whole population.

> ### *Example 5.2 Sample statistics*
>
> Consider a random sample X_1, X_2, ..., X_n. Some frequently used sample statistics are
>
> 1. **sample mean,** $\overline{X} := \dfrac{X_1 + X_2 + \cdots + X_n}{n}$,
>
> 2. **sample variance,** $S^2 := \dfrac{1}{n-1}\sum_{i=1}^{n}\left(X_i - \overline{X}\right)^2$, and
>
> 3. **sample standard deviation,** $S = \sqrt{S^2}$.

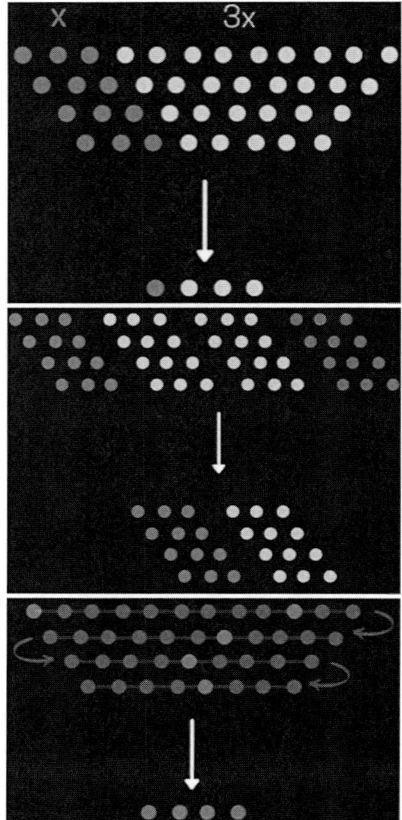

Figure 5.5 Probabilistic sampling strategies: stratified sampling (*top*), clustered sampling (*middle*), and systematic sampling (*bottom*). See color plate on page 297.

It turns out that dividing by $(n-1)$ makes the sample variance an *unbiased* estimator of the population variance σ^2.[4] In fact, out of n degrees of freedom that is available from the n observables $x_1, x_2, ..., x_n$, one unit of information is already used in the computation of $\overline{X}$. Hence, we are left with $(n-1)$ degrees of freedom; therefore, using $(n-1)$ as the normalization factor in the denominator makes sense.

5.3.3 *Sampling strategies*

Different sampling strategies may be employed depending on the experimental context. Broadly speaking, sampling strategies may be classified into two categories, namely, *non-probabilistic sampling* (e.g., voluntary participation and convenience sampling based on availability of experimental participants) and *probabilistic sampling*. Some frequently used probabilistic sampling strategies are summarized below (see Figure 5.5).[5]

1. **Random sampling**: Here, every member of the population has an equal probability of being included in the sample set whose size may be predefined.

2. **Stratified sampling**: Here, the entire population is classified into sub-populations based on certain characteristics. Sampling is performed randomly from within the sub-populations such that the groups are maintained in the same ratio in the sample set as they were in the original population. This type of sampling is most commonly used in pattern classification and machine learning applications.

3. **Clustered sampling**: In this strategy, the entire population is classified into sub-groups in a manner such that each of these sub-groups encompasses all the features of the whole population. Thereafter, the clusters are randomly selected as a group, rather than individually, to form the sample set.

4. **Systematic sampling**: A pre-defined strategy is used to pick the members of the sample set from the whole population; e.g., every 10th (randomly decided) member of the population is sampled sequentially.

5.3.4 *Definition: Sampling distribution*

The probability distribution of a statistic is called a *sampling distribution*.

The named probability distributions such as Poisson, Bernoulli, and exponential, which were introduced in Chapter 3, are consequences of theoretical considerations.[6] However, the sampling distributions are consequences of real outcomes of statistical experiments. We will study different types of sampling distributions and their applications in this chapter. We list here some of the most frequently used sampling distributions.

- Standard normal distribution
- χ^2 distribution
- t distribution
- F distribution

It is necessary to emphasize that the sampling distribution of a statistic depends on the following:

---

[4] S^2 is an unbiased estimator of σ^2, i.e., $E(S^2) = \sigma^2$.

[5] In many applications, *re-sampling* techniques like *jackknife* or *bootstrap* may be necessary.

[6] For example, the probability density function of the exponential random variable can be derived theoretically from the Poisson distribution. We will consider the random variables: $X \sim$ Poisson (λ) and $T \sim exp(\lambda)$, where λ is the rate parameter. Since T is the inter-arrival time between two Poisson arrivals, $P(T > t) \equiv P(X =$ 0 in t time units) $= \underbrace{e^{-\lambda} \times e^{-\lambda} \times \cdots \times e^{-\lambda}}_{t \text{ times}}$ $= e^{-\lambda t}$. Therefore, $P(T \leq t) = 1 - P(T > t)$ $= 1 - e^{-\lambda t}$. Further, $f_T(t) = \dfrac{d}{dt} P(T \leq t)$ $= \lambda e^{-\lambda t}$. Thus, we see that the exponential probability distribution can be entirely constructed from the Poisson distribution based solely on theoretical arguments.

$\rightarrow$ distribution of the population,

$\rightarrow$ sample size, and

$\rightarrow$ sampling strategy.

5.4   Sampling distributions

We will begin with the sampling distributions of the sample mean $\overline{X}$ and the sample variance S^2. If the random sample is drawn from the normal distribution, i.e., $X_1, X_2, ..., X_n \sim N\left(\mu, \sigma^2\right)$, then clearly by the linearity principle, we have $\overline{X} \sim N\left(\mu, \dfrac{\sigma^2}{n}\right)$.[7] Notably, $\overline{X}$ and S^2 are independent random variables.[8] It turns out that this independence of $\overline{X}$ and S^2 is a unique property of the normal distribution. We will return to the distribution of the sample variance later in this section. Now, we will state our first result on the asymptotic distribution of $\overline{X}$ without assuming the distribution of the samples X_1, X_2, ..., X_n. This is one of the most fundamental results in the theory of statistics.

5.4.1   *Sampling distribution of $\overline{X}$: Central Limit Theorem (CLT)*

> Consider a random sample of size n denoted by X_1, X_2, ..., X_n that is drawn from a population of mean μ (i.e., $E\left(X_i\right) = \mu$, $i = 1, 2, ..., n$) and variance σ^2 (i.e., $Var\ (X_i) = \sigma^2$). Then the asymptotic distribution of $Z = \dfrac{\overline{X} - \mu}{\sigma / \sqrt{n}}$ as $n \rightarrow \infty$ is the standard normal distribution $N\ (0, 1)$.

Most importantly, X_1, X_2, ..., X_n need not be normally distributed. Many random variables encountered in several science and engineering applications are normally distributed due to the effect manifested by the CLT.

5.4.2   *How large a sample suffices for the CLT result to take effect?*

This is a rather subjective question. If the underlying population distribution is much alike the standard normal distribution, then only a few (say 3 to 5) samples are required to get a decent normal approximation of the sample mean. However, if the population distribution is very different from the standard normal distribution, then about 30 samples will suffice to obtain a good normal approximation for the sample mean.

5.4.3   *Computer illustration of CLT*

In order to fully comprehend the meaning of the above result, we provide a computer simulated random sampling of exponentially distributed random variables with rate parameter $\mu = 3$. We will notice that with an increasing number of samples/realizations of the simulated experiment, the distribution of the sample means (mean subtracted and appropriately normalized)[9] resembles the standard normal distribution.

The Matlab routine for conducting the aforementioned computer experiment is presented below.

---

[7] This result can be easily deduced by using the method of moment generating functions that we studied earlier in Chapter 2.

[8] This result is a consequence of *Basu's theorem* because $\overline{X}$ is a complete sufficient statistic for estimating the model parameter μ, and S^2 is an ancillary statistic whose distribution does not depend on μ. The independence of $\overline{X}$ and S^2 can also be proven by using *Cochran's theorem*. A detailed study of these techniques and concepts can be found in more advanced textbooks on statistics (see pp. 289 and 572 in George Casella and Roger L. Berger, *Statistical Inference*, 2nd ed. [Duxbury-Thomson Learning, 2002]).

[9] The computer simulation shows that $\frac{\overline{X}-\mu}{\sigma/\sqrt{n}} \sim N\left(0,1\right)$ as $n \rightarrow \infty$. Here $X_i \sim exp(\mu)$, for all $i = 1, 2, ..., n$.

```matlab
mu = 3; var = mu^2; k=1; numsamples = [2 5 25 2000];
for N = numsamples
  ni = [];
  for i=1:N
      ni = [ni; exprnd(mu,1,10000)];
  end
  n = (mean(ni)-mu)/(sqrt(var)/sqrt(N)); x = -max(n):0.1:max(n);
  subplot(length(numsamples),1,k);
  h=histogram(n,'Normalization','probability'); hold on;
  h.BinWidth = 1.0;
  [yn,xn]=ksdensity(n);
  plot(xn,yn,'k','LineWidth',2); hold on;
  pd = makedist('Normal'); pdf_normal = pdf(pd,x);
  plot(x,pdf_normal,'rx-','LineWidth',2);
  xlabel('observables','FontSize',16); ylabel('probability distribution', 'FontSize',16);
  legend('histogram of samples from exponential distribution',...
      'pdf of distribution with N samples','standard normal distribution','FontSize',16);
  k = k+1;
end
```

It is necessary to comment on some of the Matlab commands used in the above routine.

→ `numsamples = [2 5 25 2000]`: Here, `numsamples` is a variable array which is assigned the values of the size of the experiments (number of realizations) that will be considered.

→ `exprnd(mu,1,10000)`: It returns an array of random numbers of dimension 1×10000 chosen from the exponential distribution with mean parameter `mu`.

→ `ni = [ni; exprnd(mu,1,10000)]`: `ni` is a two-dimensional array of dimension $N \times 10000$ that stores the N realizations of the exponentially distributed random variables. Each realization (sample) is an array of 10000 observables.

→ `histogram(n,'Normalization','probability')`: This function returns a histogram of the elements in the array n; the resulting histogram is normalized in such a way that the cumulative area covered by the histogram is normalized to unity.

→ `h.BinWidth = 1.0`: In conjunction with the normalization mode mentioned in the above paragraph, the width of each bin in the histogram must be set to unity to ensure compliance with the second axiom of probability (axiom of unitarity).

→ `ksdensity(n)`: It computes a probability density estimate of the sample stored in the array n.

→ `pdf(makedist('Normal'),x)`: This function generates the probability density function of a standard normal distribution for the observable variables in the range specified by x.

It is essential to emphasize that irrespective of the type of the probability distribution of the population, the sampling distribution of $Z := \dfrac{\overline{X} - \mu}{\sigma / \sqrt{n}}$ tends to the standard normal distribution. This can be easily verified by replacing the Matlab function `exprnd` with any other probability distribution, e.g., `Poisson` and `Geometric`. For more details, readers may check out the Matlab documentation page for `random`[10] for generating random variables (random samples) drawn from different probability distributions. The results of the simulation experiment are shown in Figure 5.6.

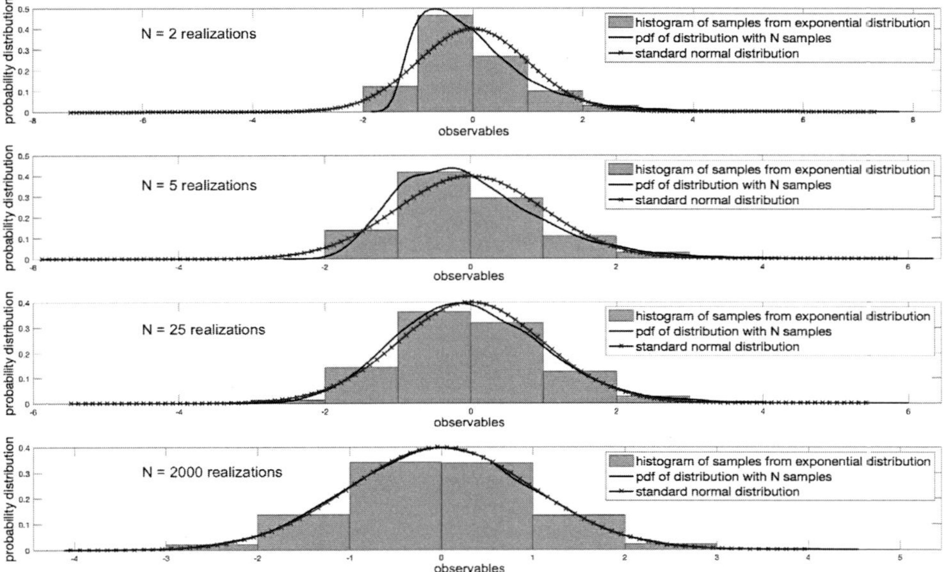

[10] Type `>> doc random` in the Matlab command window and hit return.

Figure 5.6 The central limit theorem (CLT) demonstrates that as the size of the samples is increased from $N = 2$ to $N = 2000$, the probability distribution of the mean subtracted and appropriately normalized sample mean (shown by solid lines) resembles the standard normal distribution (shown by cross-cut lines).

Example 5.3   Rating of frontal cameras of an autonomous vehicle (application of CLT)

Safety features of a self-driving car are of paramount importance to manufacturers. This is the reason why modern autonomous cars have anywhere between 10 and 30 cameras on board. A certain self-driving car has three frontal cameras which are activated when the car gets close to a vehicle or object in front of the car. This is especially useful in a traffic jam and/or on the highway when a certain distance must be maintained between two successive cars. During foggy winter conditions, measurement of the exact distance of the car (d meters (m)) from the one in front, by a single input from a camera, may become less reliable. Therefore, the system must have built-in redundancies whereby multiple frames per second (fps) must be obtained successively by the three frontal cameras. Each of these measurements (all made every second) may be regarded to be an independent random variable with mean d m and standard deviation 2 m based on multiple statistical tests performed by the manufacturer. Then the average of all these measurements must be taken and processed per second by the electronic computer of the car to make the autonomous technology more accurate. What must be the fps rating of the frontal camera unit of the car so that the manufacturer is at least 95% certain that the estimated information is accurate to within ±0.5 m?

Solution: Consider that the requisite fps rating of the frontal camera system is n. Then, if $\overline{X}$ is the mean measurement made every second, then it is reasonable to assume that $\overline{X} \sim N\left(d, \dfrac{4}{n}\right)$ based on the CLT (see note below). Therefore,

$$P(-0.5 < \overline{X} - d < 0.5) \;=\; P\left(\frac{-0.5}{2/\sqrt{n}} < \frac{\overline{X} - d}{2/\sqrt{n}} < \frac{0.5}{2/\sqrt{n}}\right)$$

$$\approx\; P(-\sqrt{n}/4 < Z < \sqrt{n}/4)$$

Direct consequence of CLT

$$=\; 2P(Z < \sqrt{n}/4) - 1, \tag{5.1}$$

See caption of Figure 5.7

where Z is the standard normal random variable. This means we must have $2P(Z < \sqrt{n}/4) - 1 \geq 0.95$ or equivalently $P(Z < \sqrt{n}/4) \geq 0.975$. Following the standard normal distribution table shown in Figure 5.8, since we have $P(Z < 1.96) = 0.975$; n must be chosen such that $\sqrt{n}/4 \geq 1.96$ or $n \geq 61.46$. This means that the fps rating of the frontal camera system of the autonomous car must be at least 62.

Note: In addition to the statement of the CLT, we can verify that

$$E\left(\overline{X}\right) = \frac{E\left(X_1\right) + E\left(X_2\right) + \cdots + E\left(X_n\right)}{n} = \mu, \text{ and } Var\left(\overline{X}\right) = \frac{\sum_{i=1}^{n} Var\left(X_i\right)}{n^2} = \frac{n\sigma^2}{n^2} = \sigma^2/n.$$

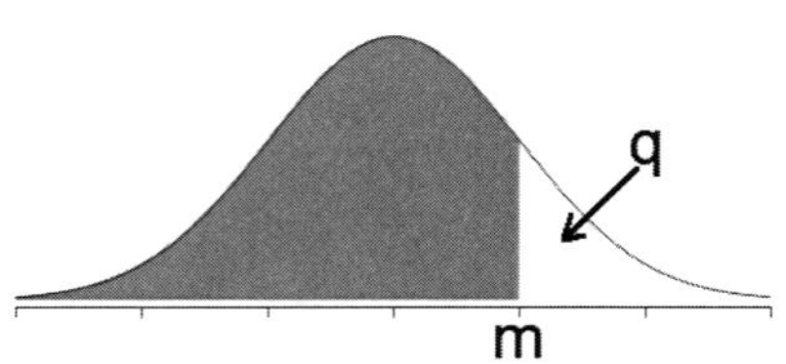

Figure 5.7 The shaded portion of the distribution $N(0, 1)$ shows $P(Z < m)$, $m > 0$. $P(-m < Z < m) = (1 - 2q) = 1 - 2\{1 - P(Z < m)\} = 2P(Z < m) - 1$.

Figure 5.8 An excerpt from the standard normal distribution table compatible with the shaded portion of the probability distribution shown in Figure 5.7. In the example discussed in section 5.4.3, $m = 1.96$. $P(Z < 1.96)$ can be computed from the table by reading off the entry corresponding to $(1.9, 0.06) = 0.9750$.

	0.00	0.01	0.02	0.03	0.04	0.05	0.06	0.07	0.08	0.09
0.0	0.5000	0.5040	0.5080	0.5120	0.5160	0.5199	0.5239	0.5279	0.5319	0.5359
0.1	0.5398	0.5438	0.5478	0.5517	0.5557	0.5596	0.5636	0.5675	0.5714	0.5753
0.2	0.5793	0.5832	0.5871	0.5910	0.5948	0.5987	0.6026	0.6064	0.6103	0.6141
0.3	0.6179	0.6217	0.6255	0.6293	0.6331	0.6368	0.6406	0.6443	0.6480	0.6517
⋮										
1.6	0.9452	0.9463	0.9474	0.9484	0.9495	0.9505	0.9515	0.9525	0.9535	0.9545
1.7	0.9554	0.9564	0.9573	0.9582	0.9591	0.9599	0.9608	0.9616	0.9625	0.9633
1.8	0.9641	0.9649	0.9656	0.9664	0.9671	0.9678	0.9686	0.9693	0.9699	0.9706
1.9	0.9713	0.9719	0.9726	0.9732	0.9738	0.9744	0.9750	0.9756	0.9761	0.9767
2.0	0.9772	0.9778	0.9783	0.9788	0.9793	0.9798	0.9803	0.9808	0.9812	0.9817
2.1	0.9821	0.9826	0.9830	0.9834	0.9838	0.9842	0.9846	0.9850	0.9854	0.9857
2.2	0.9861	0.9864	0.9868	0.9871	0.9875	0.9878	0.9881	0.9884	0.9887	0.9890

5.4.4 *Sampling distributions derived from the standard normal distribution*

In this section, we will study a few sampling distributions that either arise from the standard normal distribution or converge to the standard normal distribution asymptotically. These sampling distributions will be extensively used for performing statistical experiments and testing hypotheses.

(i) **The chi-square distribution**   The chi-square distribution, denoted by $\chi^2(v)$, has v degrees of freedom. It is the distribution of the sum of the squares of normally distributed random variables.

$$X_v = Z_1^2 + Z_2^2 + \cdots + Z_v^2; \quad \text{where} \quad Z_i \sim N(0,1); \; i = 1, 2, \cdots, v. \tag{5.2}$$

Then $X_v \sim \chi^2(v)$. The Matlab routine for simulating the χ^2 distribution based on the above definition is given below along with the results in Figure 5.9.

```
x = [0:0.1:20]; num_max = 5000;
nu = 5; mu = 0; sigma = 1;

for k=1:num_max
    X = 0;
    for i=1:nu
        Z_gauss = normrnd(mu,sigma);
        X = X + Z_gauss^2;
    end
    X_chi(k) = X;
end

h = histogram(X_chi,'Normalization','probability'); hold on;
h.BinWidth = 1.0;
[yn,xn]=ksdensity(X_chi);
plot(xn,yn,'k','LineWidth',2); hold on;

fx_chi = chi2pdf(x,nu);
plot(x,fx_chi,'m--','LineWidth',4);
legend(['histogram of samples from X_{\nu} = ` ...
    'Z_1^2 + Z_2^2 + \cdot \cdot + Z_{\nu}^2; where Z_i \sim N(0,1), i=1,2,...,\nu'],...
    'pdf of X_{\nu}','pdf of \chi^2(\nu)','FontSize',16);
xlabel('x','FontSize',16);
ylabel('f_{X_\nu}(x)','FontSize',16);
title('Number of degrees of freedom, \nu = 5','FontSize',16);
xlim([0 25]);
```

In Figure 5.9, the probability density function (pdf) of X_v based on 5000 realizations of the v-tuple random sample $\{Z_1, Z_2, \cdots, Z_v\}$ (solid lines) is compared against the χ^2 pdf generated by the Matlab function `chi2pdf` (broken lines). It may be easily verified that the accuracy of the pdf of X_v generated from multiple realizations of $\{Z_i\}_{i=1,2,\ldots,v}$ increases by considering a larger number of realizations (i.e., by increasing the value of `num_max` in the above Matlab routine).

Figure 5.9 Computer simulation of the χ^2 distribution with v degrees of freedom.

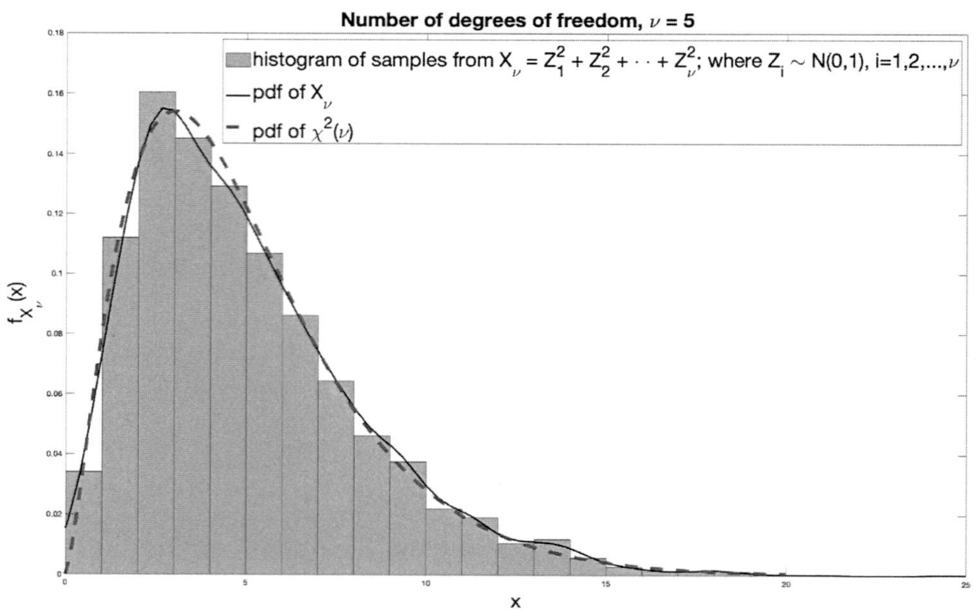

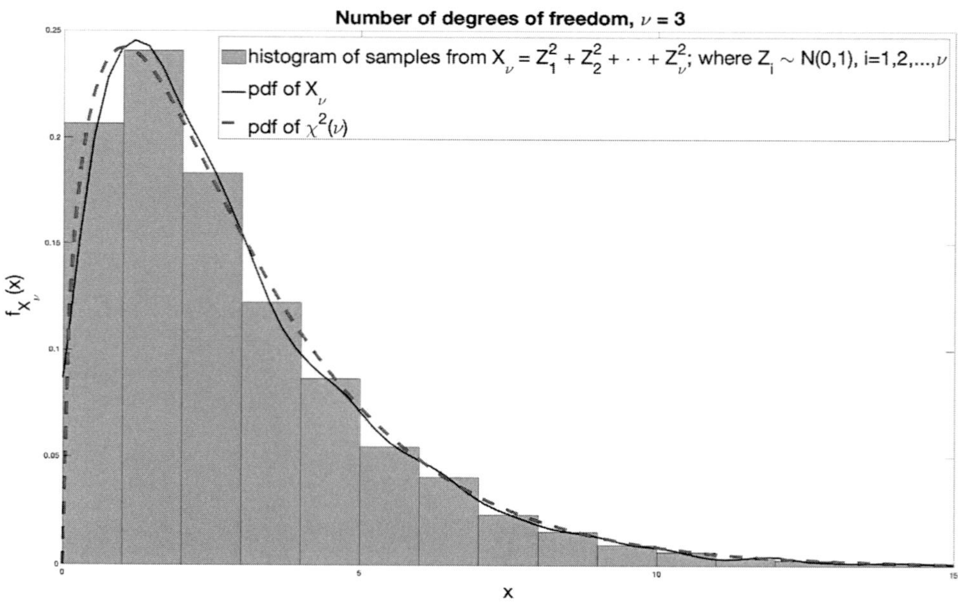

(i.i) **Properties of chi-square random variables**

1. χ^2 values are always positive (for $v > 1$) or non-negative (for $v = 1$).

2. The shape of the pdf of χ^2 random variables differs with v.

3. $E(X_v) = v$ and $Var(X_v) = 2v$.

4. $\chi^2(v) \to N(v, 2v)$ as $v \to \infty$. The approximation is very good for $v \geq 30$. This entails that for large values of v, $Z := \dfrac{X_v - v}{\sqrt{2v}} \sim N(0,1)$, where $X_v \equiv \chi_v^2 \sim \chi^2(v)$.

An illustration of this property is demonstrated in Figure 5.10 with $v = 100$ degrees of freedom. The pdf is generated based on 100000 realizations.

5. $\chi^2(n) \equiv \Gamma(\alpha, \lambda)$ when the shape parameter $\alpha = n/2$ and the rate parameter $\lambda = 1/2$.[11] Further, recall that $\Gamma(1, \lambda) \equiv exp(\lambda)$. So, when $n = 2$, the χ^2 distribution collapses to the exponential distribution with rate $\lambda = 1/2$, i.e., $\chi^2(2) \equiv exp(1/2)$. Here, $\Gamma(\alpha, \gamma)$ refers to the gamma probability distribution, whereas $\Gamma(\alpha)$ refers to the gamma function.

[11] The pdf of $\chi^2(n)$ distribution is

$$f_X(x) = \frac{\left(\frac{1}{2}\right)^{\frac{n}{2}} x^{\frac{n}{2}-1} e^{-\frac{x}{2}}}{\Gamma\left(\frac{n}{2}\right)}, x > 0.$$

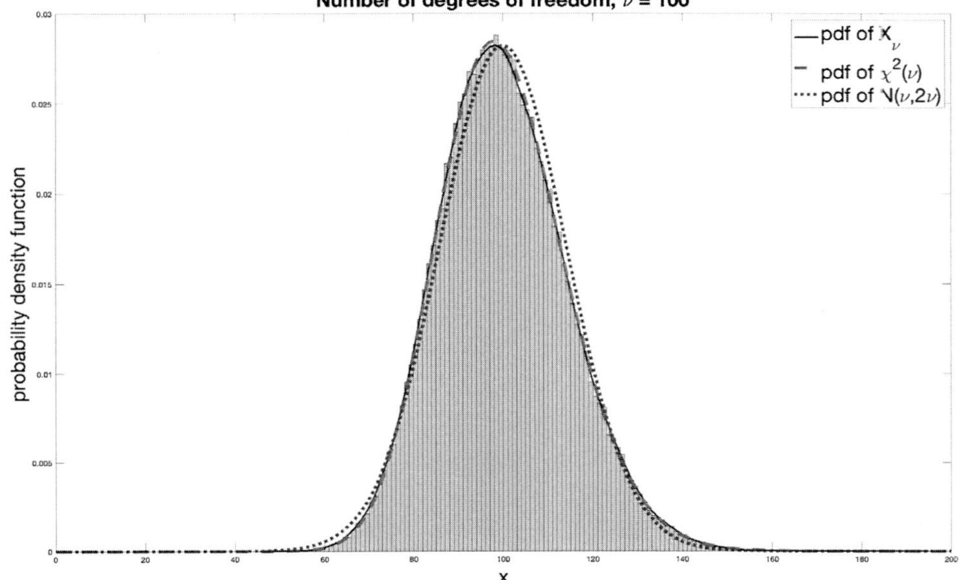

Figure 5.10 For large values of v (say 100), the $\chi^2(v)$ distribution converges to a normal distribution with mean v and variance $2v$. See color plate on page 297.

(i.ii) Application of χ^2 distribution: sampling distribution of $\dfrac{(n-1)S^2}{\sigma^2}$

Let $X_i \sim N(\mu, \sigma^2)$, $i = 1, 2, ..., n$; then $\frac{(n-1)S^2}{\sigma^2} \sim \chi^2(n-1)$, where $S^2 := \frac{\sum_{i=1}^{n}(X-\overline{X})^2}{(n-1)}$ is the *unbiased* sample variance. This result is elucidated here by simulating a large number of normally distributed random variables and, subsequently, computing the sample variance as shown in Figure 5.11. The Matlab code for this simple simulation is furbished below.

```
mu = 10; var = 2;
s = sqrt(var);
n = 5; max_realizations = 100000;
for realizations = 1:max_realizations
    y = s.*randn(n,1) + mu;
    sample_var(realizations) = (sum((y - mean(y)).^2))/(length(y)-1);
end
weighted_sample_var = (n-1)*sample_var/s^2;
h = histogram(weighted_sample_var,'Normalization','probability'); hold on;
h.BinWidth = 1.0;
[yn,xn]=ksdensity(weighted_sample_var,'Support','positive');
plot(xn,yn,'k','LineWidth',2); hold on;
x=[0:30];
fx_chi = chi2pdf(x,n-1);
plot(x,fx_chi,'m--','LineWidth',4);
legend('histogram of $\frac{(n-1)S^2}{\sigma^2}$ based on 100000 realizations',...
```

```
    'pdf of $\frac{(n-1)S^2}{\sigma^2}$','pdf of $\chi^2(n-1)$',...
    'Interpreter','latex','FontSize',25);
xlim([min(xn) max(x)]);
xlabel('x','FontSize',25);
ylabel('probability density function','FontSize',25);
title('Demonstrating $\frac{(n-1)S^2}{\sigma^2} \sim \chi^2(n-1)$ with n = 5',...
    'FontSize',25,'Interpreter','latex');
```

Figure 5.11 Sampling distribution of the sample variance.

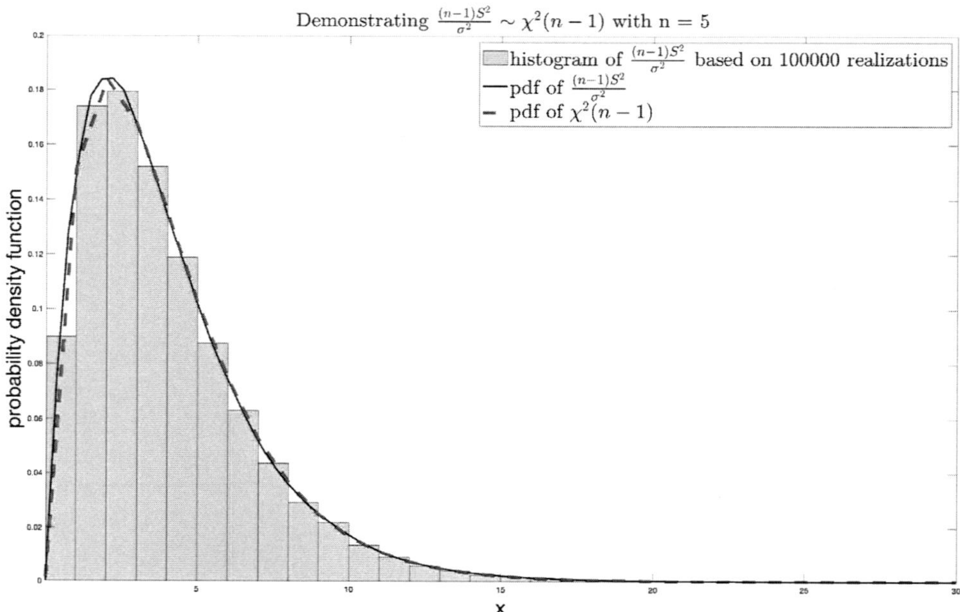

The above result can be proved more rigorously by using the method of moment generating functions. Interested readers may refer to the bibliography provided at the end of this chapter. We will omit the presentation of the proof here.

Example 5.4   Success rate of a sniper

A sniper can locate a target at a distance of 2 km. In desert conditions and/or foggy weather, the optical scope on the marksman's rifle has an imprecision in each of the horizontal and vertical coordinates that is normally distributed with mean zero and variance of 4 sq. m. What is his success rate to hit a target within a radius of 0.1 m?

Solution: Let $R^2_{err} = X^2 + Y^2$ denote the square of the error to hit the target, $X, Y \sim N(0,4)$. Consider $Z_1 = X/2$ and $Z_2 = Y/2$ whence we have $Z_i \sim N(0,1)$, $i = 1,2$. Now, following equation 5.2, we have

$$P\left(R^2_{err} < 0.01\right) = P\left(Z_1^2 + Z_2^2 < 0.01/4 = 0.0025\right) = P\left(\chi_2^2 < 0.0025\right) = 1 - e^{-\frac{0.0025}{2}} = 0.0012.$$

The penultimate equality above arises from the fact that $\chi_2^2 \sim exp(\lambda = 1/2)$ as stated above in subsection (i.i.5) of section 5.4.4. Alternatively, the equality may also be arrived at by using the Matlab command: `>> chi2cdf(0.0025,2)`.

Therefore, the success rate of the sniper under the given adverse atmospheric condition drops to 0.12%.

(ii) **The t distribution**[12]   Let us recall from the CLT that if we have random samples Y_i of size n *from* any population distribution with mean μ and variance σ^2, then $\overline{Y} \sim N\left(\mu, \dfrac{\sigma^2}{n}\right)$ as $n \to \infty$. This implies $Z := \dfrac{\overline{Y} - \mu}{\sigma / \sqrt{n}} \sim N(0,1)$.

However, in most practical cases, the population variance σ^2 is unknown. This limits the utility of Z as a suitable statistic for conducting experiments. This entails that we consider an appropriate test statistic where the population variance σ^2 is replaced by the sample variance S^2 which is a knowable (computable) quantity. From our discussion on χ^2 distribution, since we know that $\dfrac{(n-1)S^2}{\sigma^2} \sim \chi^2(n-1)$, let us consider a test statistic defined as follows.

$$T \equiv T_{n-1} := \frac{Z}{\sqrt{\dfrac{\chi^2_{(n-1)}}{n-1}}}. \tag{5.3}$$

Here, Z and $\chi^2_{(n-1)}$ are independent random variables. The above test statistic T simplifies to $T = \dfrac{\overline{Y} - \mu}{S / \sqrt{n}} \sim t(n-1)$.[13] This defines the t distribution with $(n - 1)$ degrees of freedom.[14]

(ii.i) **Properties of t distribution**

1. $t(n) \overset{n \to \infty}{\to} N(0,1)$.

2. $E(T) = 0$, for $n > 1$ (otherwise, undefined), and $Var(T) = \dfrac{n}{n-2}, for\ n > 2$ (∞ for $1 < n \leq 2$).

3. For some $\alpha \in [0,1]$, let us consider the observable quantity $t_{\alpha,\,n}$ such that $P(T_n > t_{\alpha,n}) = \alpha$. Now, given that the bell-shaped t distribution is a symmetric curve about the mean zero (see Figures 5.12 and 5.13), we can easily establish the following symmetry relation:

$$P(T > t_{\alpha,n}) = P(T < -t_{\alpha,n}), \tag{5.4}$$

that is, more succinctly,

$$-t_{\alpha,n} = t_{1-\alpha,n}. \tag{5.5}$$

The first property stated above is illustrated through a Matlab routine given below.

```
x = [-6:0.1:6];
for nu=1:1000
    plot(x,tpdf(x,nu)); hold on;
end
pd = makedist('Normal'); pdf_normal = pdf(pd,x);
plot(x,pdf_normal,'rd:','LineWidth',4);
xlabel('observables','FontSize',24);
ylabel('probability density function','FontSize',24);
set(gca,'FontSize',24)
```

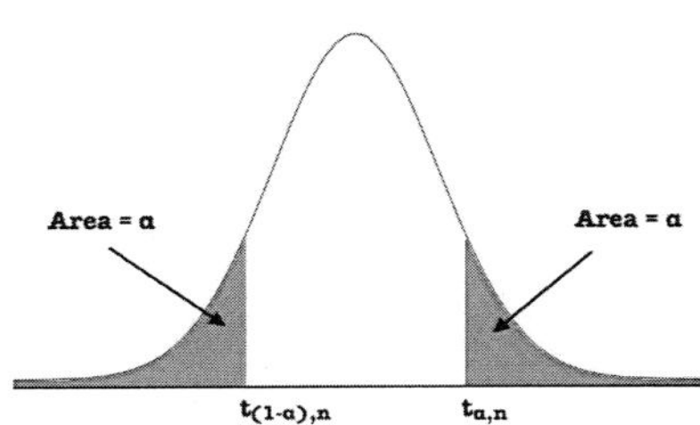

Figure 5.12 The pdf of the t distribution is symmetric about the origin.

[12] It is also known as *Student's* distribution.

[13] For clarity, here $\chi^2(v)$ refers to the chi-square distribution with v degrees of freedom and $\chi^2_v \equiv X_v$ refers to the random variable that is sampled from the $\chi^2(v)$ distribution. The random variable T follows the t distribution with $(n - 1)$ degrees of freedom.

[14] We will not present here the actual expression of the pdf of the t distribution because it is complicated. Interested readers may refer to the texts mentioned in the bibliography of this chapter.

Figure 5.13 The thin solid curves represent the pdf of the t distribution with varying degrees of freedom. The solid vertical arrow points in the direction of increasing degrees of freedom. The checkered-dotted curve represents the pdf of the standard normal distribution. $t(n) \xrightarrow{n \to \infty} N(0,1)$. See color plate on page 298.

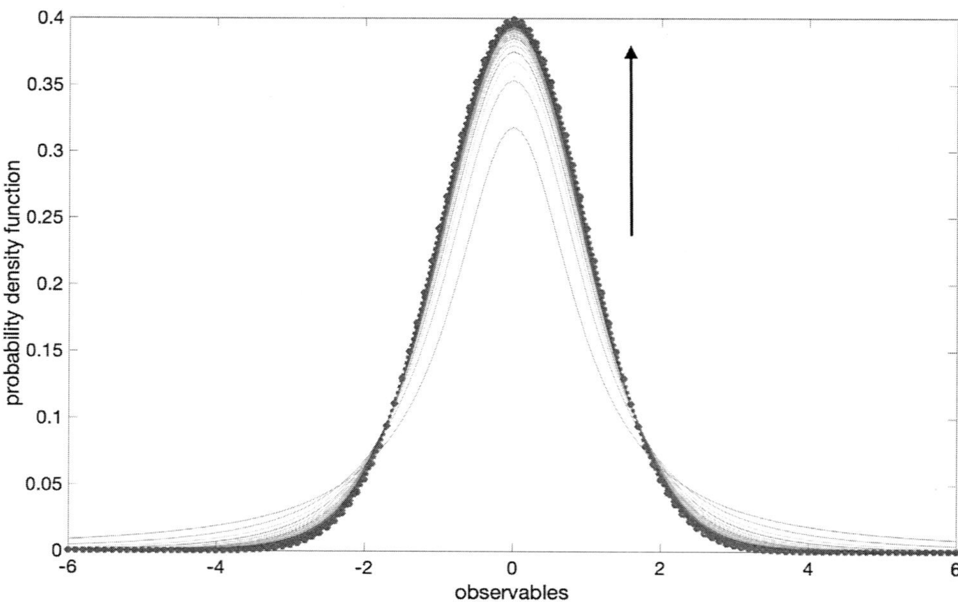

Besides the asymptotic convergence of the t distribution to the standard normal distribution, the above result also demonstrates that the t distribution is mostly suitable for data with bell-shaped distribution.

(ii.ii) **Applications of t distribution**

1. Let us consider two different samples of sizes n_1 and n_2 with means $\bar{x}_1$ and $\bar{x}_2$, respectively. We may want to know if the two means are sufficiently alike to warrant an inference that both the samples are drawn from the same population. In such a case, $T := \dfrac{\left(\bar{x}_1 - \bar{x}_2\right)\sqrt{n_1 + n_2 - 2}}{\sqrt{S_1 + S_2}} \sqrt{\dfrac{n_1 n_2}{n_1 + n_2}}$ is a suitable test statistic that follows the t distribution. Here, S_1 and S_2 are the respective sample variances.

2. The t distribution also finds applications in statistical experiments to find the significance of differences between regression coefficients obtained from different samples.

We will study the first application in greater detail in this chapter. We will revisit regression analysis in a subsequent chapter of this book. Interested readers may refer to the seminal work by R. A. Fisher to know more about the applications of the t distribution.[15]

[15] R. A. Fisher, Applications of Student's distribution, *Metron* **5** (1925): 90–104 .

Example 5.5   ***Estimating the spread of viral infection***

i. Consider that the daily number of reported influenza cases in a village is denoted by X. $X \sim N(70,9)$. What is the probability that on a given day the total number of reported cases exceeds 75?

ii. Now, consider that the actual daily mean number of influenza cases in the village is 70, i.e., $\mu_Y = 70$. It is not known if Y follows a normal distribution. Specifically, over a period of 15 days, the sample mean of number of infections $\bar{Y}$ is computed. Additionally, it is observed that the sample standard deviation is 4 reported cases. What is the probability that $\overline{Y}$ is greater than 74?

Solution:

i. X is sampled from a population that is normally distributed with mean 70 and variance 9. So $Z := \dfrac{X - \mu}{\sigma/\sqrt{n}} = \dfrac{75 - 70}{3} \approx 1.67,\ Z \sim N(0,1)$. In this case, $n = 1$. Therefore, $P(Z > 1.67) = 1 - P(Z \le 1.67) = 0.0475$. This result may be computed by using the standard normal distribution look-up table or by using the following Matlab command:

```
>> 1-cdf('Normal',1.67,0,1)
```

ii. In this case, since the population standard deviation is unknown, an appropriate test statistic is $T := \dfrac{\overline{Y} - \mu_Y}{S/\sqrt{n}} \sim t(n-1)$, which depends on the sample standard deviation S.

$$P(\overline{Y} > 74) = P\left(\frac{\overline{Y} - \mu}{S/\sqrt{n}} > \frac{74 - 70}{4/\sqrt{15}} \right) = P(T > 3.8730) = 0.00084461.$$

The above may be computed by using the Matlab command:

```
>> 1-tcdf(3.8730,14)
```

It may be carefully noted that we have used $(n - 1) = (15 - 1) = 14$ degrees of freedom to compute the probability because $T \sim t(n-1)$.

(iii) **The F distribution**[16]   If $X_n = \chi_n^2$ and $X_m = \chi_m^2$ are two independent chi-square random variables with n and m degrees of freedom, respectively, then the ratio

$$F := \frac{X_n / n}{X_m / m} \tag{5.6}$$

is a random variable which follows the F distribution. Further, if we have two independent samples of sizes n_1 and n_2 from two independent normal populations with variances σ_1^2 and σ_2^2, respectively, then the statistic

$$F := \frac{S_1^2 / \sigma_1^2}{S_2^2 / \sigma_2^2} \tag{5.7}$$

follows the $F(n_1 - 1, n_2 - 1)$ distribution.

(iii.i) **Properties of F distribution**

1. The F distribution is defined for non-negative values.

2. The pdf of the F random variables is not symmetric in shape over the range of the observables (see Figure 5.14).

3. $F(1, m) \equiv t^2(m)$ (see Figure 5.15).

   The Matlab routine to demonstrate this result computationally is given below.

[16] It is also known as the *Fisher–Snedecor distribution*.

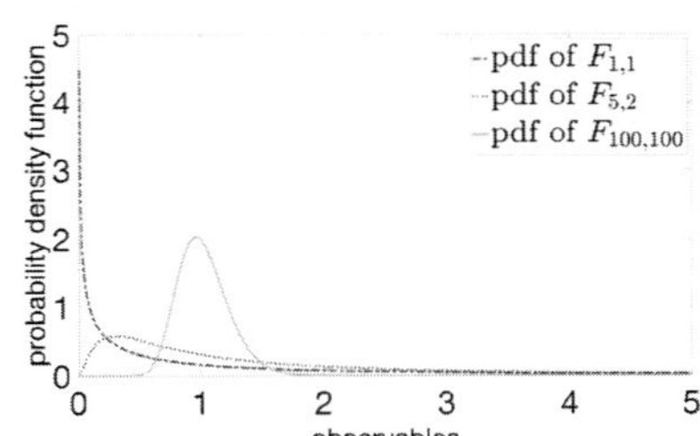

Figure 5.14 The pdfs of the F distribution with different pairs of n and m are generated using the Matlab command
```
>>plot (x, fpdf(x, n, m)),
```
where x takes the values of the observables.

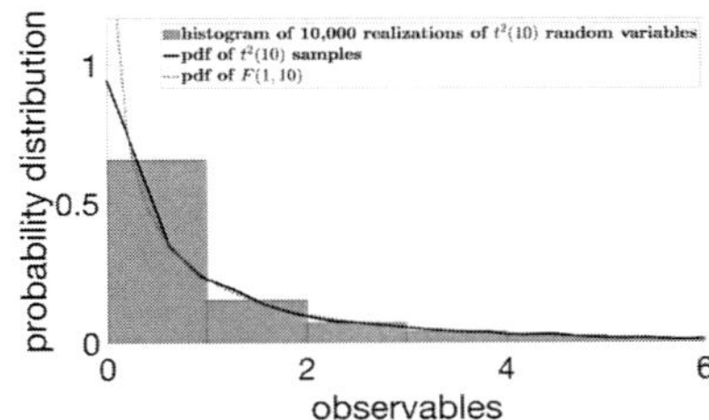

Figure 5.15 Computer simulation demonstrating $F(1, m) \equiv t^2(m)$.

```
x = [0:0.1:6];
n=1; m=10;
nu = m;
max_realizations = 10000;
for i=1:max_realizations
    t_rv = trnd(nu);
    tsq_rv(i) = (t_rv)^2;
end
h = histogram(tsq_rv,'Normalization','probability'); hold on;
h.BinWidth = 1.0;
[yn,xn]=ksdensity(tsq_rv);
plot(xn,yn,'k','LineWidth',4); hold on;
plot(x,fpdf(x,n,m),'m:','LineWidth',4);
xlim([0 max(x)]);
xlabel('observables');
ylabel('probability distribution');
set(gca,'FontSize',60);
legend('\bf histogram of 10,000 realizations of $t^2(10)$ random variables',...
    '\bf pdf of $t^2(10)$ samples',...
    '\bf pdf of $F(1,10)$','Interpreter','latex','FontSize',33);
```

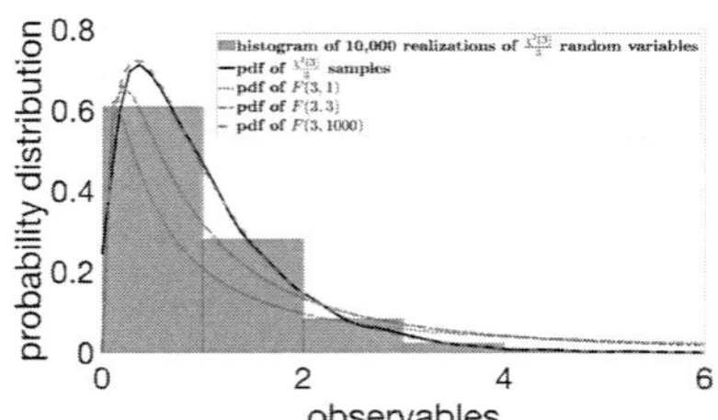

Figure 5.16 Computer simulation demonstrating $F(n,m) \overset{m\to\infty}{\equiv} \dfrac{\chi^2(n)}{n}$.

4. $F(n,m) \overset{m\to\infty}{\equiv} \dfrac{\chi^2(n)}{n}$ (see Figure 5.16).

The Matlab routine to demonstrate this result computationally is given below.

```
x = [0:0.1:6];
n = 3; m = 1000; m1 = 1; m2 = 3;
nu = n;
max_realizations = 10000;
for i = 1:max_realizations
    X_chisq(i) = chi2rnd(nu)/nu;
end
h = histogram(X_chisq,'Normalization','probability'); hold on;
h.BinWidth = 1.0;
[yn,xn]=ksdensity(X_chisq);
plot(xn,yn,'k','LineWidth',4); hold on;
plot(x,fpdf(x,n,m1),'m:',x,fpdf(x,n,m2),'m-.', 'LineWidth',4); hold on;
plot(x,fpdf(x,n,m),'m--','LineWidth',4);
xlim([0 max(x)]);
xlabel('observables');
ylabel('probability distribution');
set(gca,'FontSize',60);
legend('\bf histogram of 10,000 realizations of $\frac{\chi^2(3)}{3}$ random variables',...
    '\bf pdf of $\frac{\chi^2(3)}{3}$ samples','\bf pdf of $F(3,1)$',...
    '\bf pdf of $F(3,3)$','\bf pdf of $F(3,1000)$','Interpreter','latex','FontSize',30);
```

5. $E(F)=\frac{m}{m-2}$ and $Var(F)=\frac{2m^2(n+m-2)}{n(m-2)^2(m-4)}$, where n is the degrees of freedom corresponding to the numerator and m is the degrees of freedom corresponding to the denominator.

6. Consider $P\left(F > f_{\alpha,n,m}\right)=\int_{f_{\alpha,n,m}}^{\infty} f_F(x)dx = \alpha$ and $P(F < f_{1-\alpha,n,m}) = \int_{0}^{f_{1-\alpha,n,m}} f_F(x)$ $dx = \alpha$. Here, $f_{\alpha,n,m}$ is the *upper-tailed α-percentage point* and $f_{1-\alpha,n,m}$ is the *lower-tailed $(1 - \alpha)$-percentage point*. The lower-tailed percentage point can be found in terms of the upper-tailed percentage point as follows:

$$f_{1-\alpha,n,m} = \frac{1}{f_{\alpha,m,n}}.$$ Here, α is known as the *level of significance*.[17] $f_{\alpha,n,m}$ is the *critical value*. We will revisit α in a subsequent section of this chapter when we discuss methods to test hypotheses.

[17] α corresponds to the *rejection region*. The area under the curve $f_F(x)$ to the right of $f_{\alpha,n,m}$ is equal to α; the area under the curve $f_F(x)$ to the left of the curve $f_{\alpha,n,m}$ is equal to $1 - \alpha$ and corresponds to the *acceptance region*.

(iii.ii) Applications of *F* distribution

1. The test statistic used for performing an *analysis of variance* (ANOVA) experiment to test the difference between means of different populations is an F random variable that follows the F distribution.

2. *F* distribution is also used to test the existence of any significant difference between the variances of two different groups of population. For example, (i) a university academic policy may prefer that two instructors, co-teaching a course, grade exams in such a way so as to have the same variation in their grading; (ii) to ensure a tight fit, a manufacturing unit may require the variation in the lid and the container to be similar (see definition of F in equation (5.6) and Figure 5.17).

Figure 5.17 The variation in the pattern of grooves in the inner lining of the lid and the outer lining of the container must be similar. A manufacturing unit making covered containers must perform a statistical experiment to ensure quality check of its products. Such an experiment would rely on the application of the F distribution.

Example 5.6 Inverse symmetry of the upper- and lower-tailed percentage points of F(n, m)

Given that the upper 5-percentage point of F (5,10) can be found by using the Matlab command `>> finv(1-0.05,5,10)`,[*] compute the lower 95-percentage point of F (5, 10) both analytically and using an appropriate Matlab command.

Solution: Since the Matlab command `>> finv(1-0.05,5,10)` gives ans = `3.3258`, and the command `>> finv(1-0.05,10,5)` gives ans = `4.7351`, we have $f_{0.05,5,10} = 3.3258$ and $f_{0.05,10,5} = 4.7351$ whence we must have $f_{0.95,5,10} = \frac{1}{f_{0.05,10,5}} = \frac{1}{4.7351} = 0.2112$. Indeed using the Matlab command `>> finv(1-0.95,5,10)`, we get ans = `0.2112`, which means that we would observe the value greater than 0.2112 about 95% of the time by chance (when the numerator degrees of freedom is 5 and the denominator degrees of freedom is 10).

---

[*] The convention used in the Matlab command `>> finv(1-α,n,m)` may seem antithetical to the corresponding notation $f_{\alpha,n,m}$; so we warn the readers here to be mindful of this.

5.5 Chapter project: Prioritizing post-disaster reconstruction measures

5.5.1 *Interlude: Ratings of the field managers on bottlenecks in project implementation*

The ratings of the field managers on the issues mentioned in the project prologue section are printed below.

1. Community participation

Rating (scale: 1–10, 1: strongly disagree, 10: strongly agree)						
Cities	Mgr1	Mgr2	Mgr3	Mgr4	Mgr5	Mgr6
Port-au-Prince (Haiti)	$y_{11} = 3$	$y_{12} = 2$	$y_{13} = 9$	$y_{14} = 8$	$y_{15} = 9$	$y_{16} = 9$
Tacloban City	$y_{21} = 5$	$y_{22} = 9$	$y_{23} = 10$	$y_{24} = 5$	$y_{25} = 8$	$y_{26} = 9$
Latur	$y_{31} = 6$	$y_{32} = 7$	$y_{33} = 10$	$y_{34} = 5$	$y_{35} = 7$	$y_{36} = 8$
New Orleans	$y_{41} = 8$	$y_{42} = 9$	$y_{43} = 9$	$y_{44} = 8$	$y_{45} = 2$	$y_{46} = 8$
Kathmandu	$y_{51} = 3$	$y_{52} = 8$	$y_{53} = 7$	$y_{54} = 10$	$y_{55} = 10$	$y_{56} = 4$
Bagh City	$y_{61} = 2$	$y_{62} = 7$	$y_{63} = 9$	$y_{64} = 10$	$y_{65} = 6$	$y_{66} = 7$

2. Funding

Rating (scale: 1–10, 1: strongly disagree, 10: strongly agree)						
Cities	Mgr1	Mgr2	Mgr3	Mgr4	Mgr5	Mgr6
Port-au-Prince (Haiti)	$y_{11} = 3$	$y_{12} = 2$	$y_{13} = 9$	$y_{14} = 8$	$y_{15} = 9$	$y_{16} = 7$
Tacloban City	$y_{21} = 5$	$y_{22} = 4$	$y_{23} = 4$	$y_{24} = 5$	$y_{25} = 3$	$y_{26} = 2$
Latur	$y_{31} = 5$	$y_{32} = 2$	$y_{33} = 4$	$y_{34} = 5$	$y_{35} = 1$	$y_{36} = 2$
New Orleans	$y_{41} = 3$	$y_{42} = 1$	$y_{43} = 1$	$y_{44} = 2$	$y_{45} = 6$	$y_{46} = 2$
Kathmandu	$y_{51} = 3$	$y_{52} = 8$	$y_{53} = 7$	$y_{54} = 10$	$y_{55} = 10$	$y_{56} = 4$
Bagh City	$y_{61} = 3$	$y_{62} = 1$	$y_{63} = 9$	$y_{64} = 8$	$y_{65} = 6$	$y_{66} = 7$

3. Land ownership

Rating (scale: 1–10, 1: strongly disagree, 10: strongly agree)						
Cities	Mgr1	Mgr2	Mgr3	Mgr4	Mgr5	Mgr6
Port-au-Prince (Haiti)	$y_{11} = 9$	$y_{12} = 9$	$y_{13} = 10$	$y_{14} = 8$	$y_{15} = 7$	$y_{16} = 8$
Tacloban City	$y_{21} = 5$	$y_{22} = 4$	$y_{23} = 4$	$y_{24} = 5$	$y_{25} = 3$	$y_{26} = 2$
Latur	$y_{31} = 4$	$y_{32} = 6$	$y_{33} = 7$	$y_{34} = 2$	$y_{35} = 8$	$y_{36} = 9$
New Orleans	$y_{41} = 3$	$y_{42} = 1$	$y_{43} = 5$	$y_{44} = 2$	$y_{45} = 6$	$y_{46} = 2$

(Contd)

(Contd)

Rating (scale: 1–10, 1: strongly disagree, 10: strongly agree)					
Kathmandu	$y_{51} = 7$	$y_{52} = 4$	$y_{53} = 5$	$y_{54} = 1$	$y_{55} = 2$	$y_{56} = 3$
Bagh City	$y_{61} = 3$	$y_{62} = 2$	$y_{63} = 9$	$y_{64} = 8$	$y_{65} = 6$	$y_{66} = 7$

4. Shortage of technical staff

Rating (scale: 1–10, 1: strongly disagree, 10: strongly agree)						
Cities	Mgr1	Mgr2	Mgr3	Mgr4	Mgr5	Mgr6
Port-au-Prince (Haiti)	$y_{11} = 6$	$y_{12} = 9$	$y_{13} = 5$	$y_{14} = 5$	$y_{15} = 7$	$y_{16} = 6$
Tacloban City	$y_{21} = 6$	$y_{22} = 4$	$y_{23} = 6$	$y_{24} = 5$	$y_{25} = 7$	$y_{26} = 8$
Latur	$y_{31} = 4$	$y_{32} = 6$	$y_{33} = 7$	$y_{34} = 2$	$y_{35} = 8$	$y_{36} = 9$
New Orleans	$y_{41} = 4$	$y_{42} = 6$	$y_{43} = 6$	$y_{44} = 1$	$y_{45} = 8$	$y_{46} = 9$
Kathmandu	$y_{51} = 10$	$y_{52} = 7$	$y_{53} = 8$	$y_{54} = 1$	$y_{55} = 5$	$y_{56} = 6$
Bagh City	$y_{61} = 3$	$y_{62} = 2$	$y_{63} = 9$	$y_{64} = 8$	$y_{65} = 6$	$y_{66} = 7$

The data set above comprises ratings of construction managers on issues plaguing post-disaster reconstruction (PDR) projects. In each group (city), there are six different observations from six construction engineers who have been involved in PDR projects over the last many years. For each of the above-mentioned issues, you have to perform a one-way ANOVA calculation and test if the data provided in the tables presents a statistically significant difference in the mean rating of the construction engineers between the six different cities. This will reveal if the issues plaguing the implementation of PDR projects are affected in an identical manner across the six different cities around the world. Once the issues that are universally relevant have been identified, the total mean rating score across all the cities for a given issue should be computed and a decision on necessary corrective measures should be taken if this grand mean is greater than 5 (on a scale of 10).

The one-way ANOVA analysis is a specific statistical experiment where we test whether there is a significant difference in means (μ) of different populations. In the next few sections of this chapter, we will study the fundamental principles of performing the test of hypothesis that will aid us to complete the chapter project.

5.6   Test of hypothesis and statistical inference

The primary objective of performing a test of hypothesis is to use data from a sample (random) to make inferences about a population. Such statistical experiments rely on the use of test statistics and sampling distributions such as the standard normal distribution, χ^2 distribution, t distribution and F distribution.

Such a statistical experiment involves (i) a statement (*hypothesis*) about the parameter (characterizing the concerned population) and (ii) a *measure* of reliability of that statement in terms of probability.

5.6.1   *What is a hypothesis?*

It is a statement about a parameter(s) characterizing a population. A hypothesis usually results from speculation concerning an observed behavior, a natural phenomenon, or an established theory. If a hypothesis is stated in terms of population parameters such as the mean and variance, then it is called a *statistical hypothesis*. Data from the sample is used to test the validity of the hypothesis.

5.6.2   *Components of an experiment to test hypothesis*

The key components are itemized below.

- Construct the statement to be tested: *null* (H_0) versus *alternate* (H_1 or H_a) hypothesis.

- Identify the *rejection* (or *critical*) region to enable a decision about the hypothesis (e.g., the evidence based on the test statistic may prompt us to either reject or fail to reject the null hypothesis).

- Quantify the likely error in the aforementioned decision in terms of a probability measure. There are generally two types of errors, namely, *type-1 error* (with a probability of occurrence denoted by α) and *type-2 error* (with a probability of occurrence denoted by β). The objective is to reduce these errors while making a decision. However, in many circumstances, reducing one type of error can lead to an increase in the other type of error.

5.6.3   *Steps involved in performing a test of hypothesis*

The steps involved in performing a test of hypothesis are listed below.

1. Step 1: Specify H_0 and H_a and an acceptable level of significance α.

2. Step 2: Define a sample based test statistic (e.g., $\overline{X}$ and S^2) and a rejection (or critical) region for H_0 that is most suitable for the experiment.

3. Step 3: Collect the sample data and calculate the test statistic.

4. Step 4: Make a decision to either reject or fail to reject H_0.

5. Step 5: Interpret the result in the language of the problem at hand (e.g., provide confidence intervals) and provide an estimate of the error in the decision.

5.6.4   *Errors in inference*

The type and measure of the error are captured succinctly in the following table:

Decision/Truth condition	H_0 is true	H_0 is not true
H_0 is not rejected	Decision is correct (with probability $1 - \alpha$)	type-2 error (with probability β)
H_0 is rejected	type-1 error (with probability α)	Decision is correct (with probability $1 - \beta$)

5.6.5 *What might determine our choice of α?*

Deciding on the level of significance of a test is as much an art as it is guided by the context of the problem. This may be best illustrated through an example which we study in this section.

Example 5.7 Quality control in a packaging industry

A company that packages salted peanuts in 8 kg jars is interested in maintaining control on the amount of peanuts put in the jars by one of the machines in its packaging units. *Control* is defined as averaging 8 kg per jar and not consistently over- or under-filling the jars. To monitor this control, a sample of 16 jars is taken from the packaging line at random time intervals and their contents weighed. The mean weight of peanuts in these 16 jars will be used to test the null hypothesis that the machine is indeed working properly. If it is found not to be doing so, an expensive adjustment will be required. What may be a suitable level of significance for this test? For convenience, let us suppose the population standard deviation $\sigma = 0.2$. kg of the weight of the jars is known to us.*

Figure 5.18 A sample of 16 peanut jars from a packaging unit is subjected to statistical tests to check for discrepancies in weights.

Solution:

i. Step 1: The hypothesis is stated as follows.

$$H_0: \quad \mu = 8 \tag{5.8}$$

$$H_a: \quad \mu \neq 8 \tag{5.9}$$

In many problems, we may want to decide on α at this point. But what if we are not sure about how to choose α? This example will attempt to address this issue.

ii. Steps 2 and 3: The most appropriate test statistic for the case in hand is the sample mean, $\overline{X} = \dfrac{\sum_{i=1}^{16} X_i}{16}$.

iii. Step 4: A suitable rejection criteria may be selected as $\overline{X} < 7.9$ or $\overline{X} > 8.1$.

iv. Step 5: The identification of the rejection criteria must aid us in assigning the level of significance of the test α.** This may be estimated as follows.

$$\alpha = \mathrm{Prob}\left(\overline{X} < 7.9 \text{ or } \overline{X} > 8.1 \text{ when } \mu = 8\right). \tag{5.10}$$

$$P\left(\overline{X} < 7.9\right) = P\left(\frac{\overline{X} - \mu}{\sigma/\sqrt{n}} < \frac{7.9 - 8}{0.2/\sqrt{16}}\right) \tag{5.11}$$

$$= P\left(Z < -2.0\right) \text{ (here } Z \text{ is a standard normal random variable)} \tag{5.12}$$

$$= 0.0228. \tag{5.13}$$

Likewise, $P\left(\overline{X} > 8.1\right) = P\left(Z > 2.0\right) = 0.0228$. Since the events $\overline{X} < 7.9$ and $\overline{X} > 8.1$ are *disjoint*, we have

$$\alpha = \text{Prob}\left(\text{type-1 error}\right) = P\left(\overline{X} < 7.9\right) + P\left(\overline{X} > 8.1\right) = 0.0228 + 0.0228 = 0.0456.$$

---

*In most practical cases, σ will not be known, and we may have to base our analysis on the t distribution as opposed to the standard normal distribution that is used in this problem.

**We may also interpret α as the maximum allowable type-1 error.

5.6.6   *Additional comments on the level of significance α*

1. Notwithstanding the insight that may be gleaned from the previous example, we may not have a clear idea of what an appropriate maximum allowable type-1 error should be for a typical statistical experiment. There is no general rule of thumb to choose α.

2. α may also be sensitive to minor changes in the sample statistic and may conflate matters in appropriately testing the veracity of the hypothesis.

3. There is always a trade-off between α and β[18] because any effort to reduce one may likely increase the other.

[18] β is the probability of making type-2 error. We will not discuss much about β in this introductory text. Interested readers may refer to more advanced texts listed in the bibliography.

5.6.7   *Two-sample test for means*

In this section, we will discuss another test of hypothesis which uses the t distribution. Consider a case where we have two different populations that are normally distributed with the same variance. A random variable sampled from each population is denoted by $X_1 \sim N\left(\mu_1, \sigma^2\right)$ and $X_2 \sim N\left(\mu_2, \sigma^2\right)$. Further, let there be n_1 samples taken from the first population, $X_{1i} \sim N\left(\mu_1, \sigma^2\right)$, $i = 1, 2, 3, ..., n_1$, and let there be n_2 samples taken from the second population, $X_{2j} \sim N\left(\mu_2, \sigma^2\right)$, $j = 1, 2, 3, ..., n_2$.

1. Step 1: Construct the hypothesis.

$$H_0: \quad \mu_1 = \mu_2$$
$$H_1: \quad \mu_1 \neq \mu_2 \quad \left(\text{double sided test}\right)$$
$$\left(\text{alternatively, } \mu_1 > \mu_2 \text{ or } \mu_1 < \mu_2\right) \quad \left(\text{single sided test}\right)$$

Further, choose and set α.

2. Steps 2 and 3: The test statistic is $t := \dfrac{\left(\overline{X_1} - \overline{X_2}\right) - \left(\mu_1 - \mu_2\right)}{S\sqrt{\frac{1}{n_1} + \frac{1}{n_2}}}$, where $S^2 = \dfrac{(n_1-1)S_1^2 + (n_2-1)S_2^2}{n_1 + n_2 - 2}$ and S_j^2 is the sample variance corresponding to the samples taken from the jth population set. Here, $j = 1, 2$.

3. Step 4: Identify a rejection criteria.[19] There may be three distinct cases depending on the type of test (double or single sided alternate hypothesis).

- Reject H_0 in favor of $H_1\left(\mu_1 \neq \mu_2\right)$ if

$$\left|t\right| \geq t\left(\alpha/2, n_1 + n_2 - 2\right).$$

- Reject H_0 in favor of $H_1\left(\mu_1 > \mu_2\right)$ if

$$t \geq t\left(\alpha, n_1 + n_2 - 2\right).$$

[19] In the parlance of statistics, one never accepts a null hypothesis. One either rejects or fails to reject the null hypothesis against an alternate hypothesis.

- Reject H_0 in favor of $H_1 \left(\mu_1 < \mu_2 \right)$ if

$$t \le -t \left(\alpha, n_1 + n_2 - 2 \right).$$

The RHS term refers to the t observable value from the t distribution with a given significance level ($\alpha/2$ or α as stated above) and ($n_1 + n_2 - 2$) degrees of freedom.

Figure 5.19 My vintage car Rocky-X3 is likely to derive better mileage when powered by gasoline brand Gusto.

Example 5.8   Choosing between two gasoline brands for optimal mileage and performance

While comparing two different gasoline brands, a consumer survey reveals the following:

- a full tank of brand *Gusto* requires 4 cans and covers 546 km with a standard deviation of 31 km,

- a full tank of brand *Jiva* requires 4 cans and covers 492 km with a standard deviation of 26 km.

Assume that the performance parameters (mentioned above) of both brands are sampled from normal distributions with equal variances; test if there is a significantly better value in terms of mileage (km per tank) offered by Gusto over Jiva or if the mileage of both brands are statistically similar. Choose $\alpha = 0.05$.

Solution: Let us define the hypothesis at the outset as follows. We will use the subscript 1 for the brand Gusto and the subscript 2 for the brand Jiva.

$$H_0: \qquad \mu_1 = \mu_2$$
$$H_1: \qquad \mu_1 > \mu_2$$

For samples from brand Gusto, we have: $\overline{X}_1 = 546$, $S_1 = 31$, $n_1 = 4$.

For samples from brand Jiva, we have: $\overline{X}_2 = 492$, $S_2 = 26$, $n_2 = 4$.

The sample variance $S^2 = \dfrac{\left(4-1\right)31^2 + \left(4-1\right)26^2}{4+4-2} \Rightarrow S = 28.609$. Therefore,

$$t \text{ (under } H_0) = \frac{546 - 492}{28.609\sqrt{1/4 + 1/4}} = 2.67.$$

The observable value $t\,(0.05,6) = 1.9432$ may be computed either from the t distribution look-up table or using Matlab `>> tinv(1-0.05,6)`, where ($n_1 + n_2 - 2$) = (4 + 4 -2) = 6. Since this is a single tailed test,

$$H_1 : \mu_1 > \mu_2,$$

and because

$$t_{\text{calculated}} = 2.67 > t\left(0.05,6\right) = 1.9432,$$

we reject H_0 in favor of H_1. In other words, brand Gusto will likely give us better mileage than brand Jiva at the level of significance $\alpha = 0.05$ (or with 95% confidence level).

Compare the technique presented above with the one using a Z-test statistic when the sample variances are known and the sample size is sufficiently large. We illustrate the Z-test with the help of an example below.

Example 5.9 Z-test for difference of means – estimating average population of villages

A random sample of 200 villages was taken from a district and an average population per village was found to be 485 with a standard deviation of 50. Another random sample of 200 villages from the district gave an average population of 510 per village with a standard deviation of 40. Is the difference between the averages of the two samples statistically significant?

Note: A test statistic (Z) is said to be significant if $|Z| > 3$.

Solution:

$$n_1 = 200 = n_2$$
$$\overline{x}_1 = 485, \ \overline{x}_2 = 510$$
$$s_1 = 50, \ s_2 = 40$$

Null hypothesis: $\mu_1 = \mu_2$

$$Z = \frac{\overline{x}_1 - \overline{x}_2}{\sqrt{\dfrac{s_1^2}{n_1} + \dfrac{s_2^2}{n_2}}} = -5.5.$$

Since $|Z| > 3$, null hypothesis is rejected. Hence, the difference of the mean of the samples is highly significant.

Example 5.10 Z-test for difference of means – assessing brand quality by statistical comparison

A potential buyer of light bulbs bought 50 bulbs each of two brands. Upon testing these bulbs he found that brand A has mean life of 1282 hours with a standard deviation of 80 hours, whereas brand B has a mean life of 1280 hours with a standard deviation of 94 hours. Can the buyer be quite certain that the two brands differ in quality?

Solution:

$$n_1 = 50 = n_2$$
$$\overline{x}_1 = 1282, \ \overline{x}_2 = 1280$$
$$s_1 = 80, \ s_2 = 94$$

Null hypothesis: $\mu_1 = \mu_2$, that is the two brands do not differ in quality.

$$Z = \frac{\overline{x}_1 - \overline{x}_2}{\sqrt{\dfrac{s_1^2}{n_1} + \dfrac{s_2^2}{n_2}}} = 0.1146.$$

Since $|Z| \leq 3$, null hypothesis cannot be rejected. Hence, there is no significant difference in the quality of the two brands.

5.6.8 *Analysis of variance (ANOVA)*

It must be emphasized that the t statistic based test for two means cannot be generalized for more than two different population sets. For multi-population tests, we may have to resort to ANOVA. In order to appreciate the machinery of ANOVA, it is crucial to understand some notation and conventions. We will start here with a note on data representation.

ONE-WAY ANOVA

Data collated from survey samples is denoted by y_{ij}, where the first subscript represents the ith population group (i = 1, 2, ..., t) and the second subscript represents the jth observation (data point; j = 1, 2, ..., n). We will consider n_1, n_2, ..., n_t observations for the t population groups. If $n_1 = n_2 = ... = n_t = n$, then we have a *balanced* data set. The total number of observations is $\sum_{i=1}^{t} n_i$ (= nt in the case of balanced data).

Null hypothesis:

$$H_0 : \mu_1 = \mu_2 = \cdots = \mu_t$$

H_1 : one of the above equality is *not* satisfied $\hspace{2em}$ (5.14)

Assumption: Data from each population group is normally distributed $N(\mu_i, \sigma^2)$, where σ^2 is the same across population groups.

Organization of data:

Population group	Observations/ data	$\Sigma_j y_{ij}$ (totals)	$\dfrac{Y_{i.}}{n_i}$ (means)	Sum of squares
1	$y_{11} y_{12} \cdots y_{1n_1}$	$Y_{1.}$	$\overline{y}_{1.}$	SS_1
2	$y_{21} y_{22} \cdots y_{2n_2}$	$Y_{2.}$	$\overline{y}_{2.}$	SS_2
3	$y_{31} y_{32} \cdots y_{3n_3}$	$Y_{3.}$	$\overline{y}_{3.}$	SS_3
.	.	.	.	.
.	.	.	.	.
.	.	.	.	.
t	$y_{t1} y_{t2} \cdots y_{tn_t}$	$Y_{t.}$	$\overline{y}_{t.}$	SS_t
	Overall	$Y_{..}$	$\overline{y}_{..}$	SS_p

We have used the convention whereby the position of the . (dot) in the subscript represents which of the two indices (in the subscript) is being summed. For example, for some variable α_{ij}, we will use the following convention for summation: $\sum_i \alpha_{ij} = \alpha_{.j}$, where the summation is performed over the first index i.

Sum of squares: $SS_i = \sum_j \left(y_{ij} - \overline{y}_{i.} \right)^2 \equiv \sum_j y_{ij}^2 - \dfrac{Y_{i.}^2}{n_i}.$

Pooled sum of squares: $SS_p = \sum_{i=1}^{t} SS_i.$

Pooled degrees of freedom: $\sum_{i=1}^{t} n_i - t = t(n-1).$[20]

[20] The last equality is true for balanced data.

The pooled variance s_p^2 can now be defined as $s_p^2 = \dfrac{SS_p}{\sum_{i=1}^{t} n_i - t}$. Now another estimate of the sample variance is possible by considering the mean data across the population groups (factor levels). This sample variance estimate is formulated as follows: $s_{means}^2 = \dfrac{\sum_i \left(\overline{y}_{i.} - \overline{y}_{..} \right)^2}{t-1}$. Under the null hypothesis and based on the discussions under the first paragraph of the section on sampling distributions, we may deduce that the factor level means have a distribution with mean μ and variance $\dfrac{\sigma^2}{n}$. Thus, we have an estimate for the population variance $\sigma^2 = ns_{means}^2$ with $(t-1)$ degrees of freedom. Of course, an alternate estimate of the population variance is s_p^2 with $t(n-1)$ degrees of freedom. We know from the definition of the F statistic that the F value represents the ratio of two independent estimates of a common variance. Therefore,

$$F_{cal} = \dfrac{ns_{means}^2}{s_p^2}. \tag{5.15}$$

If $F_{cal} > F_\alpha\left(t-1, (n-1)t\right)$, then we *reject* H_0.

ALTERNATE FORMULATION OF ONE-WAY ANOVA

This alternate formulation leads to the same inference. In this formulation, we have the following definitions:

$$SSB\left(\text{sum of sqs. between groups}\right) = \sum_i \dfrac{\left(Y_{i.}\right)^2}{n_i} - \dfrac{Y_{..}^2}{\sum_i n_i}$$

$$\text{with } (t-1) \text{ degrees of freedom.} \tag{5.16}$$

$$SSW\left(\text{sum of sqs. within groups}\right) = \sum_j \sum_i y_{ij}^2 - \sum_i \dfrac{Y_{i.}^2}{n_i}$$

$$\text{with } \left(\sum_i n_i - t \right) \text{ degrees of freedom.} \tag{5.17}$$

Consequently, the total sum of squares is $TSS = SSB + SSW$. The one-way ANOVA table can now be reformulated as follows.

Source	d.o.f. [†]	SS [†]	$MS^{†} = \dfrac{SS}{d.o.f.}$	F_{cal}
Between groups	$t-1$	SSB	MSB	$\dfrac{MSB}{MSW}$
Within groups	$(\Sigma_i n_i - t)$	SSW	MSW	
Total	$\Sigma_i n_i - 1$	TSS		

Note: † d.o.f. means degrees of freedom, SS means sum of squares, MS means mean sum of squares.

Inference: If $F_{cal} > F_\alpha\big(t-1,(n-1)t\big)$, then we *reject* H_0 in favor of H_1.

Example 5.11   Rice yield across varieties

An experiment to compare the yield of four varieties of rice is conducted. Each of the plots on a test farm where soil fertility is fairly homogeneous is treated alike relative to water and fertilizer. Four plots are randomly assigned each of the four varieties of rice. The yield in kg/acre is recorded for each plot for this randomized experiment. Does the data presented in the following table indicate a difference in the mean yield between the four varieties? Choose $\alpha = 0.01$.

Figure 5.20 Rice plantation in multiple plots of land to test the productivity across different rice varieties.

Variety	Yield
1	934 1041 1028 935
2	880 963 924 946
3	987 951 976 840
4	992 1143 1140 1191

Solution: The hypothesis is stated as follows.

H_0 : $\mu_1 = \mu_2 = \mu_3 = \mu_4$
H_1 :   not all varieties have the same mean.

Here, μ_i denotes the mean yield of the ith variety. $n = 4$, $t = 4$. The one-way ANOVA table is printed below.

Rice variety	Yield data	$Y_{i.}$ (totals)	$\bar{y}_{i.}$ (means)	SS_i
1	934 1041 1028 935	3938	984.50	10085
2	880 963 924 946	3713	928.25	3868.75
3	987 951 976 840	3754	938.50	13617
4	992 1143 1140 1191	4466	1116.5	22305
	Overall	15871	991.94	49875.75

$$ns^2_{means} = n\frac{\sum_i\left(\bar{y}_{i.} - \bar{y}_{..}\right)^2}{t-1} = 29977.06. \text{ Further, } s^2_p = \frac{\sum_i SS_i}{t(n-1)} = 4156.31.$$

$$F_{cal} = \frac{29977.06}{4156.31} = 7.21.$$

Using the F distribution table or using Matlab, we can find $F_{0.01}(3,12) = 5.95$. Since $F_{cal} > F_{0.01}(3,12)$, we reject H_0 and infer that there is a significant difference in the yield between the different rice varieties.

We would arrive at the same conclusion if we used the alternate approach to perform the calculation. In the alternate approach, the one-way ANOVA table is as follows.

Source	d.o.f.	SS	MS $= \dfrac{SS}{d.o.f.}$	F_{cal}
Between varieties	3	89931.19	29977.06	$\dfrac{MSB}{MSW} = 7.21$
Within varieties	12	49875.75	4156.31	
Total	15	139806.94		

5.6.9   *Advanced ANOVA techniques*

An extension of the one-way ANOVA is a two-way ANOVA technique where we may consider the influence of two independent factors (e.g., we may want to test the influence of rice variety and temperature of the environment in our aforementioned experimental example) on the dependent observable or outcome (e.g., yield of the rice). In such a scenario, certain combinations of the two factors may *interact* differently from a simple additive approach. This interaction phenomena must necessarily be captured in the analysis. There may be problems where more than two factors may be involved, and we may have to consider what is known as a 2^k *factorial experimental design*. We will not elaborate on these multi-factor ANOVA designs here. However, an illustrative exercise problem is included at the end of the chapter for the interested audience to explore this technique.

5.7   Chapter project: Prioritizing post-disaster reconstruction measures

5.7.1   *Epilogue: Identifying issues that are universally plaguing reconstruction efforts by performing ANOVA on the survey ratings of the managers*

Now that we know the basic principles involved in performing ANOVA, let us use the survey ratings of the managers to populate the following ANOVA table:

Source	Degree of freedom	Sum of sqs.	Mean of sqs.	F_{cal}
Between groups	$dfB = t - 1$	SSB	MSB	$F_{cal} = \dfrac{MSB}{MSW}$
Within groups	$dfW = \sum_i n_i - t$	SSW	MSW	
Total	$\sum_i n_i - 1 = n - 1$	TSS = SSB+SSW		

Here, the number of groups $= t = 6$ and the number of observations in group i $= n_i = 6$ and $n = 36$.

Software Implementation

Use Matlab to construct the one-way ANOVA table and implement the following algorithm:

- Compute the entries of the ANOVA table. One-way ANOVA computation involves the following calculations:

  1. Compute $Y_{i.} = \sum_{j=1}^{n_i} y_{ij}, Y_{..} = \sum_{i,j} y_{ij}$.

  2. Compute $SSB = \sum_i \dfrac{Y_{i.}^2}{n_i} - \dfrac{Y_{..}^2}{\sum_i n_i}, SSW = \sum_{i,j} y_{ij}^2 - \sum_i \dfrac{Y_{i.}^2}{n_i}, TSS = SSB + SSW$.

  3. Set $MSB = \dfrac{SSB}{t-1}$ and $MSW = \dfrac{SSW}{\sum_i n_i - t}$.

  4. Compute $F_{cal} = \dfrac{MSB}{MSW}$.

- Next, compare F_{cal} and $F_{tab} = F_\alpha\left(dfB, dfW\right)$

  (F_{tab} is found from F distribution table).

- If $F_{tab} > F_{cal}$, then fail to reject H_0 (i.e., possibly all means [mean rating values] are *statistically* equal); else if $F_{tab} < F_{cal}$, then reject H_0 (and abandon dwelling on the issue for the time being).

- If $F_{tab} > F_{cal}$ and if $\mu = \dfrac{\sum_{i,j} y_{ij}}{n} > 5.0$, then the issue is universally relevant across cities and demands corrective measures to successfully implement PDR projects.

Questions

1. State the assumptions of one-way ANOVA. Comment if these assumptions seem reasonable in the context of the PDR data provided here.

2. Implement the algorithm prescribed above in Matlab. Specifically, write a Matlab script to construct and display the ANOVA table for each of the datasets provided in the interlude section 5.5.1 of the project.

3. Compute F_{tab} for the given problem from the F distribution table corresponding to a level of significance of test $\alpha = 0.01$.

4. Based on the strategy prescribed in the algorithm, select and mention the issues that are universally relevant across different cities and which need immediate redressal.

5. Mention at least two drawbacks of one-way ANOVA test.

5.8   Selected bibliography

Billingsley, P. *Probability and Measure* (2nd edition). John Wiley and Sons, 1990.

Gupta, S. C. and V. K. Kapoor. *Fundamentals of Mathematical Statistics* (11th edition). Sultan Chand & Sons, 2019.

Hogg, R. V., E. A. Tanis, and D. L. Zimmerman. *Probability and Statistical Inference* (9th edition). Pearson, 2020.

Montgomery, D. C. and G. C. Runger. *Applied Statistics and Probability for Engineers* (6th edition). Wiley, 2019.

Ross, S. M. *Introduction to Probability and Statistics for Engineers and Scientists* (6th edition). Academic Press, Elsevier, 2021.

5.9   Exercise problems

1. (*Service time of a cashier in a shopping mall: application of CLT*) A cashier in a shopping mall serves customers standing in a queue one by one. Suppose that the service time X_i for the ith customer has a mean $E(X_i) = 2$ min and $Var(X_i) = 1$ min. $\{X_i\}$ for all i customers are independent random variables. Let W be the total time spent by the cashier to serve 50 customers. Find $P(80 < W < 120)$.

Figure 5.21 It's a long queue!

2. (*Testing efficacy of students' health and welfare programs in schools*) A middle school conducted a survey of its students to collect health information for future planning purposes. The survey revealed that a substantial proportion of students were not engaging in regular exercise, many did not have access to optimal nutrition, and a substantial number were exposed to environmental hazards like pollution. In response to a question on regular exercise, 60% of all parents of the students reported that their children were not regularly exercising, 25% reported their wards were exercising sporadically, and 15% reported that their children were exercising regularly. The next year the school launched a health promotion campaign on campus in an attempt to increase healthy behaviors among students. The program included modules on exercise, nutrition, and environmental awareness. To evaluate the impact of the program, the school again surveyed the students and their parents. The survey was completed by 470 students and the following data were collected on the exercise question:

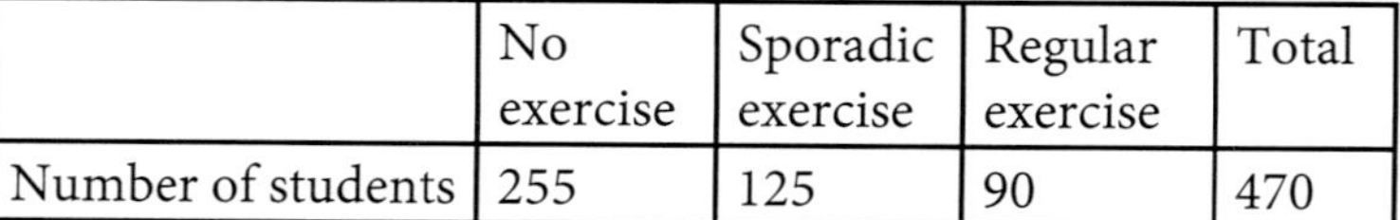

	No exercise	Sporadic exercise	Regular exercise	Total
Number of students	255	125	90	470

Based on the data, is there evidence of a shift in the distribution of responses to the exercise question following the implementation of the health promotion campaign on campus? In order to answer this question, design a suitable statistical experiment and frame a hypothesis. Consider a 5% level of significance for testing your hypothesis.

3. (***Testing level of bacterial contamination in shipped packages***) Matlab provides an inbuilt database called *hogg* that is sourced from the work of Hogg and Ledolter.[21] This dataset has a record of total bacteria count in randomly selected cartons of milk packed in different shipping boxes. The relevant data may be obtained by typing the following commands in the Matlab command window:

```
>> load hogg
>> hogg
hogg =

    24    14    11     7    19
    15     7     9     7    24
    21    12     7     4    19
    27    17    13     7    15
    33    14    12    12    10
    23    16    18    18    20
```

[21] R. V. Hogg and J. Ledolter, *Engineering Statistics* (Macmillan USA, 1987).

Here, the columns represent the randomly selected shipping boxes and the rows constitute the bacterial count in randomly picked milk cartons from the respective shipping box. Perform a one-way ANOVA to test if the bacterial contamination is the same, on average, across the five different shipping boxes. Use the level of significance of test $\alpha = 0.01$. Compare your analysis with the Matlab inbuilt `anova1` function that outputs the one-way ANOVA table and the p-value of the test.[22]

[22] You may use the following Matlab commands:
`>>[p,tbl,stats] = anova1(hogg);` and
`>>doc anova1` to learn about testing the validity of the null hypothesis from the p-value of the ANOVA experiment.

4. (***Consistency of caffeine content in Coke across beverage counters***) Coke is available as a fountain soft drink from different vendors and beverage counters across the world. In order to attain consistency in quality, it is desired that the caffeine content in mg per 12 oz does not exceed 34 mg. The data collected from 50 randomly selected beverage counters from across the world is available in a consolidated manner through this link here: Coke-1.csv. Perform a suitable test of hypothesis to validate the null hypothesis (mean caffeine content across counters is 34 mg per 12 oz) against the alternate hypothesis that the mean caffeine content is greater than 34 mg per 12 oz. In this context, answer the following questions:

   i.   Clearly state the null and the alternate hypotheses in mathematical terms.

   ii. Identify the appropriate statistical test for this experiment.

   iii. Verify that the assumptions for using this aforementioned test is met.[23]

   iv. Calculate the sampling distribution of mean under the null hypothesis.

   v.   Conduct the test of hypothesis and report your inference at $\alpha = 0.05$ significance level.

Figure 5.22 Watch out for the extra dose of caffeine in your drink!

[23] You may want to use the Matlab inbuilt function `kstest` to check one of these assumptions.

5. (***Prognosis of pulmonary infection***) The change in the amount of carbon monoxide transfer, which is an indicator of improved pulmonary function in smokers with chickenpox, over a one-week time frame is recorded as follows: 33, 2, 24, 17, 4, 1, −6 (units are in ml). Is there any evidence of significant improvement in pulmonary function at a 95% confidence level

   i.   if the data is normally distributed with variance $\sigma^2 = 100$?

   ii. if the data is normally distributed with unknown variance σ^2?

   You may use $\alpha = 0.05$.

6. (***Efficacy of calcium channel blockers among hypertensive patients of age group 45–59***) The efficacy of a treatment for hypertension is to be studied using a randomized clinical trial. Thirty-eight hypertensive patients in the age group 45–59 were randomly allocated to either a placebo group (who were administered over-the-counter potassium tablets) or an intervention group (who were administered a calcium channel blocker) and a two-month follow-up study was performed at the clinic. At the end of the trial phase, the difference in systolic blood pressure was measured for the patients in each group and recorded. The summary of the results is given below.

Group	Number of patient participants	Mean difference in systolic blood pressure	Sample variance
Placebo	21	−0.108	2.101^2
Intervention	17	3.753	4.630^2

   Is there any evidence of significant improvement in the treatment group? Use $\alpha = 0.01$.

7. (***How long will you keep your first car?***) An economist wishes to investigate whether people are keeping cars longer now than in the past. He knows that five years ago, 38% of all passenger vehicles in operation were at least ten years old. He commissions a study in which 325 automobiles are randomly sampled. Of them, 132 are ten years old or older.

   i.   Find the sample proportion.

   ii. Find the probability that, when a sample of size 325 is drawn from a population in which the true proportion is 0.38, the sample proportion will be as large as the value you computed in part (i). You may assume that the normal distribution applies.

   iii. Give an interpretation of the result in part (ii). Is there strong evidence that people are keeping their cars longer than was the case five years ago?

8. (***Cholesterol content in eggs***) Suppose the mean amount of cholesterol in eggs labeled "large" is 186 mg, with standard deviation 7 mg. Find the probability that the mean amount of cholesterol in a sample of 144 eggs will be within 2 mg of the population mean.

9. (***Simulating Cochran's theorem***) Write a Matlab code to simulate and establish the veracity of the following theorem: *if* $Z_1, \ldots, Z_k$ *are independent*

Figure 5.23 Estimating cholesterol in eggs using a sampling distribution.

and identically distributed (i.i.d.) standard normal random variables, then

$$\sum_{i=1}^{k}\left(Z_i - \overline{Z}\right)^2 \sim \chi^2_{k-1}, \text{ where } \overline{Z} \text{ is the sample mean.}$$

10. (**Two-way ANOVA**) Aircraft primer paints are applied to aluminium surfaces by two methods: *dipping* and *spraying*. The purpose of using the primer is to improve paint adhesion, and some parts can be primed using either method. The process engineering group responsible for this operation is interested in knowing whether three different primers differ in their adhesion properties. A factorial experiment was performed to investigate the effect of paint primer type and application method on paint adhesion. For each combination of primer type and application method, three specimens were painted, then a finish paint was applied, and the adhesion force was measured. The data from the experiment are shown in the following table. Perform a two-way ANOVA and identify the most effective primer type and the better of the two paint application methods.

Primer type	Dipping	$y_{ij.}$	Spraying	$y_{ij.}$	$y_{i..}$
1	4.0, 4.5, 4.3	12.8	5.4, 4.9, 5.6	15.9	28.7
2	5.6, 4.9, 5.4	15.9	5.8, 6.1, 6.3	18.2	34.1
3	3.8, 3.7, 4.0	11.5	5.5, 5.0, 5.0	15.5	27.0
$y_{.j.}$	40.2		49.6		$y_{...} = 89.8$

6

Prediction and Time Series Modeling

Empirical techniques rely on abstracting meaning from observable phenomena by constructing relationships between different observations. This process of abstraction is facilitated by appropriate measurements (experiments), suitable organization of data generated by measurements, and, finally, rigorous analysis of the data. The latter is a functional exercise that synthesizes information (data) and theory (model) and enables *prediction* of hitherto unobserved phenomena.[1] It is important to underscore that a good theory (model) that explains a certain phenomenon well by appealing to a set of laws and conditions is expected to be a good candidate for predicting the same using reliable data. For example, a good model for the weight of a normal human being is $w = m * h$, where w and h refer to weight and height of the person, and m can be set to unity if appropriate units are chosen. A rational explanation of such a formula for weight based on anatomical considerations is perhaps very reasonable. From an empirical standpoint, if we collect height and weight data of normal humans, we will notice that a linear model of the form $w = m * h$ represents the data reasonably well and may be used to predict the weight of the person based on the height of the person. This fact ascertains a functional *symmetry* between *explanation* and *prediction*. Therefore, a good predictive model must automatically be able to explain the data (and related events) well.

A very pertinent question in the context of model prediction must be addressed here. *How can a predictive model determine what data (forecast) is observable?* The possibility of obtaining realizable (observable) forecasts is determined by the hypotheses and conditions to which the model must adhere. In this chapter, we will not only discuss different models that forecast and explain observations but also lay down the conditions and assumptions governing the usability of the models. This will also in turn shed light on whether certain predictions of the model(s) are realizable and, if yes, under what circumstances.

Often in the field of probability and statistics, one is able to deduce the most likely event conditioned by other known events. Further, it may also be possible to estimate correlation metrics between two (or more) distinct data sequences (or time series signals). But as the well-known phrase goes, *correlation does not imply causation*. In order to establish a causal signature of events, it may be essential to quantify the causal interdependencies between different time series data using appropriate metrics. The last section of this chapter will elaborate on one such classical technique using the *Granger causality index* (GCI) to infer causal relations between seemingly random independent sequences of events.

Figure 6.1 An excerpt from Shakespeare's *Macbeth* – 1.3.59–61.

If you can look into the seeds of time,
And say which grain will grow and which will not,
Speak then to me, …

[1] Prediction can only be validated if the unobserved phenomenon under investigation is actually observable. Prediction is a function of our senses (measurements) and memory (accessibility to organized data). It is a theoretical construct guided by real experiences (data).

6.1   Chapter objectives

The chapter objectives are listed as follows.

1. Students will learn to estimate correlation between data and investigate statistical properties like stationarity and ergodicity of a time series data.

2. Students will develop the technical skills to construct a curve of regression to fit a given data.

3. Students will learn to construct different time series models like autoregressive (AR), moving average (MA), and autoregressive moving average (ARMA) models to forecast trends and make predictions.

4. Students will learn to estimate the order and coefficients of time series models based on the partial auto-correlation functions (PACFs) and the Yule–Walker equations.

5. Students will train to infer causal relationships between disparate time series data by using the Granger Causality Index (GCI).

6.2   Chapter project: Indicators of a federal government's policy effectiveness index

6.2.1   *Prologue: Measuring efficacy of a federal policy intervention in terms of trends in employment generation*

The COVID pandemic of 2020 has exposed the vulnerability of governments across the globe that have struggled to bring forth effective economic policy interventions to arrest the loss of jobs en masse. Further, several macroeconomic shocks have effected additional dents to the labor markets. In such a scenario, predicting trends in employment generation is not only imperative to calibrate appropriate counter-measures but also essential to understand the dynamic response of economies to macroeconomic impulses and policy decisions.

In this project, you will learn to select an appropriate auto-regressive $(AR(p))$ model where p is the order of the model determined by available data from a certain labor market. You will use the partial auto-correlation functions (PACFs) and the Yule–Walker equations to estimate the order p of the $AR(p)$ model and determine its coefficients. You will then use this model to predict the future trend of monthly jobs created based on the data from the preceding many years.

As an introductory exercise, plot the monthly employment data provided in the link JOBS.csv and comment on the trend in job creation based on visual inspection. Do you observe any noteworthy statistical attributes in this plot?

6.3   Introduction: Statistical attributes of time series data

We will begin this chapter by first introducing a few elementary concepts that elucidate important statistical properties of a time series data. These properties are essential to construct stochastic models that forecast and explain trends. Many scientific experiments are designed to generate data, often spatio-temporal data, to glean information about the natural laws that govern the underlying processes.

Figure 6.2 Policy interventions, economic recovery, and job creation are essential indicators of a government's longevity.

Therefore, it is crucial that the method of sampling and generating data from measurements is carefully crafted in a manner that this data can be subsequently subjected to rigorous mathematical analyses and used for intelligent designing of predictive models. The process of sampling data from laboratory and observational experiments must be designed with the objective of eliminating any statistical bias. One of the elementary techniques to ensure that we have good and reliable data free of bias is to conduct repeated experiments and sample data from multiple realizations of an experiment. Formally, we will represent this process as follows. Consider that we are interested in collecting information about the average monthly rainfall in a certain district. We choose to set up k different stations in the district strategically so as to cover all geographically significant terrains. From each of these k stations, we collect average rainfall data every month over several months. This data is represented below.

$$\{\mathbf{y}^{(1)}\}_{t \geq 0} = \{y_0^{(1)}, y_1^{(1)}, y_2^{(1)}, ..., y_\tau^{(1)}, ...\}$$
$$\{\mathbf{y}^{(2)}\}_{t \geq 0} = \{y_0^{(2)}, y_1^{(2)}, y_2^{(2)}, ..., y_\tau^{(2)}, ...\}$$

$$\cdot$$
$$\cdot$$
$$\cdot$$

$$\{\mathbf{y}^{(k)}\}_{t \geq 0} = \{y_0^{(k)}, y_1^{(k)}, y_2^{(k)}, ..., y_\tau^{(k)}, ...\}$$

The above data corresponds to k realizations of monthly average rainfall. Each realization is a time series data. We may now choose to sample this data at the τth instant and generate a sample of k realizations of the random variable Y_τ as follows.

$$Y_\tau = \{y_\tau^{(1)}, y_\tau^{(2)}, ..., y_\tau^{(k)}\}. \tag{6.1}$$

Subsequently, we may be interested to know the probability distribution of this random variable Y_τ and compute its statistical moments. In a similar manner, we may generate k realizations of the random variables Y_1, Y_2, ..., and so on. We may want to know if these random variables are *correlated*. We may also want to find the joint probability distribution of these random variables. A whole family of statistical tools are available at our disposal to know more about the underlying natural phenomena that are responsible for the observed monthly rainfall patterns.

6.3.1 *Gaussian white noise*

Another example of a time series data is Gaussian white noise. It is often denoted as $\varepsilon_t = \{\varepsilon_1, \varepsilon_2, \varepsilon_3, ...\}$ which follows a Gaussian distribution, $\varepsilon_t \sim N(0, \sigma^2)$. It has the following properties:

$$E[\varepsilon_t] = 0, \tag{6.2}$$

$$E\left[\varepsilon_t^2\right] = \sigma^2, \text{ and} \tag{6.3}$$

$$E\left[\varepsilon_t \varepsilon_\tau\right] = 0, \forall t \neq \tau. \tag{6.4}$$

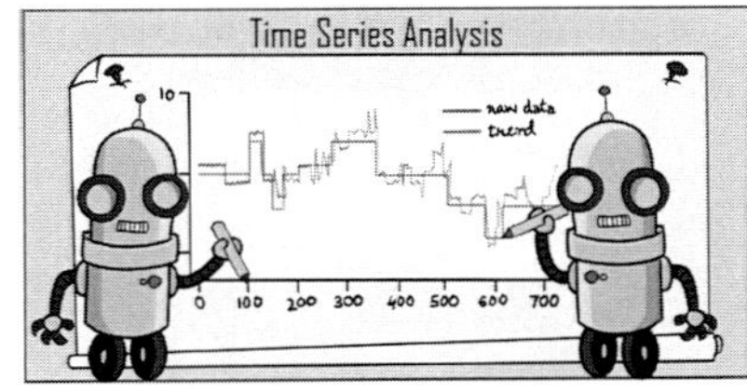

Figure 6.3 Robots evaluating trends in data in the Internet of Things (IoT) age.

6.3.2   *Correlation and Pearson's coefficient of correlation*

Often, in order to understand the co-relationship between two or more distinct random processes (time series data), it is essential to formulate an appropriate metric. We will first begin by considering two random variables and define the correlation coefficient.

Recall that in Chapter 1 we have defined covariance of two random variables X and Y as $Cov(X,Y) = E(X - \mu_X)(Y - \mu_Y)$, where $\mu_X = E(X)$, $\mu_Y = E(Y)$. In order to obtain a standard metric, it may be useful to normalize the measurement of this co-variation to a scale between -1 and $+1$. Correlation between two random variables X and Y is thus defined as follows.

$$\rho \equiv \rho_{XY} = \frac{Cov(X,Y)}{\sqrt{\sigma_X^2 \sigma_Y^2}} = \frac{Cov(X,Y)}{\sigma_X \sigma_Y}. \tag{6.5}$$

Here $\sigma_X^2 = Var(X)$ and $\sigma_Y^2 = Var(Y)$. ρ is also known as the *correlation coefficient* or *Pearson's correlation coefficient of **population***.

In many other scenarios, we may have access to data in pairs of the form (x_i, y_i), $i = 1, 2, 3, ..., n$ and we may be interested to know the co-relationship between the data variables x and y. For instance, x may be the amount of rainfall in a year and y may be the corresponding yearly rice productivity. In such a case, we may resort to the *Pearson's **sample** correlation coefficient*.

$$r \equiv r_{XY} = \frac{\sum_{i=1}^{n}(x_i - \bar{x})(y_i - \bar{y})}{\sqrt{\sum_{i=1}^{n}(x_i - \bar{x})^2 \sum_{i=1}^{n}(y_i - \bar{y})^2}}. \tag{6.6}$$

Both ρ and r determine the extent of *linear* relationship between the variables.

Example 6.1   Relationship between per capita gross domestic product (GDP) and life expectancy

Consider a demographic survey of different nations from 2005 as tabulated below. Compute the Pearson's sample correlation coefficient, r, and comment on your answer.

Country	GDP per person, x_i (USD)	Life expectancy, y_i (years)
USA	41890	78
Japan	31267	82
Germany	29461	79
Russia	10845	65
Mexico	10751	76
Brazil	8402	72
China	6757	72
India	3452	64
Bangladesh	2053	63
Nigeria	1128	47

Solution: $\overline{x} = \frac{\sum_{i=1}^{10} x_i}{10} = 14600.6$ and $\overline{y} = \frac{\sum_{i=1}^{10} y_i}{10} = 69.8$. In order to compute r, we will use the following table:

$(x_i - \overline{x})$	$(x_i - \overline{x})^2$	$(y_i - \overline{y})$	$(y_i - \overline{y})^2$	$(x_i - \overline{x})(y_i - \overline{y})$
27289.4	744711352.36	8.2	67.24	50074391332.6865
16666.4	277768888.96	12.2	148.84	41343121432.8064
14860.4	220831488.16	9.2	84.64	18691177157.8624
−3755.6	14104531.36	−4.8	23.04	324968402.534400
−3849.6	14819420.16	6.2	38.44	569658510.950401
−6198.6	38422641.96	2.2	4.84	185965587.086400
−7843.6	61522060.96	2.2	4.84	297766775.046401
−11148.6	124291281.96	−5.8	33.64	4181158725.13440
−12547.6	157442265.76	−6.8	46.24	7280130368.74239
−13472.6	181510950.76	−22.8	519.84	94356652643.0784
Total	1835424882.40000	–	971.6000	984246.2

Finally, $r = \dfrac{\sum_{i=1}^{n}(x_i - \overline{x})(y_i - \overline{y})}{\sqrt{\sum_{i=1}^{n}(x_i - \overline{x})^2 \sum_{i=1}^{n}(y_i - \overline{y})^2}} = 0.737$, which implies that life expectancy of

people of a nation is directly correlated to its per capita GDP (also refer to the illustration in Figure 6.4).

We can use Matlab to compute the same result. Let us store the (x_i, y_i) $i =$ 1, 2, ..., 10 data in the Matlab variable `data`. We then use the following Matlab routine to compute r.

```
>> corr_mat = corrcoef(data)

corr_mat =

    1.0000    0.7370
    0.7370    1.0000
```

The diagonal entries are the self-correlations r_{xx} and r_{yy} and hence equal to unity. The off-diagonal entries are the Pearson's sample correlation coefficients $r_{xy} = r_{yx} = 0.7370$ as calculated above.

Figure 6.4 A long and fulfilling life of a citizen is a true reflection of the health of a nation and the potency of its economic state–life expectancy vis–à–vis GDP.

6.3.3 *Variation in time series data*

As mentioned earlier, we could define a random vector $\mathbf{Y} := \{Y_1, Y_2, Y_3, ..., Y_t, ..., Y_n\}$ of n random variables. It is inconsequential whether the number is finite or infinite. $E(Y_t) = \mu_t$, $\forall t$. We may be interested in estimating two different types of variations:

(i) variation of Y_t itself and (ii) co-variation between the lagged variables Y_t and Y_{t-j}, $j \neq 0$. The former is known as the variance (often denoted by σ^2) and the latter is known as the *auto-covariance* γ_{jt}.

$$\sigma^2 \equiv \gamma_{0t} := E\left(Y_t - \mu_t\right)^2,$$ (6.7)

$$\gamma_{jt} := E\left(Y_t - \mu_t\right)\left(Y_{t-j} - \mu_{t-j}\right).$$ (6.8)

We may think of γ_{jt} in a way analogous to the covariance between two random variables X and Y that is defined as $Cov\left(X,Y\right) := E\left(X - \mu_x\right)\left(Y - \mu_Y\right)$. Similarly, the auto-correlation function (ACF) ρ_j of a weakly stationary stochastic process is defined as follows.

$$\rho_j = \frac{\gamma_j}{\gamma_0}.$$ (6.9)

It follows that $\rho_0 \equiv 1$ and $|\rho_j| < 1$ for all $j = 1, 2, 3, \ldots$ Clearly, $\rho_j = \rho_{-j}$ owing to a symmetry of the auto-covariance function.

The auto-covariance, along with the mean, is an important parameter that determines the *stationarity* and *ergodicity* of a time series data. We will discuss these ideas in the following sections.

6.3.4   *Stationarity*

Generally speaking, when the statistical properties of a temporal data remain invariant over time, without the actual values necessarily staying the same, we say that the random process generating the data is *stationary*. Statistical theories (models) require that the underlying phenomena are stationary in order for them to be useful. So if a certain data exhibits *trend*[2] or *seasonality*,[3] then it is important to stabilize these attributes first before employing any stochastic model on the time series data. The reason for this is that the co-dependency structure between any two time series data can be meaningfully discovered by models if the temporal and/or seasonal trends are first discounted (and removed) and if the respective processes are stationary.

It may be helpful to refer to the example in Figure 6.5 to fully appreciate how de-trending the original time series data appropriately results in a residual stationary stochastic process. This residual system may be subjected to time series analysis. The Matlab code for generating the plot of this data is given below.

[2]Trend may be linear if $Y_t = a + \beta t + \varepsilon_t$, where a, β are constants and ε_t is Gaussian white noise. This trend may be stabilized by the method of differencing, $\Delta Y_t := Y_t - Y_{t-1} = \varepsilon_t - \varepsilon_{t-1} + \beta$, which is a stationary process owing to the stationarity of the Gaussian white noise process. This method of stabilization of the trend may not work if differencing results in only mean invariance without arresting the growth of variance concurrently (such as in a random walk). In this particular example of the linear trend, differencing results in finitely bounded values of mean and variance.

[3]Seasonality is when we have periodic fluctuations in the data and *may* thereby render it non-stationary if the error in the fluctuations depends on time.

```
load Data_GDP
t = [1:length(Data)];
Data_detrend_2 = detrend(Data,2);
plot(t,Data,'-.b',t,Data_detrend_2,'-r',t,Data-Data_detrend_2,':k','linewidth',2,...
    'MarkerSize',10);
legend('Input data','Detrended data','Quadratic trend',...
    'Location','northwest','FontSize',16)
xlim([0,234]);
set(gca,'xtick',[]);
```

```
title('Quarterly GDP of USA in units of billion dollars. 1947-2005','FontSize',14);
xlabel('quarters (1947-2005)','FontSize',14);
ylabel('GDP in billion USD','FontSize',14);
```

Matlab has a default file named `Data_GDP` that contains the quarterly GDP of USA from 1947 to 2005.[4] This information is stored in the variable `Data`. The function `detrend(Data,2)` produces the residual data after subtracting the trend (modeled by a quadratic interpolating polynomial) from the original data. This residual temporal data appeals to our eyes as statistically stationary. Next, we will validate this rigorously. We must now present a mathematical definition of a stationary process.

[4] This will require an active Econometrics Toolbox linked to your Matlab application. For details refer to the following link: https://in.mathworks.com/help/econ/data-sets-and-examples.html (accessed January 24, 2024).

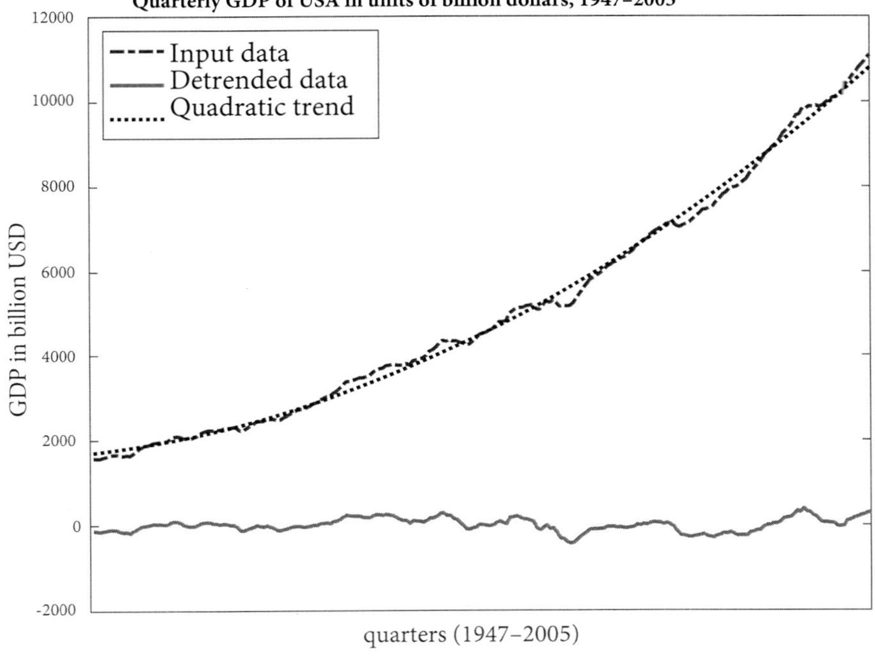

Figure 6.5 The input data comprises quarterly GDP of USA from the year 1947 to 2005 and shows a quadratic trend (dotted lines). The residual data (trend subtracted input data) is stationary.

There are two notions of stationarity, namely, weak (covariance) stationarity and strong stationarity.

1. *Weak or covariance stationary process.* A stochastic process Y_t is *weakly stationary* if the following conditions are satisfied:

a) $E(Y_t) = \mu$ is independent of time,

b) $E\big(Y_t - \mu_t\big)\big(Y_{t-j} - \mu_{t-j}\big) = \gamma_j$ is independent of time, and

c) the auto-covariance function γ_j is an even function of the lag variable j, i.e.,
$$\gamma_j = \gamma_{-j}, \ \forall j.$$

Visual inspection of the residual plot in Figure 6.5 is sufficient to validate the first condition mentioned above (temporal invariance of the mean). The third condition follows from the definition of γ_j. The invariance of the auto-covariance function is demonstrated by selecting different random sections of the time series data and computing γ_j for each of these sections. We have used the following Matlab code to perform this test and the result is plotted in Figure 6.6.

```
N = length(Data_detrend_2);
T1 = randi([floor(N/4),floor(3*N/4)]); T2 = randi([floor(N/4),floor(3*N/4)]);
T3 = randi([floor(N/4),floor(3*N/4)]);
acf1 = autocorr(Data_detrend_2(T1:end));
acf2 = autocorr(Data_detrend_2(T2:end));
acf3 = autocorr(Data_detrend_2(T3:end));
lags = [1:length(acf1)];
plot(lags,acf1,'-.k',lags,acf2,':m',lags,acf3,'-g','linewidth',2,'MarkerSize',10);
legend('data sequence 1','data sequence 2','data sequence 3','Location','northeast','FontSize',16);
xlabel('lag, j','FontSize',18); ylabel('\gamma_j','FontSize',18); xlim([0,length(lags)]);
```

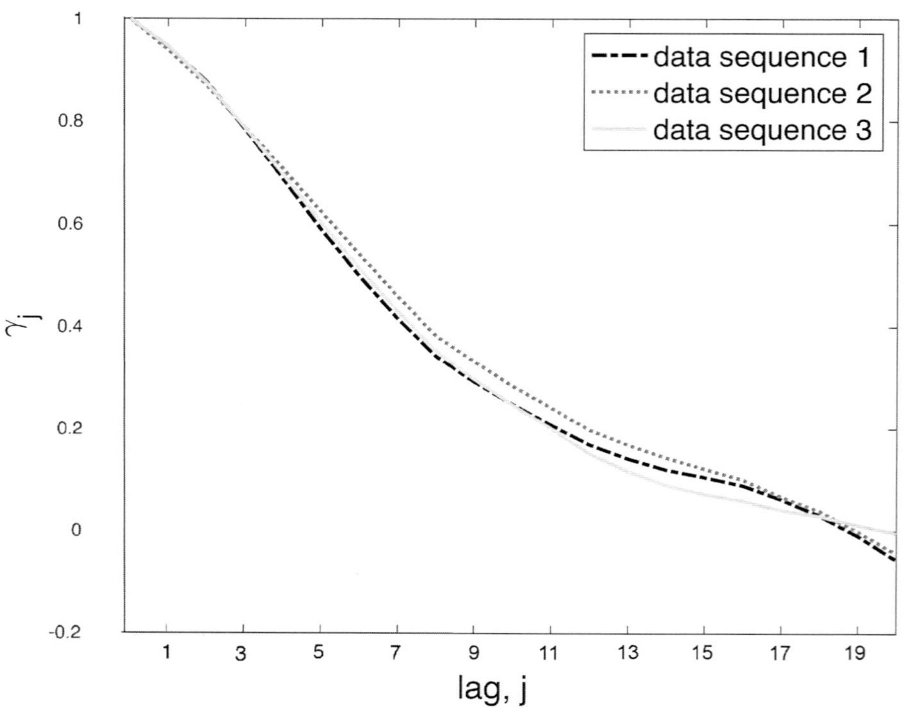

Figure 6.6 The auto covariance functions of three random sections of the residual data of Figure 6.5 demonstrating the temporal invariance of γ_j for all j.

2. *Strong or strictly stationary process.* A stochastic process Y_t is *strictly stationary* if the joint distribution of the random variables $Y_t, Y_{t+j_1}, Y_{t+j_2}, Y_{t+j_3}, ..., Y_{t+j_n}$ depends on the intervals separating the time stamps $j_1, j_2, ..., j_n$ but does not depend on the time instant t.

In this chapter, we will restrict our discussion to weakly stationary processes unless otherwise mentioned.

6.3.5 *Ergodicity*

A stochastic process is *ergodic* when the time averages of the random realizations of the process converge, in probability, to their ensemble averages, i.e., $\overline{Y} \xrightarrow{p} E(Y_t)$, where $\overline{Y} = \frac{1}{T}\sum_{t=1}^{T} Y_t$.

The auto-covariance function may be used to test the ergodicity of a random process because $\gamma_j \to 0$ sufficiently fast as $j \to \infty$ when the system is ergodic. This is true if and only if $\sum_{j \geq 0} |\gamma_j| < \infty$.

In many applications it may be easier to compute the temporal average (sample mean) of data rather than the ensemble average as the latter will require repeated experiments and hence may be expensive. If we know that the underlying process is ergodic, we may simply use the temporal averages as a suitable replacement of ensemble averages (see Figure 6.7).[5]

6.4 Predictive models

Now that we are somewhat familiar with the statistical features of a data set, we begin our discussion on prediction with a classical model and its variants. The goal here is to consider several data points available to us over a certain range and use them to make predictions about events for which the data is not available. This missing data could be due to gaps in the data collection process or simply a matter of paucity of enough memory to store a lot of data. It may also be due to hitherto unseen scenarios. The basic idea behind these predictive models is to construct a mathematical function (e.g., a line, a curve, a surface, etc.) that describes the available data with the least error. We will hope that after constructing such a mathematical function, we may be able to use it not only to explain the trend or profile of the data but also to make predictions in the case of missing data.

6.4.1 *Least squares regression*

Given a sequence of data $(\mathbf{x},\mathbf{y})$, where $\mathbf{x} = (x_1, x_2, \ldots, x_n)$ and likewise for $\mathbf{y}$, we may be interested in acquiring knowledge about the trend in the data; e.g., how does $\mathbf{y}$ change with $\mathbf{x}$? In terms of the example discussed above, we may be interested to know the life expectancy of the people of a nation if its per capita GDP is USD 30,000. This may enable nations to calibrate their economic policies (in terms of GDP per person) for achieving a certain quality of life for their citizens in terms of life expectancy (say 80 years). Let us plot the per capita GDP versus life expectancy data from the example in section 6.3.2 and see if we can identify a trend by visual inspection. We will then attempt to plot a *line of best fit* to the data.

In Figure 6.8, either of the two lines could qualify as the line of best fit and demonstrate the linear trend for life expectancy with increasing per capita GDP. So, in a sense, fitting a line to data has a subjective element depending on how we interpret the data.[6] In any case, the cognitive exercise of drawing such a line of best fit, perhaps, involves a notion of minimizing the deviation of the line from as many of the data points as possible. We can think of the deviation as an error profile – the shortest distance between the line and each of the data points accumulated into one whole. This leads us to an expression of the form $\nabla_i = \left(y_i - M\left(x_i \right) \right)$, where $M(\cdot)$ is a mathematical model. In the case of the line of best fit or the line of regression, $M(x)$ takes the form $M(x) = a + bx$, where a and b are constants. However, we are interested in minimizing the deviation of the model from the actual data points and should not care much about the actual position of the line (above or below) with respect to the data point. So we may have considered $\nabla_i = \left| y_i - M\left(x_i \right) \right|$, but the absolute value function is a non-differential function. Since we should preempt

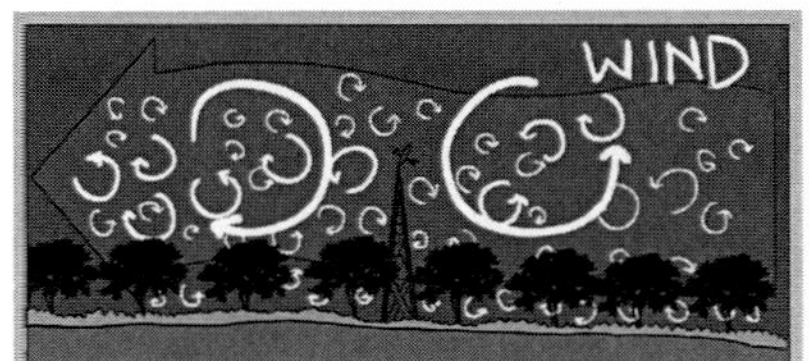

Figure 6.7 An artist's rendition of a measuring station recording wind data in the stable atmospheric boundary layer (SABL). These ground-based weather stations rely on the *ergodic* hypothesis of atmospheric turbulence in order to develop statistical models of weather patterns. The ergodic hypothesis permits scientists to approximate ensemble-averaged wind speed measurements (which would require multiple devices) with temporally averaged measurements (using only one device).

[5] Chad W. Higgins et al., Are atmospheric surface layer flows ergodic? *Geophysical Research Letters* **40** (2013): 3342–3346, doi:10.1002/grl.50642.

[6] In this example, we may ignore the data for Nigeria as an outlier and come up with the dashed line as the line of best fit. Of course, the dotted line perhaps serves as a better candidate for the linear trend if we consider all the data points.

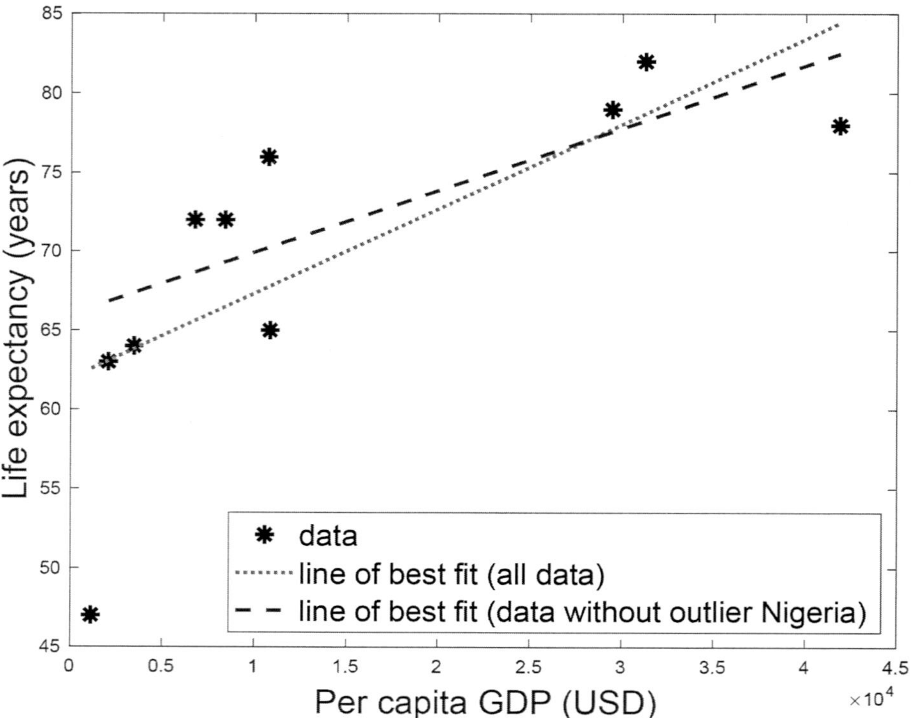

Figure 6.8 Data from the example in section 6.3.2 and the line of best fit to the data.

that the process of minimization will likely involve the operation of differentiation, it may be prudent to choose an alternative form, $e_i = \nabla_i^2 = (y_i - M(x_i))^2$, which will result in a system of linear equations whose solutions may be computed conveniently using appropriate matrix methods. Essentially, we will seek to minimize $e := \sum_{i=1}^{n} e_i$ with respect to the model parameters.

In fact, it is not necessary that the model M is a straight line; M can take the general form $M(x) = a + bf(x) + cg(x)$, where a, b, and c are real constants and f and g are model functions. With this general functional model, it is still possible to establish a system of linear equations in a, b, and c whose solution gives the least squares curve of best fit to the data. Let us first derive the general regression model M using this principle.

Consider the error functional, $e = \sum_{i=1}^{n} e_i = \sum_{i=1}^{n} \nabla_i^2 = \sum_{i=1}^{n} \left(y_i - M\left(x_i\right) \right)^2$. In order to find the critical points in (a,b,c) space, we set $\dfrac{\partial e}{\partial a}, \dfrac{\partial e}{\partial b}, \dfrac{\partial e}{\partial c} = 0$. This results in the following three equations in this order:

$$\sum_{i=1}^{n} 2\left\{ y_i - \left(a + bf_i + cg_i\right) \right\}\left(-1\right) = 0, \tag{6.10}$$

$$\sum_{i=1}^{n} 2\left\{ y_i - \left(a + bf_i + cg_i\right) \right\}\left(-f_i\right) = 0, \text{ and} \tag{6.11}$$

$$\sum_{i=1}^{n} 2\left\{ y_i - \left(a + bf_i + cg_i\right) \right\}\left(-g_i\right) = 0. \tag{6.12}$$

For brevity, we have used f_i for $f(x_i)$ and so on. It is important to note that equations (6.10)–(6.12) give us the *least* squares solution to the data or the curve of best fit that *minimizes* the cumulative error e. This can be seen easily by computing the second order derivatives and noting that $\dfrac{\partial^2 e}{\partial a^2}, \dfrac{\partial^2 e}{\partial b^2}, \dfrac{\partial^2 e}{\partial c^2} > 0$. Equations (6.10)–(6.12) can be written in matrix vector form $\Lambda \mathbf{p} = \mathbf{q}$ as follows.

$$
\begin{pmatrix}
\sum_{i=1}^{n} 1 & \sum_{i=1}^{n} f_i & \sum_{i=1}^{n} g_i \\[2mm]
\sum_{i=1}^{n} f_i & \sum_{i=1}^{n} f_i^2 & \sum_{i=1}^{n} f_i g_i \\[2mm]
\sum_{i=1}^{n} g_i & \sum_{i=1}^{n} f_i g_i & \sum_{i=1}^{n} g_i^2
\end{pmatrix}
\begin{pmatrix} a \\ b \\ c \end{pmatrix}
=
\begin{pmatrix}
\sum_{i=1}^{n} y_i \\[2mm]
\sum_{i=1}^{n} y_i f_i \\[2mm]
\sum_{i=1}^{n} y_i g_i
\end{pmatrix}.
\tag{6.13}
$$

The regression model parameters are given by $\boxed{\mathbf{p} = \begin{pmatrix} a \\ b \\ c \end{pmatrix} = \Lambda^{-1}\mathbf{q}}$ as long as Λ is an invertible matrix.[7] With such a $\mathbf{p}$ and the chosen model functions f and g, the least squares regression model M is constructed.

[7] Under what condition(s) is Λ not invertible?

Example 6.2 *Daily milk requirement of an ice-cream maker*

The sale of ice creams depends on the number of hours of bright sunshine. As an ice-cream maker, Alice needs to know how much milk she must acquire daily to make the required number of ice creams to cater to her customers. Based on her sale in the previous summer months, she has access to the following data.

Number of hours of sunshine, x_i	Number of ice creams sold, y_i
2	4
3	5
5	7
7	10
9	15

Figure 6.9 An ice-cream maker sells more ice creams on a hot sunny day.

Answer the following questions:

i. Find the linear least squares regression model that best fits the data.

ii. How many ice creams might she sell on a day with eight hours of bright sunshine?

Solution: Here the total number of sample data points $n = 5$. We seek the linear model $M(x) = a + bx$. Here $f(x) = x$. The matrix equations $\Lambda p = q$ is

$$
\begin{pmatrix}
\sum_{i=1}^{5} 1 & \sum_{i=1}^{5} x_i \\[2mm]
\sum_{i=1}^{5} x_i & \sum_{i=1}^{5} x_i^2
\end{pmatrix}
\begin{pmatrix} a \\ b \end{pmatrix}
=
\begin{pmatrix}
\sum_{i=1}^{5} y_i \\[2mm]
\sum_{i=1}^{5} y_i x_i
\end{pmatrix}.
\tag{6.14}
$$

$$\Lambda = \begin{pmatrix} 5 & 26 \\ 26 & 168 \end{pmatrix},\ \mathbf{q} = \begin{pmatrix} 41 \\ 263 \end{pmatrix},\ \Lambda^{-1} = \begin{pmatrix} 1.0244 & -0.1585 \\ -0.1585 & 0.0305 \end{pmatrix}. \text{ The least squares solution}$$

$$\text{is } \mathbf{p} = \Lambda^{-1}\mathbf{q} = \begin{pmatrix} 0.3049 \\ 1.5183 \end{pmatrix}.$$

i. Therefore, the linear model is $M(x) = 0.305 + 1.518x$.

ii. $M(8) = 0.305 + 1.518(8) = 12.45$. Therefore, roughly 13 ice creams will sell in a day with eight hours of sunshine.

Example 6.3 *Tracing the trajectory of a particle*

A physicist performed a few experiments and obtained the following data on the displacement (from the origin) versus time for a particle of interest. The particle's displacements from origin at time $t = 0$, $\frac{\pi}{2}$, π, $\frac{3\pi}{2}$, and 2π were 4 μm, 3 μm, $-2\ \mu m$, $-1\ \mu m$ and 4 μm, respectively.

i. If it is assumed that the particle moves like a parabola over time, find the best fit curve of the particle's trajectory.

ii. If it is assumed that the particle has wave-like displacement over time, find the best fit curve of the particle's trajectory.

iii. Compare the two results and identify the most suitable curve that gives the minimum error to the experimental data and find the displacements (from the origin) at time instants $t = \pi/4$ and $7\pi/4$.

Solution: In the problem, we are given the following data of the particle's location.

Time t_i	Displacement y_i (in μm)
0	4
$\pi/2$	3
π	-2
$3\pi/2$	-1
2π	4

i. Here $n = 5$. Now if the curve is parabolic, we assume $f(t) = t$ and $g(t) = t^2$.

$$y(t) = a + bt + ct^2.$$

We use the following table to construct our regression model.

t_i	y_i	f_i	g_i	$f_i g_i$	$f_i y_i$	$g_i y_i$	f_i^2	g_i^2
0	4	0	0	0	0	0	0	0
$\pi/2$	3	$\pi/2$	$\pi^2/4$	$\pi^3/8$	$3\pi/2$	$3\pi^2/4$	$\pi^2/4$	$\pi^4/16$
π	-2	π	π^2	π^3	-2π	$-2\pi^2$	π^2	π^4
$3\pi/2$	-1	$3\pi/2$	$9\pi^2/4$	$27\pi^3/8$	$-3\pi/2$	$-9\pi^2/4$	$9\pi^2/4$	$81\pi^4/16$
2π	4	2π	$4\pi^2$	$8\pi^3$	8π	$16\pi^2$	$4\pi^2$	$16\pi^4$
Total	8	5π	$30\pi^2/4$	$100\pi^3/8$	$12\pi/2$	$50\pi^2/4$	$30\pi^2/4$	$354\pi^4/16$

The least squares coefficients a, b, and c can be determined by the following system of linear equations. This can be written in matrix form as $\Lambda\alpha = \chi$, that is,

$$\begin{pmatrix} \sum_{i=1}^{n}1 & \sum_{i=1}^{n}f_i & \sum_{i=1}^{n}g_i \\ \sum_{i=1}^{n}f_i & \sum_{i=1}^{n}f_i^2 & \sum_{i=1}^{n}f_i g_i \\ \sum_{i=1}^{n}g_i & \sum_{i=1}^{n}f_i g_i & \sum_{i=1}^{n}g_i^2 \end{pmatrix} \begin{pmatrix} a \\ b \\ c \end{pmatrix} = \begin{pmatrix} \sum_{i=1}^{n}y_i \\ \sum_{i=1}^{n}y_i f_i \\ \sum_{i=1}^{n}y_i g_i \end{pmatrix}.$$

Assuming that the matrix is non-singular, the solution is given by

$$\alpha = \Lambda^{-1}\chi.$$

From the table, we get

$$\Lambda = \begin{bmatrix} 5 & 5\pi & \dfrac{30\pi^2}{4} \\ 5\pi & \dfrac{30\pi^2}{4} & \dfrac{100\pi^3}{8} \\ \dfrac{30\pi^2}{4} & \dfrac{100\pi^3}{8} & \dfrac{354\pi^4}{16} \end{bmatrix} \text{ and } \chi = \begin{bmatrix} 8 \\ \dfrac{12\pi}{2} \\ \dfrac{50\pi^2}{4} \end{bmatrix}.$$

The inverse of Λ is

$$\Lambda^{-1} = \begin{bmatrix} 0.8857 & -0.4911 & 0.0579 \\ -0.4911 & 0.5037 & -0.0737 \\ 0.0579 & -0.0737 & 0.0117 \end{bmatrix}.$$

Hence,

$$\alpha = \Lambda^{-1}\chi = \begin{bmatrix} 4.9714 \\ -3.5287 \\ 0.5211 \end{bmatrix}.$$

Therefore, the parabolic curve of best fit is

$$y(t) = 4.9714 - 3.5287t + 0.5211t^2.$$

ii. If the curve follows a wave-like pattern, we assume $f(t) = \sin t$ and $g(t) = \cos t$.

$$y(t) = a + b \sin t + c \cos t.$$

We consider the following table in this case.

t_i	y_i	f_i	g_i	f_i^2	g_i^2	$f_i\,g_i$	$y_i\,f_i$	$y_i\,g_i$
0	4	0	1	0	1	0	0	4
$\pi/2$	3	1	0	1	0	0	3	0
π	-2	0	-1	0	1	0	0	2
$3\pi/2$	-1	-1	0	1	0	0	1	0
2π	4	0	1	0	1	0	0	4
Total	8	0	1	2	3	0	4	10

From this table, we obtain

$$\Lambda = \begin{bmatrix} 5 & 0 & 1 \\ 0 & 2 & 0 \\ 1 & 0 & 3 \end{bmatrix} \text{ and } \chi = \begin{bmatrix} 8 \\ 4 \\ 10 \end{bmatrix}.$$

The inverse of Λ is

$$\Lambda^{-1} = \begin{bmatrix} 3/14 & 0 & -1/14 \\ 0 & 0.5 & 0 \\ -1/14 & 0 & 5/14 \end{bmatrix}.$$

Hence,

$$\alpha = \Lambda^{-1}\chi = \begin{bmatrix} 1 \\ 2 \\ 3 \end{bmatrix}.$$

Therefore, the curve of best fit is

$$y(t) = 1 + 2\sin(t) + 3\cos(t).$$

iii. The error function is defined as follows.

$$\text{Error} = e(a,b,c) = \sum_{i=1}^{n} \left\{ y_i - \left(a + bf_i + cg_i \right) \right\}^2.$$

For the parabolic curve, the error is

$$\text{Error} = \sum_{i=1}^{5} \left\{ y_i - \left(4.9714 - 3.5287 t_i + 0.5211(t_i)^2 \right) \right\}^2 > 0.$$

And for the wavy curve, the error comes out to be

$$\text{Error} = \sum_{i=1}^{5} \left\{ y_i - \left(1 + 2\sin\left(t_i\right) + 3\cos\left(t_i\right) \right) \right\}^2 = 0.$$

Therefore, the error is minimum for the wavy curve; thus,

$$y(t) = 1 + 2\sin(t) + 3\cos(t)$$

is a better model for the given data. At time $t = \pi/4$ and $7\pi/4$, we have

$$y\left(\frac{\pi}{4}\right) = 1 + \frac{5}{\sqrt{2}} = \frac{2 + 5\sqrt{2}}{2} = 4.5355\ \mu m,$$

and

$$y\left(\frac{7\pi}{4}\right) = 1 + \frac{1}{\sqrt{2}} = \frac{2 + \sqrt{2}}{2} = 1.7071\ \mu m.$$

6.4.2 *Matrix formulation of least squares regression: normal system*

We can generalize the model $M(x)$ defined in the previous section to include m model functions λ_j and model parameters p_j, where $j = 1, 2, ..., m$. So the model now takes the form $M(x) = p_1\lambda_1(x) + p_2\lambda_2(x) + ... + p_m\lambda_m(x)$ with $\lambda_1(x) \equiv 1, \forall x$. Ideally, $M(x)$ would fit the data perfectly, i.e., $y_i \equiv M(x_i)$, whence we would arrive at a simple linear system written in matrix form as follows.

$$\Lambda p = q. \tag{6.15}$$

$$\begin{pmatrix} 1 & \lambda_{12} & \lambda_{13} & \cdots & \lambda_{1m} \\ 1 & \lambda_{22} & \lambda_{23} & \cdots & \lambda_{2m} \\ . & . & . & \cdots & . \\ . & . & . & \cdots & . \\ 1 & \lambda_{n2} & \lambda_{n3} & \cdots & \lambda_{nm} \end{pmatrix} \begin{pmatrix} p_1 \\ p_2 \\ . \\ . \\ p_m \end{pmatrix} = \begin{pmatrix} y_1 \\ y_2 \\ . \\ . \\ y_m \end{pmatrix}. \tag{6.16}$$

Here we have used the notation: $\lambda_{ik} = \lambda_k(x_i)$. These λ_{ij}s populate the matrix Λ. There are n rows in Λ corresponding to n equations for n data points (x_i, y_i), $i = 1, 2, ..., n$ and m columns corresponding to m model parameters.

However, like we observed in Figure 6.8, it is unlikely for a finite dimensional model to exactly fit a dataset. Our goal, as always, must be to minimize the two-norm (squared) of the error vector: $e = \left\|\Lambda p - q\right\|^2$, for which we must set the derivatives $\dfrac{\partial e}{\partial p_k}$ equal to 0 for all $k = 1, 2, ..., m$ following the arguments presented in the previous section. This leads to the following set of equations.

$$\sum_{i=1}^{n}\sum_{j=1}^{m}\lambda_{ik}\lambda_{ij}p_j = \sum_{i=1}^{n}\lambda_{ik}q_i, \tag{6.17}$$

where $k = 1, 2, ..., n$ designates the equation for the kth data point. The system of equations (6.17) may be written more compactly in matrix-vector form as follows.

$$\left(\Lambda^T \Lambda\right)p = \Lambda^T q. \tag{6.18}$$

Equation (6.18) is known as the *normal system* whose solution p is the *least squares solution* to the linear system $\Lambda\, p = q$.[8]

[8] If Λ is of dimensions $(n \times m)$ and if $rank(\Lambda) = m$ (i.e., Λ is full column rank), then $\Lambda^T\Lambda$ is always invertible and there exists a *unique* least squares solution of the normal system (6.18), namely, $(\Lambda^T\Lambda)^{-1}\Lambda^T q$.

Example 6.4   Quadratic least squares model

Consider the following data from an experiment. Find a quadratic least squares model that best fits this data.

x	y
-1	0
0	1
1	3
2	9
3	19

Solution: Our objective here is to fit a model $M(x) = a + bx + cx^2$. We will begin by constructing the entries of the matrix Λ. $\lambda_{12} = x_1 = -1$, $\lambda_{13} = 1^2$. In fact, $\lambda_{i2} = x_i$ and

$$\lambda_{i3} = x_i^2 \text{ for all } i = 1, 2, ..., n = 5. \text{ Therefore, } \Lambda = \begin{pmatrix} 1 & -1 & 1 \\ 1 & 0 & 0 \\ 1 & 1 & 1 \\ 1 & 2 & 4 \\ 1 & 3 & 9 \end{pmatrix} \text{ and } \mathbf{q} = \begin{pmatrix} 0 \\ 1 \\ 3 \\ 9 \\ 19 \end{pmatrix}. \text{ The}$$

matrix Λ has rank $m = 3$ and hence there exists a unique least squares solution

$$\mathbf{p} = \left(\Lambda^T \Lambda\right)^{-1} \Lambda^T \mathbf{q} = \begin{pmatrix} 0.2286 \\ 1.4571 \\ 1.5714 \end{pmatrix}. \text{ This model and the data are plotted in Figure 6.10.}$$

Figure 6.10 Quadratic least squares fit to the data in the example of section 6.4.2.

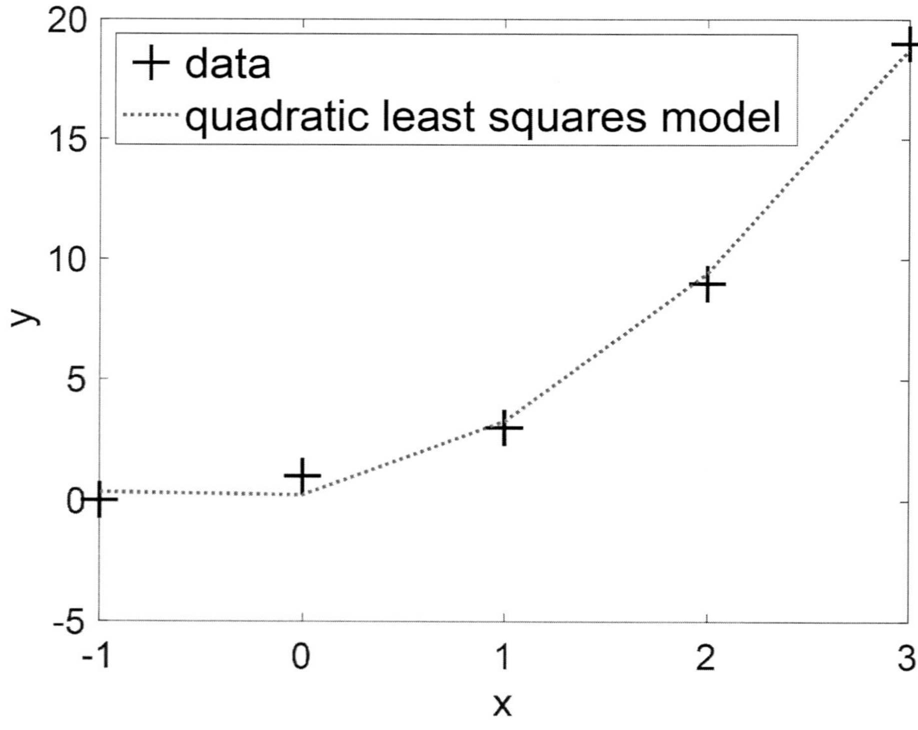

6.4.3   *Multiple linear regression*

So far we have considered datasets where the dependent variable y is a function of only one independent variable x. This functional dependence could be either linear or non-linear. However, in most practical applications, y may have a functional dependence on multiple independent attributes, i.e., each data variable x_i may be a $k = (m - 1)$ dimensional vector $(x_{i1}, x_{i2}, ..., x_{ik})$ for each of the $i = 1, 2, ..., n$ data points. In this section, we will seek a *multiple linear regression* model $M(x) = p_0 + p_1 x_1 + ... + p_k x_k$. This is similar in form to the model in section 6.4.2 if we consider the individual model functions to have a linear dependence on the different components of the independent data vector $\mathbf{x}$. The least squares solution is given by

$$\mathbf{p} = \begin{pmatrix} p_0 \\ p_1 \\ p_2 \\ \cdot \\ \cdot \\ p_k \end{pmatrix}_{m \times 1} \tag{6.19}$$

$$= \left(\Lambda^T \Lambda \right)^{-1} \Lambda^T \mathbf{q}, \tag{6.20}$$

where the matrix $\Lambda = \begin{pmatrix} 1 & x_{11} & x_{12} & \cdot & \cdot & x_{1k} \\ 1 & x_{21} & x_{22} & \cdot & \cdot & x_{2k} \\ \cdot & \cdot & \cdot & \cdot & \cdot & \cdot \\ \cdot & \cdot & \cdot & \cdot & \cdot & \cdot \\ 1 & x_{n1} & x_{n2} & \cdot & \cdot & x_{nk} \end{pmatrix}_{n \times m}$.

Example 6.5 Multiple linear regression for predicting multi-dimensional data

Consider the following data sourced from the Eurosat public repository. Here the GDP of Belgium (y) is a function of two independent indicators: (i) educational expenditure by the state (x_1) and (ii) employee compensation (x_2). All figures below are in million euros and have been rounded to the nearest thousand. Construct a linear regression model of this data.

Figure 6.11 Literacy and purchasing power are known to propel GDP growth.

GDP (y)	Educational expenditure (x_1)	Employee compensation (x_2)
256000	14000	129000
264000	15000	136000
273000	16000	142000
281000	17000	145000
296000	17000	149000
310000	18000	161000

(Contd)

(Contd)

344000	20000	180000
351000	22000	185000
363000	23000	193000
376000	24000	200000
386000	25000	204000
393000	26000	208000
417000	27000	212000
445000	28000	219000
461000	29000	226000

Solution: We will use a variable `data` to store the values of x_1, x_2, and y in each of the first three columns in that order.

$$\texttt{data = [x(:,1) x(:,2) y]}$$

Next, we will construct the matrix Λ (we will call it A in the routine below) and solve the normal system (6.20) in Matlab as follows.

```
A(:,1)  = ones(length(y),1);
A(:,2)  = data(:,1);
A(:,3)  = data(:,2);

p = (inv(A'*A))*(A')*(data(:,3));
mdl = fitlm([data(:,1) data(:,2)], data(:,3) );
% fitlm is the Matlab in-built routine for multiple linear regression model

x1fit = min(data(:,1)):100:max(data(:,1));
x2fit = min(data(:,2)):100:max(data(:,2));
[X1, X2] = meshgrid(x1fit,x2fit);
LS = p(1) + p(2).*X1 + p(3).*X2;
plot3(data(:,1),data(:,2),data(:,3),'ko',...
    'linewidth',4,'MarkerSize',20,'MarkerFaceColor',[1 1 1]); hold on;
%scatter3(data(:,1),data(:,2),data(:,3),'linewidth',2,'MarkerSize',20);
hold on;
LS_plot = mesh(X1,X2,LS,'FaceAlpha',0.85);
xlabel('Education expenditure','FontSize',20);
ylabel('Employee compensation','FontSize',20);
zlabel('GDP','FontSize',20);
legend('data','multiple linear regression plane',...
    'Location','northwest','FontSize',16);
set(LS_plot,'edgecolor','none','facecolor',[0.35 0.35 0.35])
set(gca,'FontSize',20);
```

The result is plotted in Figure 6.12.

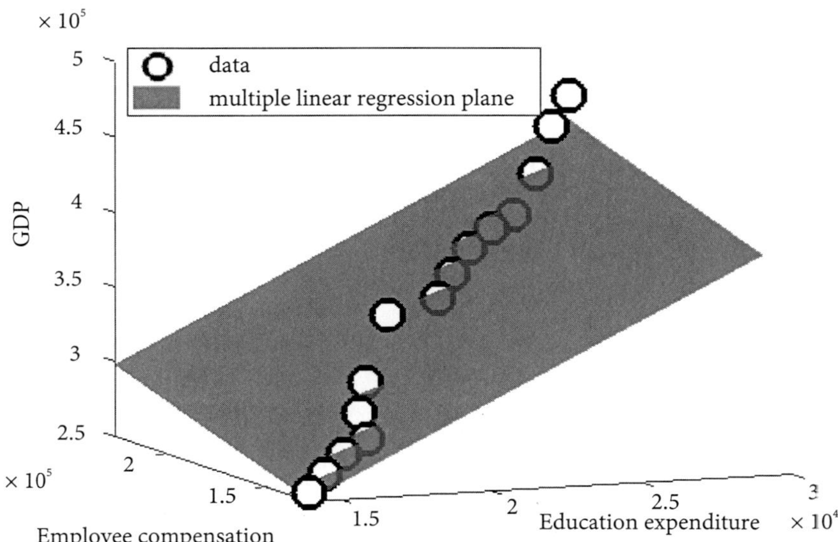

Figure 6.12 Multiple linear regression model for multi-dimensional data from the example in section 6.4.3. All figures are in million euros.

6.5 Time series models

In this section, we will learn to construct and apply a few classical models to time series data. We will comment on the stochastic properties of the underlying time series data (process) to which the model is applied. The scope of the current textbook will not allow us to explore this area of study in depth; our objective here is to introduce the readers to the most commonly used time series models and some of the basic ideas related to the determination of order of such a model.

6.5.1 *Moving average model*

MA(1) model For a stationary time series data Y_t with continuous stochastic variations without trend, it should make sense to capture the essential aspects of this process by considering a weighted sum of a finitely correlated white noise process ε_t offset by a constant term. The moving average model of order one is constructed as follows.

$$Y_t = \mu + \varepsilon_t + \theta\varepsilon_{t-1}, \tag{6.21}$$

where μ and θ are constants. The order of the MA(1) model is determined by the maximum lag value, for which the auto-correlation of Y_t is non-trivial and turns out to be one. This will be demonstrated below along with other statistical properties of this model. A simulation of an $MA(1)$ process is shown in Figure 6.13.

1. *Auto-correlation*: The first few statistical moments of this system are computed here. $E(Y_t) = \mu + 0 + 0 = \mu$, which is constant and independent of time t. $\gamma_{0t} = Var(Y_t) = 0 + \sigma^2 + \theta^2\sigma^2 = (1 + \theta^2)\sigma^2$ is also constant over time. $\gamma_{jt} = E(Y_t - \mu_t)(Y_{t-j} - \mu_{t-j}) = E(\varepsilon_t + \theta\varepsilon_{t-1})(\varepsilon_{t-j} + \theta\varepsilon_{t-j-1}) = \theta\sigma^2$ for $j = 1$. $\gamma_{jt} = 0$

for all $j \geq 2$. This result follows directly from the third statistical property of Gaussian white noise. Consequently, the ACFs are $\rho_0 = 1$, $\rho_1 = \frac{\gamma_1}{\gamma_0} = \frac{\theta}{\left(1+\theta^2\right)}$, and $\rho_j = 0$, $\forall j \geq 2$.

Figure 6.13 Simulation of an *MA* (1) stochastic process with $\theta_1 = 0.1$ and $\varepsilon_t \sim N(0, 50)$.

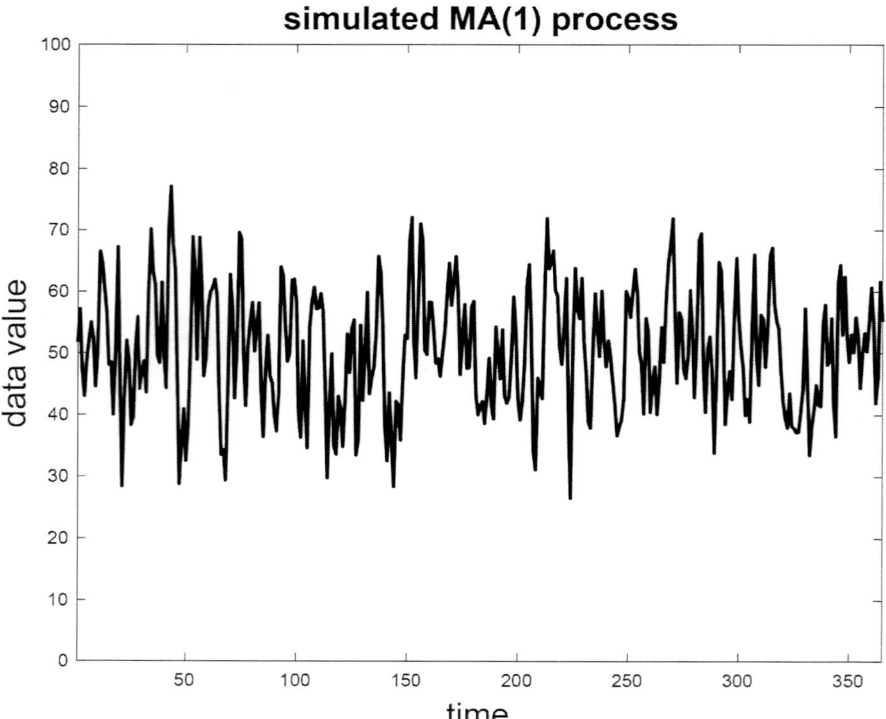

Since γ_{0t} and γ_{1t} are non-zero, but all higher order auto-correlations γ_{jt}, $j = 2, 3, \ldots$, are zero, it follows that the model defined by equation (6.21) is an order one model. See Figure 6.14 as a reference.

2. *Stationarity:*   Since the mean and ACFs, as per the calculation above, are independent of time, the $MA(1)$ system is statistically weakly stationary.

3. *Ergodicity:* Since $\sum_{j=1,2,\ldots} \left|\gamma_j\right| = \left|\gamma_0\right| + \left|\gamma_1\right| + 0 = \left(1+\theta^2\right)\sigma^2 + \theta\sigma^2 < \infty$, the $MA(1)$ is an ergodic time series model.

The Matlab routines corresponding to Figures 6.13 and 6.14 are furbished below.

```
model = arima('Constant',50,'MA',{1 0.1},'Variance',50); % simulate MA(1) process
sim_data = simulate(model,365);
figure, plot([1:length(sim_data)],sim_data, 'k','linewidth',2, 'MarkerSize',10);
title('simulated MA(1) process','FontSize',18);
xlabel('time','FontSize',18); ylabel('data value','FontSize',18);
xlim([1, 365]); ylim([0,100]);
[acf2, lags2] = autocorr(sim_data);
figure, stem(lags2,acf2,'k','linewidth',2,'MarkerSize',10);
xlabel('lag','FontSize',18); ylabel('auto-correlation', 'FontSize',18);
```

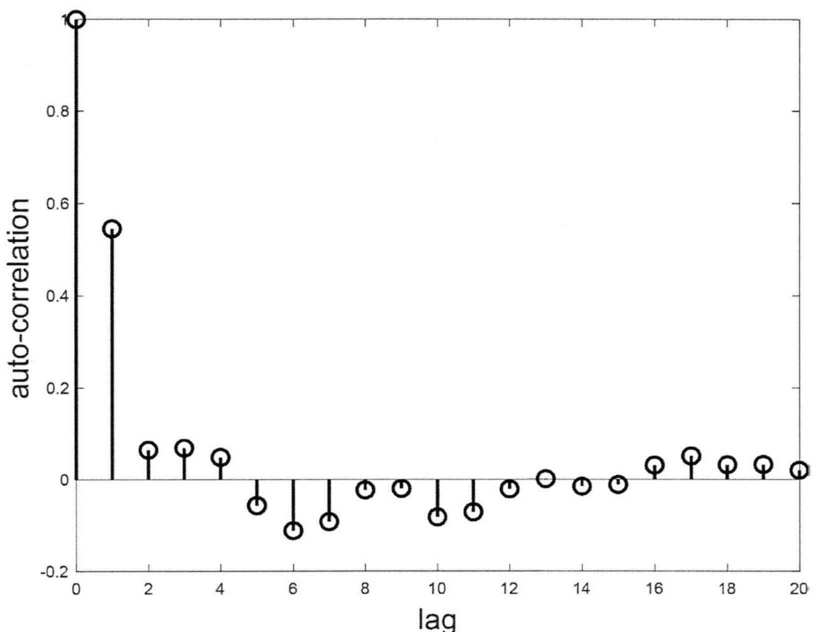

Figure 6.14 Auto-correlation profile of the simulated $MA(1)$ system shown in Figure 6.13. The profile demonstrates that for the $MA(1)$ system, the higher order ACFs $\gamma_{jt} \approx 0,\ \forall j \geq 2$.

MA(q) model   Consider a stochastic time series model $Y_t = \mu + \varepsilon_t + \theta_1\varepsilon_{t-1} + \dots + \theta_q\varepsilon_{t-q}$, where ε_t is Gaussian white noise and $\theta_1, \theta_2, \dots, \theta_q$ are real constants. For this model, $E(Y_t) = \mu$ and $\gamma_0 = \left(1 + \theta_1^2 + \theta_2^2 + \dots + \theta_q^2\right)\sigma^2$. Further, we may easily check that $\gamma_j = 0,\ \forall j > q$. The $MA(q)$ model is stationary and ergodic.

Example 6.6   *Moving average model of second order*

Consider the $MA(2)$ model $Y_t = \mu + \varepsilon_t + \theta_1\varepsilon_{t-1} + \theta_2\varepsilon_{t-2}$ with $\theta_1 = \frac{2}{5}$, $\theta_2 = -\frac{1}{5}$. Calculate all the ACFs. Comment on the stationarity and ergodicity of the model.

Solution:

$$
\begin{aligned}
\gamma_{jt} &= E(Y_t - \mu)(Y_{t-j} - \mu)\\
&= E\left(\varepsilon_t + \theta_1\varepsilon_{t-1} + \theta_2\varepsilon_{t-2}\right)\left(\varepsilon_{t-j} + \theta_1\varepsilon_{t-j-1} + \theta_2\varepsilon_{t-j-2}\right)\\
&= E\left(\varepsilon_t\varepsilon_{t-j}\right) + \theta_1 E\left(\varepsilon_t\varepsilon_{t-j-1}\right) + \theta_2 E\left(\varepsilon_t\varepsilon_{t-j-2}\right) + \theta_1 E\left(\varepsilon_{t-1}\varepsilon_{t-j}\right) + \theta_1^2 E\left(\varepsilon_{t-1}\varepsilon_{t-j-1}\right)\\
&\quad + \theta_1\theta_2 E\left(\varepsilon_{t-1}\varepsilon_{t-j-2}\right) + \theta_2 E\left(\varepsilon_{t-2}\varepsilon_{t-j}\right) + \theta_2\theta_1 E\left(\varepsilon_{t-2}\varepsilon_{t-j-1}\right) + \theta_2^2 E\left(\varepsilon_{t-2}\varepsilon_{t-j-2}\right).
\end{aligned}
$$

$$\tag{6.22}$$

This allows us to deduce the following results:

$$\gamma_0 = \left(1 + \theta_1^2 + \theta_2^2\right)\sigma^2. \tag{6.23}$$

Since $E\left(\varepsilon_t^2\right) = E\left(\varepsilon_{t-1}^2\right) = \sigma^2$ and so on

$$\gamma_1 = \theta_1\left(1 + \theta_2\right)\sigma^2. \tag{6.24}$$

$$\gamma_2 = \theta_2\sigma^2. \tag{6.25}$$

$$\gamma_j = 0,\ \forall j > 2. \tag{6.26}$$

Normalization by γ_0 gives us the ACFs: $\rho_0 = 1$, $\rho_1 = \dfrac{\gamma_1}{\gamma_0} = \dfrac{4}{15}$, $\rho_2 = \dfrac{\gamma_2}{\gamma_0} = -\dfrac{1}{6}$. Further, the $MA(2)$ model is also stationary and ergodic for reasons already mentioned earlier in this section.

MA(∞) model   This model is defined as $Y_t = \mu + \sum_{j \geq 0} \psi_j \varepsilon_{t-j}$, where μ, ψ_j are constants.[9] An $MA(\infty)$ model is stationary if $\sum_{j \geq 0} |\psi_j| < \infty$ because the absolute summability condition ensures that all statistical moments $\gamma_0, \gamma_1, \gamma_2, \dots$ are finite and constant in time. This may be easily checked by performing simple algebraic calculations similar to the calculations in the example above. The absolute summability of the coefficients ψ_j also guarantees that the model is ergodic.

[9] **Wold's decomposition theorem:** Any weakly stationary process can be represented as an $MA(\infty)$ process.

6.5.2 *Auto-regressive model*

Unlike the $MA(1)$ process, where the stochastic variable Y_t is correlated with the variable Y_{t-1} but not with any other variable $Y_{t-j,\, j \geq 2}$ in the past, certain processes may be correlated up to past j time instances from Y_t to Y_{t-j}, although the correlation may be diminishing in magnitude with an increasing span between the time dates and may become infinitesimally negligible for a large separation of time dates ($j \to \infty$). In this section, we will discuss models that capture such phenomena.

AR(1) model   Consider a stochastic time series model $\boxed{Y_t := c + \phi Y_{t-1} + \varepsilon_t}$, where c, ϕ are constants and ε_t is Gaussian white noise. We will deduce a condition on ϕ which will ensure that this model is covariance stationary and ergodic. In order to do so, let us begin with the $MA(\infty)$ process, where the auto-correlation between two instances is non-trivial even for a large time separation (perhaps negligibly small for a very large time separation).

Recall that the $MA(\infty)$ process is defined as $Y_t = \mu + \sum\limits_{j=0,1,2,\dots}^{\infty} \psi_j \varepsilon_{t-j}$, which converges if the power series $\sum\limits_{j=0,1,2,\dots}^{\infty} \psi_j \varepsilon_{t-j}$ converges absolutely, i.e., if $\sum\limits_{j=0,1,2,\dots}^{\infty} |\psi_j| < \infty$. This process is non-trivially auto-correlated up to an infinite order as desired.[10] Now, since μ is a finite constant, we will rewrite μ as $\mu = \dfrac{c}{1-\phi}$ for some constants c and ϕ. This can be written as an infinite geometric series $\sum\limits_{j=0}^{\infty} c\phi^j$ if $|\phi| < 1$. Now let us define $\psi_j := \phi^j$ and a stochastic variable $W_t := (c + \varepsilon_t)$. If we use these definitions of ψ_j and W_t in the definition of the $MA(\infty)$ process, then we will arrive at the following reformulation of the stochastic time series $Y_t = \mu + \varepsilon_t + \phi \varepsilon_{t-1} + \phi^2 \varepsilon_{t-2} + \dots..$ This can be further simplified as follows using a certain lag operator $\mathbb{L}$.

[10] In fact, the ACF tends to zero as the lag tends to infinity. So the gist of the point here is that the process has a long-range correlation.

$$Y_t = \frac{c}{1-\phi} + \varepsilon_t + \phi \varepsilon_{t-1} + \phi^2 \varepsilon_{t-2} + \cdots$$

$$= c\left(1 + \phi + \phi^2 + \cdots\right) + \varepsilon_t + \phi \varepsilon_{t-1} + \phi^2 \varepsilon_{t-2} + \cdots$$

$$= \left(c + \varepsilon_t\right) + \phi\left(c + \varepsilon_{t-1}\right) + \phi^2\left(c + \varepsilon_{t-2}\right) + \cdots$$

$$= W_t + \phi W_{t-1} + \phi^2 W_{t-2} + \cdots$$

$$= W_t + \phi \mathbb{L} W_t + \phi^2 \mathbb{L}^2 W_t \cdots$$

$\mathbb{L}\, y_t := y_{t-1}$ defines the lag operator

$$= \left(1 + \phi \mathbb{L} + \phi^2 \mathbb{L}^2 + \cdots\right) W_t$$

$$= \left(1 - \phi \mathbb{L}\right)^{-1} W_t.$$

Maclaurin series of $\left(1 - \phi \mathbb{L}\right)^{-1}$

Thereafter, multiplying the operator $\left(1 - \phi \mathbb{L}\right)^{11}$ to both sides of the above equation results in $Y_t = c + \phi Y_{t-1} + \varepsilon_t$, $|\phi| < 1$, which is the $AR(1)$ model. This completes the derivation of the $AR(1)$ model from the $MA(\infty)$ model. Let us now compute the statistical moments of Y_t for the $AR(1)$ model.

[11] We have omitted certain technical details about the asymptotic behavior of the lag operator $\mathbb{L}$ as this would be beyond the scope of this text.

$$E\left(Y_t\right) \quad = \quad \frac{c}{1-\phi}, \text{ because } Y_t = \mu + \varepsilon_t + \phi \varepsilon_{t-1} + \phi^2 \varepsilon_{t-2} + \cdots \qquad (6.27)$$

$$\gamma_0 \quad = \quad E\left(Y_t - \mu\right)^2$$

$$= \quad E\left(\varepsilon_t + \phi \varepsilon_{t-1} + \phi^2 \varepsilon_{t-2} + \cdots\right)^2$$

$$= \quad \left(1 + \phi^2 + \phi^4 + \phi^6 + \cdots\right)\sigma^2$$

$$= \quad \frac{\sigma^2}{1-\phi^2}. \qquad (6.28)$$

$$\gamma_j \quad = \quad \frac{\phi^j}{1-\phi^2}\sigma^2. \qquad (6.29)$$

Therefore, the ACF for the $AR(1)$ model is $\rho_j := \frac{\gamma_j}{\gamma_0} = \phi^j \to 0$ as $j \to \infty$ because $|\phi| < 1$. A simulation of an $AR(1)$ process is demonstrated in Figure 6.15 using the following Matlab routine.

```
arima('Constant', 0, 'ARLags',[1],'AR',{0.5},'Variance',1);
```

AR(p) model An $AR(p)$ model of order p is defined as $Y_t := c + \phi_1 Y_{t-1} + \phi_2 Y_{t-2} + \cdots + \phi_p Y_{t-p} + \varepsilon_t$, where c and the ϕs are model constants. The statistical moments of the $AR(p)$ model are listed as follows.

$$\mu \quad = \quad \frac{c}{1 - \phi_1 - \phi_2 - \cdots - \phi_p}, \qquad (6.30)$$

$$\gamma_0 \quad = \quad \phi_1 \gamma_1 + \phi_2 \gamma_2 + \cdots + \phi_p \gamma_p + \sigma^2, \qquad (6.31)$$

$$\gamma_j \quad = \quad \sum_{m=1}^{p} \phi_m \gamma_{j-m}, \quad \forall j = 1, 2, \ldots. \qquad (6.32)$$

Further note that $\gamma_j = \gamma_{-j}$ holds true for all j. Dividing equations (6.31) and (6.32) by γ_0 results in the *Yule–Walker equations* which we shall use in the chapter project to estimate the coefficients of the auto-regressive model. For completion, we simply write this system of equations here as follows.

$$\rho_j = \sum_{m=1}^{p} \phi_m \rho_{j-m}, \quad \forall j = 1, 2, \ldots. \tag{6.33}$$

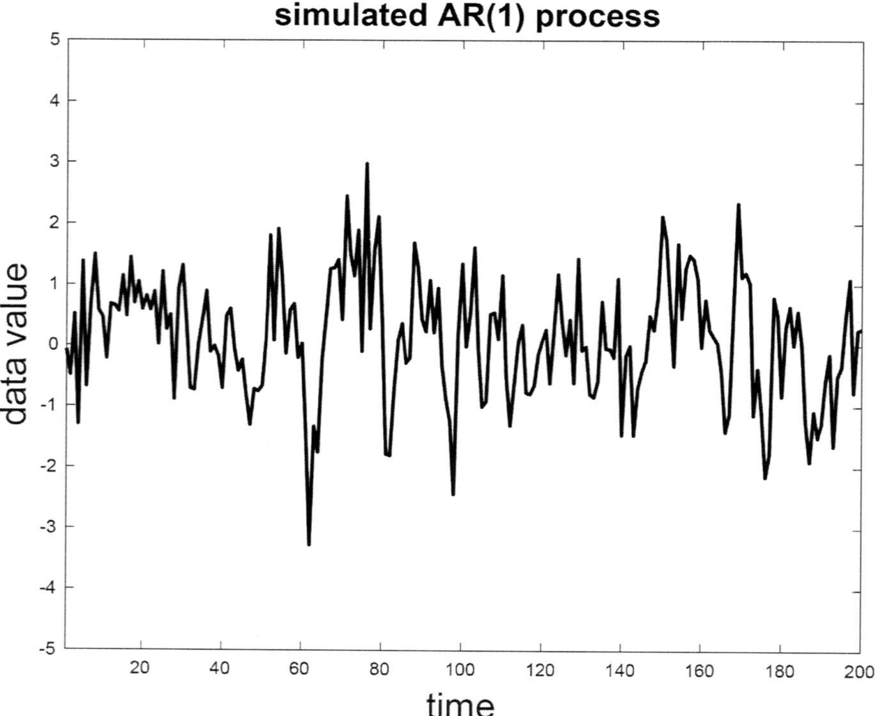

Figure 6.15 Simulation of an $AR(1)$ stochastic process $Y_t = 0.5 Y_{t-1} + \varepsilon_t$, where $\varepsilon_t \sim N(0, 1)$.

The stationarity of the $AR(p)$ model is determined by the roots of the characteristic polynomial of the $AR(p)$ model which is given here as follows.

$$\Phi(\mathbb{L}) := (1 - \alpha_1 \mathbb{L})(1 - \alpha_2 \mathbb{L}) \cdots (1 - \alpha_p \mathbb{L}), \tag{6.34}$$

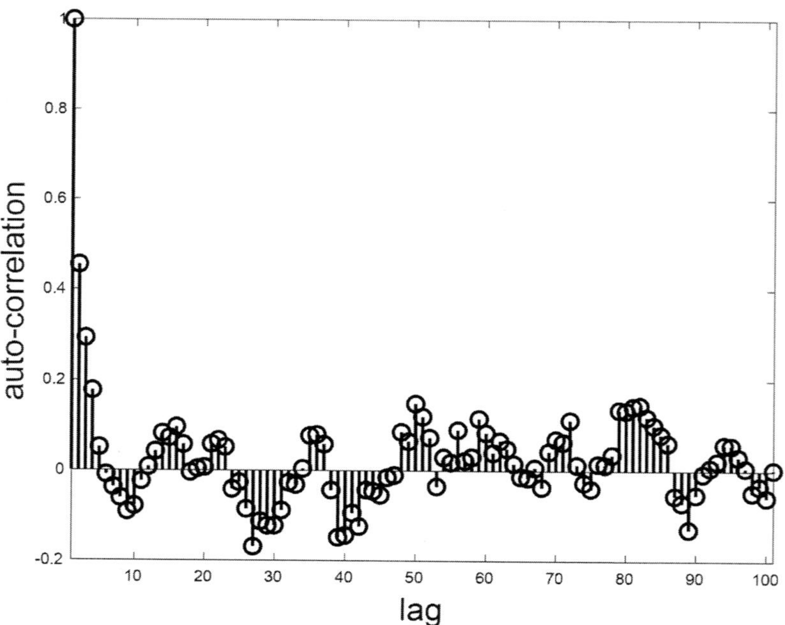

Figure 6.16 Autocorrelation profile of the simulated $AR(1)$ system shown in Figure 6.15 demonstrating that $\rho_j \to 0$ as $j \to \infty$.

where the α_is are the *reciprocal* roots of the polynomial $\Phi(L)$, i.e., if the roots of $\Phi(L)=0$ are $\beta_1,\beta_2,...,\beta_p$, then $\alpha_i=\dfrac{1}{\beta_i}$ for all $i=1,2,...,p$. The stochastic $AR(p)$ model is covariance stationary if the roots $|\beta_i|>1$, $\forall i=1,2,...,p$, or equivalently, if the reciprocal roots $|\alpha_i|<1$, $\forall i=1,2,...,p$. We illustrate this technique with the help of an example below.

Example 6.7   Stationarity of an AR(2) process

Consider the $AR(2)$ process $Y_t=0.5Y_{t-1}+0.25Y_{t-2}+\varepsilon_t$, where ε_t is Gaussian white noise. Find whether this $AR(2)$ process is covariance stationary or not.

Solution: The characteristic polynomial is $\Phi(\mathbb{L})=\left(1-0.5\mathbb{L}-0.25\mathbb{L}^2\right)$ whose roots are $\beta_1=\dfrac{-\phi_1+\sqrt{\phi_1^2+4\phi_2}}{2\phi_2}=1.2361$ and $\beta_2=\dfrac{-\phi_1-\sqrt{\phi_1^2+4\phi_2}}{2\phi_2}=-3.2361$. Since $|\beta_1|,|\beta_2|>1$, the $AR(2)$ process is covariance stationary.

In fact, it can be checked by performing similar calculations that an $AR(2)$ process is covariance stationary if $|\phi_1|<2$ and $|\phi_2|<1$.

6.5.3   Auto-regressive moving average model – ARMA(p, q)

We may combine the features of the auto-regressive and moving average models to construct a hybrid model as follows.

$$Y_t = c+\sum_{m=1}^{p}\phi_m Y_{t-m}+\varepsilon_t+\sum_{k=1}^{q}\theta_k\varepsilon_{t-k},\tag{6.35}$$

where c, the ϕs, and the θs are model parameters and ε_t is Gaussian white noise.

The first statistical moment $\mu=\dfrac{c}{1-\phi_1-\phi_2-\cdots-\phi_p}$. The higher order moments, the ACFs, are more complicated to derive and will be omitted from discussion here. The interested reader is referred to the textbook on time series analysis by James D. Hamilton mentioned in the bibliography section of this chapter. Further, the stationarity of the $ARMA(p, q)$ model depends entirely on the auto-regressive part of the model, i.e., on ϕ_js, and not on the moving average parameters.

6.5.4   Estimating the coefficients of an auto-regressive model: The Yule–Walker equations

As illustrated in the above sections, auto-regressive models are used for time series forecasting when the underlying process is dependent (linearly) on values from previous time instants but may be offset by random shocks. Consequently, the formal structure of such a process takes the form

$$X_{t+1}=\phi_1^{(p)}X_t+\phi_2^{(p)}X_{t-1}+...+\phi_p^{(p)}X_{t-p+1}+\varepsilon_{t+1},\tag{6.36}$$

[12] A p-dimensional process constitutes the following p random variables stacked here as a vector: $\begin{pmatrix} X_t \\ \vdots \\ X_{t-p+1} \end{pmatrix}$. These p random variables determine the outcome of the random variable X at the $(t+1)^{th}$ time stamp in a "linear" fashion accurate up to some random perturbative noise (ε).

where the $\phi_s^{(p)}$ are the AR coefficients of a p-dimensional process[12] and ε_t is Gaussian white noise that prescribes the effects of random shocks. Here, the effect of the linear dependence on previous states is significant up to lag p and truncated thereafter. We will consider a sample time series data that lists the total number of jobs filled in a public sector enterprise every month over many years. Our goal is to fit an auto-regressive (AR) model to this time series data of length N. This will essentially require us to estimate the order (lag) of the AR process by using the *Yule–Walker* equations.

We begin by multiplying equation (6.36) by X_t and, subsequently, compute the expected value, $\langle \cdot \rangle$, of the terms in the resulting equation to obtain the following:

$$\langle X_t X_{t+1} \rangle = \sum_{j=1}^{p} \phi_j^{(p)} \langle X_t X_{t-j+1} \rangle + \underbrace{\langle X_t \varepsilon_{t+1} \rangle}_{0}. \tag{6.37}$$

> ### *Change of notation:* $r_k \leftrightarrow \rho_k$ *and* $c_k \leftrightarrow \gamma_k$
>
> In this section, we will make a slight change in notation here for denoting the autocovariance and ACFs. This modification is desirable so that the notation matches seamlessly with the programming variables used in the Matlab routine and the pseudocode discussed below. Therefore, we will use r_k, instead of ρ_k, to represent the ACF. Likewise, we will use c_k, instead of γ_k, to represent the autocovariance function. This way we can also use an uppercase letter R to denote the matrix of r_ks as shown below and use the same letter to refer to either the matrix entries or the matrix as a whole. This change of notation applies to this section of the chapter and we will revert to the original notation elsewhere in the textbook.

Reconsidering equation (6.37), the last term above cancels to zero because the process X_t is independent of and uncorrelated with the random shocks ε_t. Upon division by $(N-1)$, the equation reduces to $c_1 = \sum_{j=1}^{p} \phi_j^{(p)} c_{j-1}$, which is the first auto-covariance function. Normalization by c_0 gives the first ACF $r_1 = \sum_{j=1}^{p} \phi_j^{(p)} r_{j-1}$. Likewise, to compute the second ACF, we multiply the terms in equation (6.36) by X_{t-1} and follow the same steps as above to obtain $r_2 = \sum_{j=1}^{p} \phi_j^{(p)} r_{j-2}$. Similarly, the kth ACF is $r_k = \sum_{j=1}^{p} \phi_j^{(p)} r_{j-k}$. Putting all this together, we have the *Yule–Walker* equations:

$$\begin{aligned}
r_1 &= \phi_1^{(p)} r_0 + \phi_2^{(p)} r_1 + \phi_3^{(p)} r_2 + \cdots + \phi_p^{(p)} r_{p-1}, \\
r_2 &= \phi_1^{(p)} r_1 + \phi_2^{(p)} r_0 + \phi_3^{(p)} r_1 + \cdots + \phi_p^{(p)} r_{p-2}, \\
&\quad\vdots \\
&\quad\vdots \\
r_{p-1} &= \phi_1^{(p)} r_{p-2} + \phi_2^{(p)} r_{p-3} + \phi_3^{(p)} r_{p-4} + \cdots + \phi_p^{(p)} r_1, \\
r_p &= \phi_1^{(p)} r_{p-1} + \phi_2^{(p)} r_{p-2} + \phi_3^{(p)} r_{p-3} + \cdots + \phi_p^{(p)} r_0.
\end{aligned}$$

The above set of equations can be written succinctly in matrix form as follows.

$$\Phi = \mathbf{R}^{-1}\mathbf{r},$$

(6.38)

where $\Phi^{(p)} = \begin{pmatrix} \phi_1^{(p)} \\ \phi_2^{(p)} \\ \cdot \\ \cdot \\ \phi_{p-1}^{(p)} \\ \phi_p^{(p)} \end{pmatrix}$, $R = \begin{pmatrix} r_0 & r_1 & r_2 & \cdots & r_{p-2} & r_{p-1} \\ r_1 & r_0 & r_1 & \cdots & r_{p-3} & r_{p-2} \\ & & & \cdot & & \\ & & & \cdot & & \\ r_{p-2} & r_{p-3} & r_{p-4} & \cdots & r_0 & r_1 \\ r_{p-1} & r_{p-2} & r_{p-3} & \cdots & r_1 & r_0 \end{pmatrix}$ and $\mathbf{r} = \begin{pmatrix} r_1 \\ r_2 \\ \cdot \\ \cdot \\ r_{p-1} \\ r_p \end{pmatrix}$.

The matrix $\mathbf{R}$ is always invertible because it is full rank and symmetric and hence the Yule–Walker equations are well posed and can be used to solve for the unknown AR coefficients (Φ). Here the ACFs are given by $\mathbf{r}$. It must be noted that $r_0 := 1$ and $c_{-k} = c_k$.

6.5.5 *Dependency structure of the correlated random variables: Partial auto-correlation function (PACF) and estimating the order of a suitable time series model*

The correlation between X_t and X_{t+h} comprises both *direct* and *indirect* dependencies. The indirect dependency between X_t and X_{t+h} arises due to the linear dependency between X_t and X_{t+1}, X_{t+1} and X_{t+2}, X_{t+2} and X_{t+3}, and so on, all the way through X_{t+h-1} and X_{t+h}. The PACFs prescribe the direct dependency between X_t and X_{t+h} with the effects of all the intermediary variables, X_{t+1} through X_{t+h-1}, removed. The PACFs $\Psi^{(p)}$ are negative of the last coefficients $\phi_k^{(k)}$s computed using the Yule–Walker equations for every lag (order) k starting with 1 through p, i.e., $\Psi^{(p)} = \begin{pmatrix} \Psi_1^{(p)} \\ \Psi_2^{(p)} \\ \cdot \\ \cdot \\ \Psi_{p-1}^{(p)} \\ \Psi_p^{(p)} \end{pmatrix}$

$= \begin{pmatrix} -\phi_1^{(1)} \\ -\phi_2^{(2)} \\ \cdot \\ \cdot \\ -\phi_{p-1}^{(p-1)} \\ -\phi_p^{(p)} \end{pmatrix}$. Consequently, the following table prescribes a strategy to estimate the

order of various time series models using the ACFs and PACFs. Specifically, for the AR model, the rapid (abrupt) decay of the PACFs prescribes the order of the model.

Model	ACF (r_h)	PACF (ψ_h)
AR($\tilde{p}$)	decays infinitely as $r_h \overset{h\to\infty}{\to} 0$	truncates abruptly as $\psi_h = 0, \;\; \forall h > \tilde{p}$
MA($\tilde{q}$)	truncates abruptly as $r_h = 0, \;\; \forall h > \tilde{q}$	decays infinitely as $\psi_h \overset{h\to\infty}{\to} 0$
ARMA($\tilde{p},\tilde{q}$)	decays as AR($\tilde{p}$), $\forall h > \tilde{q}$	decays as MA($\tilde{q}$), $\forall h > \tilde{p}$

6.6   Chapter project: Indicators of a federal government's policy effectiveness index

6.6.1   *Interlude: Model construction (AR(p)) and forecasting trends in job creation*

The algorithm to construct the AR(p) model constitutes the following steps.

1. *Importing data and dimensions of the Yule–Walker system*: Use the Matlab function `readtable` to import the tabular data from the spreadsheet JOBS.csv. The table has two columns, namely, *Month* and *TotalFilledJobs*. Plot the entries of the second column against the entries of the first column (or simply in sequence) in *blue* and then plot, in the same figure, the mean subtracted entries of the second column. Visually inspect the plotted mean subtracted time series data and check if it is periodic with period ($p + 1$). If so, use a system of p equations to construct the Yule–Walker system (prescribed by equation (6.38)), else use some arbitrarily large enough number (say $N/4$, where N is the number of data points) to construct the Yule–Walker system.

2. *Computing AR coefficients using the Yule–Walker equations*: Here we will use two strategies:

- First, write your own Matlab function, `myArYule`, to compute Φ as prescribed by equation (6.38). The input to your function should be the mean subtracted time series data and the dimension of the Yule–Walker system selected as above. The output of the function should be the AR coefficients, Φ, and the PACFs, Ψ.

- Second, use the Matlab function `aryule`, to compute the AR coefficients and the PACFs. Compare the results of both these strategies for consistency.

3. *Select the order of the AR model using PACFs*: Practically, the PACF is considered zero (at a 5% significance level) if the value of ψ_h falls within the critical region defined by the upper and lower limits given by $\pm 1.96/\sqrt{N}$.[13] In your software routine, plot the PACFs in conjunction with the critical region set at 5% significance level. Finally, choose the order $\tilde{p}$ of the AR model in such a way that $\psi_h \neq 0$ for $h = \tilde{p}$ but $\psi_h = 0, \; \forall h > \tilde{p}$.

[13] This approximation relies on the assumption that $N > 30$ and that the underlying process has a finite second moment (finite variance).

Pseudocode of the myArYule algorithm

```
INPUT: data, max_order.
 for current_order from max_order to 1
    initialize R to identity matrix
    compute ACFs r from data
    for i from 1 to current_order
        k=1;
        for j from i+1 to current_order
            R(i,j) = r(k)
            k = k + 1
        end
    end
compute remaining entries of R by using symmetry of R
compute  Φ_temp = R⁻¹ * r
update  Φ_temp by concatenating 1 and Φ_temp
if current_order == max_order
    Φ = Φ_temp
else
    -- Are we having fun ☺ or what? ☹ --
end
pacf(current_order) = negative of last entry of Φ_temp
end
OUTPUT: Φ, pacf.
```

Some useful Matlab functions you may want to consider are: `xcorr`, `eye`, `inv`, `aryule`, `readtable`, `plot`, `subplot`, `xlabel`, `ylabel`, `length`, `stem`, `sqrt`, `title`, `grid`, `xlim`, `hold on`. Use the command `doc <function_name>` to study more about them.

Questions

Answer the following questions:

1. Implement the algorithm of section 6.6.1 outlined above in Matlab.

2. Estimate the order $\tilde{p}$ of the AR model for the given time series data in JOBS.csv.

3. Based on the AR coefficients computed by your Matlab routine, comment whether the AR($\tilde{p}$) model has a stable solution. Explain your answer in detail.

4. Use this model to forecast the additional jobs filled each month in the year 2012 for months March through December.

6.7   Causation from time series data

Consider two covariance-stationary time series data $\{X_t\}_{t\geq0}$ and $\{Y_t\}_{t\geq0}$ representing two different stochastic events, e.g., the annual rate of foreign direct investment (FDI) and the rate of increase of GDP on a yearly basis. Often, both these data may be correlated. This itself may not reveal any causal dependence because correlation

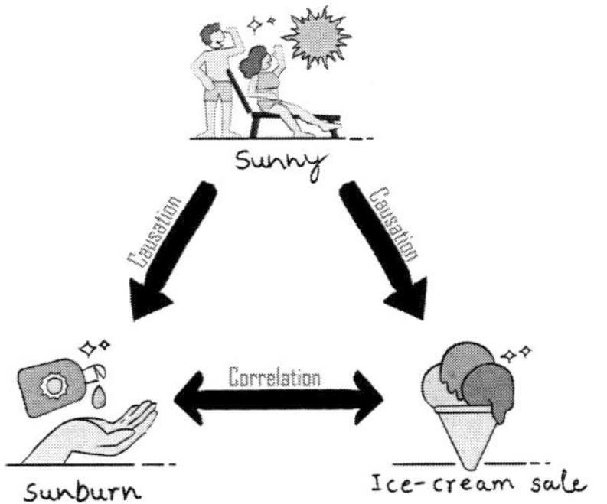

Figure 6.17 Correlation does *not* imply causation!

between the data X_t and Y_t may not automatically imply causation. A natural point of interest may be to investigate if there is any causal dependency between the two data. In this section, we will study a classical technique using *Granger causality* to establish any causal relationship between two different time series signal.

6.7.1   *Granger causality*

Let us first begin with the definition of Granger causality (GC). Consider that we have two time series signal $X_t = \{X_t, X_{t-1}, X_{t-2}, ..., X_{t-k}, ...\}$ and $Y_t = \{Y_t, Y_{t-1}, Y_{t-2}, ..., Y_{t-k}, ...\}$. In order to forecast the value of Y at time $(t + 1)$ using all historical information Y_t, we may incur an error whose variance is $\sigma^2_{Y_{t+1}|Y_t}$. This is typical of time series forecasting methods like auto-regressive models as seen earlier in this chapter. The GC may then be defined as follows.

1. *Granger causality*: X_t is Granger causal to Y_t, i.e., $X_t \to Y_t$, if and only if the application of an optimal linear predictor results in

$$\sigma^2_{Y_{t+1}|\{Y_t, X_t\}} < \sigma^2_{Y_{t+1}|Y_t}.$$

2. *Instantaneous Granger causality*: X_t is instantaneously Granger causal to Y_t, i.e., $X_t \Rightarrow Y_t$[14] if and only if the application of an optimal linear predictor results in

$$\sigma^2_{Y_{t+1}|\{Y_t, X_t, X_{t+1}\}} < \sigma^2_{Y_{t+1}|\{Y_t, X_t\}}.$$

3. *Feedback*: There exists a *feedback* between X_t and Y_t $(X_t \leftrightarrow Y_t)$ if $X_t \to Y_t$ and $Y_t \to X_t$.

6.7.2   *Granger causality test*

The Granger causality test (GCT)[15] is based on the linear auto-regressive model. If GC $X_t \to Y_t$ holds, we say X **Granger causes** Y as opposed to X causes Y. The distinction entails that X might be causing Y as opposed to X is surely causing Y.

GCT for a simple ARDL model:[16] At the outset, we will use a simple ARDL model to elucidate the concept of GC. Consider the following ARDL model where both X and Y are covariance stationary:

$$Y_t = \mu + \phi_1 Y_{t-1} + \beta_1 X_{t-1} + \varepsilon_t, \tag{6.39}$$

where ε_t is Gaussian white noise and other parameters are as defined earlier in this chapter. Unless $\beta_1 = 0$, the past value of X, i.e., the value of X_{t-1} has explanatory power for the present value of Y, i.e., the value of Y_t. Several standard techniques can be employed to estimate the lag weights of an ARDL model. We may then use the framework of a hypothesis test to reject (or fail to reject) the null hypothesis $H_0 : \beta_1 = 0$[17] against the alternate hypothesis $H_1 : \beta_1 \neq 0$. If H_1 holds true, i.e., the estimator $\widehat{\beta}_1$ is statistically significant (say $p < 0.05$), then X Granger causes Y.

---

[14] Note that $X_t \Rightarrow Y_t$ if and only if $Y_t \Rightarrow X_t$.

[15] C. W. J. Granger, Investigating Causal Relations by Econometric Models and Cross-spectral Methods, *Econometrica* 37 no. 3 (1969): 424–438.

[16] ARDL refers to auto-regressive distributed lag.

[17] The null hypothesis, in this case, is that X does *not* Granger cause Y.

The relevant test statistic is $T_s = \sqrt{\dfrac{(T-2)S_{XX}}{SS_R}}\left|\hat{\beta}_1\right|$,[18] where SS_R is the *sum of squares residual*, $SS_{XX} = \sum_{t=1}^{T}\left(X_t - \overline{X}\right)^2$, and $\hat{\beta}_1$ is the ordinary least squares estimate of section 6.4.2. An α significance level test is as follows.

$$\text{reject} \quad H_0 \quad \text{if} \quad T_s > t_{\frac{\alpha}{2},(T-2)},$$
$$\text{fail to reject} \quad H_0 \quad \text{otherwise.} \tag{6.40}$$

Alternatively (and equivalently), if $p = P\left(\left|\mathfrak{T}_{T-2}\right| > T_s\right) < 0.05$,[19] then reject H_0 in favor of H_1 and thereby establish that **X** does Granger cause **Y**. This is the GCT for the simple ARDL model.

GCT for an ARDL(p, q) model: Now consider the model with a linear dependence between p-lagged dependent process $\{Y_{t-m}\}$, $m = 1, 2, \ldots, p$ and q-lagged independent process $\{X_{t-m}\}$, $m = 1, 2, \ldots, q$. Let us say we have a total of $t = 1$ through T sample points.

$$Y_t = \mu + \phi_1 Y_{t-1} + \phi_2 Y_{t-2} + \cdots + \phi_p Y_{t-p} + \beta_1 X_{t-1} + \beta_2 X_{t-2} + \cdots \beta_q X_{t-q} + \varepsilon_t, \tag{6.41}$$

where ε_t is the Gaussian white noise. The above model defined by equation (6.41) is known as the *unrestricted model (UR)*. We say that **X** does *not* Granger cause **Y** if each of the β_i, $\forall i = 1, 2, \ldots, q$, is zero (null hypothesis H_0). This is obvious from visual inspection of the terms in equation (6.41). On the other hand, if any of the estimated values $\hat{\beta}_1$ are statistically significant (non-zero), then the alternative hypothesis H_1 is true and **X** Granger causes **Y**. Generally, an ordinary least squares regression analysis will suffice to estimate the *restricted model (R)* under the null hypothesis whence we have

$$Y_t = \mu + \phi_1 Y_{t-1} + \phi_2 Y_{t-2} + \cdots + \phi_p Y_{t-p} + \varepsilon_t. \tag{6.42}$$

If the models (6.41) and (6.42) are the same, i.e., $SSR_{UR} \approx SSR_R$, then the null hypothesis is not rejected. This test is conducted by considering a suitable F statistic akin to a regression analysis by comparing the calculated value $F_{cal} = \dfrac{\frac{SSR_R - SSR_{UR}}{q}}{\frac{SSR_{UR}}{T-q-p-2}}$ and the critical tabulated value $F_{tab}(q, T - q - p - 2)$. If $F_{cal} > F_{tab}$, then we reject H_0 in favor of H_1 and, consequently, infer that **X** Granger causes **Y**.

We will revisit the GCT for validating causal relations between two data sets relevant to measuring a nation's economic growth in this chapter's project epilogue. Before we move to the concluding section of the chapter project, we will discuss a few more important points regarding GC.

Bi-directional Granger causality test using the VAR model: In many situations, it may not be very clear which of the two (multiple) variables **X** and **Y** is independent and which one is dependent variable. Hence, it may become necessary to perform

---

[18] For detailed explanation of this choice of test statistic T, you may refer to p. 378 of the textbook by Sheldon M. Ross listed in this chapter's bibliography section.

[19] Here $\mathfrak{T}_{T-2}$ is a t-random variable with $T - 2$ degrees of freedom.

a bi-directional GCT. We will swap the roles of **X** and **Y** with respect to their role as dependent or independent variables and write two equations of the same form.

$$Y_t = \mu_1 + \phi_{11}Y_{t-1} + \cdots + \phi_{1p}Y_{t-p} + \beta_{11}X_{t-1} + \cdots + \beta_{1q}X_{t-q} + \varepsilon_{1t} \tag{6.43}$$

$$X_t = \mu_2 + \phi_{21}Y_{t-1} + \cdots + \phi_{2p}Y_{t-p} + \beta_{21}X_{t-1} + \cdots + \beta_{2q}X_{t-q} + \varepsilon_{2t} \tag{6.44}$$

The above equations (6.43) and (6.44) together constitute the vector auto-regressive (VAR) model of order p. A VAR (p) model is stationary if all the roots of the equation $det\left(\mathbb{I} - \Theta_1 - \Theta_2 - \cdots - \Theta_p\right) = 0$ are *outside* a unit complex circle.[20] It can be subjected to GCT to validate if $\mathbf{X} \to \mathbf{Y}$ and/or $\mathbf{Y} \to \mathbf{X}$. For simplicity, let us consider the VAR (1) model for a 2-dimensional vector $\vec{Y}_t = \begin{pmatrix} Y_t \\ X_t \end{pmatrix}$ that can be expressed succinctly as follows.

$$\vec{Y}_t = \vec{\mu} + \Theta_1 \vec{Y}_{t-1} + \vec{\varepsilon}_t, \tag{6.45}$$

where $\vec{\mu} = \begin{pmatrix} \mu_1 \\ \mu_2 \end{pmatrix}$, $\Theta_1 = \begin{pmatrix} \phi_{11} & \beta_{11} \\ \phi_{21} & \beta_{21} \end{pmatrix}$, and $\vec{\varepsilon}_t = \begin{pmatrix} \varepsilon_{1t} \\ \varepsilon_{2t} \end{pmatrix}$.

The GCT for a 2-dimensional VAR (1) model is summarized below as an illustrative example. The null hypotheses are

$$H_0 : \beta_{11} = 0 \qquad (\mathbf{X} \text{ does not Granger cause } \mathbf{Y})$$

$$H_0 : \phi_{21} = 0 \qquad (\mathbf{Y} \text{ does not Granger cause } \mathbf{X})$$

Each null hypothesis is tested using an F statistic. H_0 is rejected when the p-value is less than 0.05. For higher order VAR models, performing a pair-wise hypothesis test such as the one above may become tedious. In the following project section, we will learn to use Matlab and perform such tests for a higher order VAR model.

6.8   Chapter project: Indicators of a federal government's policy effectiveness index

6.8.1   *Epilogue: Granger causality test for establishing causal relationship between FDI rate and internet usage rate in a country*

In this concluding section of the chapter project we will delve a little more into other economic indicators that reflect the state of a government's policy initiatives and success. Specifically, we will analyze nearly 30 years' data of annual FDI rate (% of GDP) and internet usage rate (as % of population) in India starting from 1990 to 2018.[21] This data table is provided to the readers through this link: World_Bank_Data.csv. We will use the variables X and Y for representing FDI rate and internet usage rate, respectively.

[20] Θ_j is a matrix associated with the jth lag, and its entries are ϕ_{kj} and β_{kj} for the kth equation in the relevant VAR model.

Figure 6.18 Is there a causal relationship between FDI and internet infrastructure in a country?

[21] https://data.worldbank.org/country/india.

<u>Questions</u>

1. Compute the coefficient of regression r_{XY} for X and Y.

2. Perform a Granger causality test using Matlab `gctest` routine to check whether $X \rightarrow Y$ or $Y \rightarrow X$. Use a vector auto-regression model of order $p = 5$. Use level of significance of test $\alpha = 0.1$.

3. In the above test, if an F-test statistic is used, compute F_{cal}. Compare F_{cal} and F_{tab}.

4. Comment on the findings to questions 1 and 2 above.

6.9   Selected bibliography

Enders, Walter. *Applied Econometric Times Series* (4th edition). Wiley, 2014.

Hamilton, James D. *Time Series Analysis* (1st Indian edition). Princeton University Press, 2012.

Kirchgässner, Gebhard, Jürgen Wolters, and Uwe Hassler. *Introduction to Modern Time Series Analysis* (2nd edition). Springer, 2013.

Robinson, Derek J. S. *A Course in Linear Algebra with Applications* (2nd edition). World Scientific, 2006.

Ross, Sheldon M. *Introduction to Probability and Statistics for Engineers and Scientists* (6th edition). Academic Press, 2021.

Wooldridge, Jeffrey M. *Introductory Econometrics* (7th edition). Cengage Learning Inc., 2020.

6.10   Exercise problems

1. (***Ordinary least squares (OLS) estimator***) Consider a regression model $Y_t = \beta_0 + \beta_1 x_i + \beta_2\left(3x_i^2 - 2\right) + e_t$, $i = 1,2,3$, where $x_1 = -1, x_2 = 0, x_3 = +1$, and e_t is a normal distributed error term. Find the OLS estimates of β_0, β_1, and β_2. Show that the least squares estimates of β_0 and β_1 are unchanged if $\beta_2 = 0$. Why?

2. (***Unbiased OLS estimators***) Consider a multiple regression model $y = \mathbb{X}\vec{\beta} + \vec{\varepsilon}$,

    where $\mathbb{X} = \begin{pmatrix} x_{1,0} & x_{1,1} & x_{1,2} & \cdot & \cdot & x_{1,p-1} \\ x_{2,0} & x_{2,1} & x_{2,2} & \cdot & \cdot & x_{2,p-1} \\ \cdot & \cdot & \cdot & \cdot & \cdot & \cdot \\ \cdot & \cdot & \cdot & \cdot & \cdot & \cdot \\ x_{n,0} & x_{n,1} & x_{n,2} & \cdot & \cdot & x_{n,p-1} \end{pmatrix}$ is the regression coefficient matrix

    and y, $\vec{\beta}$, and $\vec{\varepsilon}$ are p-dimensional vectors. Here n is the total number of data points and p is the levels of inputs. We have seen that the OLS estimator is given by $\vec{\beta} = \left(\mathbb{X}^T\mathbb{X}\right)^{-1}\mathbb{X}^T y$. Further, $E(y) = \mathbb{X}\vec{\beta}$. Show that $\hat{\beta}$

is an unbiased estimator of $\vec{\beta}$, i.e., $E\left(\hat{\beta}\right)=\vec{\beta}$. Additionally, find the variance of $\vec{\beta}$.

3. (***Tension in a string***) The tension T observed in a non-extensible string required to maintain a body of unknown weight w in equilibrium on a smooth inclined plane of angle $\theta \in (0, \pi/2)$ is a random variable with mean $E(T) = w \sin \theta$. If for n observations of $\theta = \theta_i$, $i = 1, 2, ..., n$, the values of T are T_i, $i = 1, 2, ..., n$, find the OLS estimate of w.

4. (***Cigarette smoking is injurious to health – what does statistics say?***) Typical research conducted in 1971 across different countries regarding mortality (related to coronary heart disease [CHD]) and cigarette consumption is tabulated below.

Figure 6.19 Are we blowing out our hearts over packs of cigarettes?

Country	Cigarette consumption per adult per year (x)	CHD mortality per 100000 (y)
USA	3900	256.9
Canada	3350	211.6
Australia	3220	238.1
New Zealand	3220	211.8
UK	2790	194.1
Switzerland	2780	124.5
Ireland	2770	187.3
Iceland	2290	110.5
Finland	2160	233.1
West Germany	1890	150.3
Netherlands	1810	124.7
Greece	1800	41.2
Austria	1770	182.1
Belgium	1700	118.1
Mexico	1680	31.9
Italy	1510	114.3
Denmark	1500	144.9
France	1410	59.7
Sweden	1270	126.9
Spain	1200	43.9
Norway	1090	136.3

i. Construct a linear regression model $y = \hat{\beta}_1 x + \hat{\beta}_0$.

ii. Perform a test of hypothesis to validate if this data supports the thesis that smoking causes CHD related mortality. Specifically, test

$$H_0: \quad \beta_1 = 0$$
$$vs$$
$$H_1: \quad \beta_1 > 0 \tag{6.46}$$

at a level of significance $\alpha = 0.05$. Clearly specify the test statistic for this test.

5. (***Forecasting energy consumption***) The file EnergyConsumptionMP_1996-2018.csv contains biennial data for energy consumption in the state of Madhya Pradesh starting from 1996 through 2018. The file has two columns, the first lists the year and the second enumerates the total energy consumption in that year in MU (million units). Use the Matlab function `readtable` to import the data into the independent variable named `xdata` and the dependent variable `ydata`. Then use the Matlab function `plot` to map the energy consumption of the respective years in a scatter plot. Answer the following questions:

Figure 6.20 The impending energy crisis has become an existential challenge for humanity in this century.

   i.   Derive the matrix form of the equation for the least squares regression parameters.

   ii.   Is the regression matrix Λ always invertible? If so, why? If not, explain the conditions when it is not invertible.

   iii.   Implement the least-squares algorithm described in this chapter in Matlab. Compare the performance of your regression model with that of the Matlab function `lsqcurvefit`. Use initial guess for the regression parameters to be $1.0e + 08*[1, -1.5, 3]$. Repeat with an initial guess of $[1, 1, 1]$. Comment on your observations.

   iv.   Implement a linear regression model of the form $y = a + bx$ and plot your predictions on a separate figure.

   v.   Use this model to predict the total energy consumption in the state of Madhya Pradesh for the years 1999, 2007, and 2021.

6. (***Moving Average MA(2) process***) Is the following ***MA(2)*** process covariance stationary?

$$Y_t = \left(1 + 2.0\mathbb{L} + 0.8\mathbb{L}^2\right)\varepsilon_t. \tag{6.47}$$

Here ε_t is a Gaussian white noise process and $\mathbb{L}$ is the lag operator such that $\mathbb{L}^k Z_t = Z_{t-k}$ for $k = 1, 2, \ldots$ If Y_t is indeed weekly stationary, then compute its auto-covariances.

7. (***Auto-regressive AR(2) process***) Is the following ***AR(2)*** stochastic process covariance stationary?

$$\left(1 - 1.0\mathbb{L} + 0.25\mathbb{L}^2\right)Y_t = \varepsilon_t. \tag{6.48}$$

Here ε_t is a Gaussian white noise process and $\mathbb{L}$ is the usual lag operator. If Y_t is indeed weekly stationary, then compute its auto-covariances.

8. (***Auto-regressive AR(∞) process***) Consider an $AR(\infty)$ process

$$\left(1 + \phi_1\mathbb{L} + \phi_2\mathbb{L}^2 + \cdots\right)\left(Y_t - \mu\right) = \varepsilon_t \tag{6.49}$$

that is associated with an invertible $MA(q)$ process $\left(Y_t - \mu\right) = \left(1 + \theta_1\mathbb{L} + \theta_2\mathbb{L}^2 + \cdots + \theta_q\mathbb{L}^q\right)\varepsilon_t$. Find a closed form expression for ϕ_j

as a function of the roots of $\left(1+\theta_1 z+\theta_2 z^2+\cdots+\theta_q z^q\right)=0$ assuming that all roots are distinct.

9. (***Auto-regressive moving average ARMA (p, p) process***) Let $\mathbf{X}_t$ follow an $AR(p)$ process and ε_t is Gaussian white noise. Show that $Y_t = X_t + \varepsilon_t$ follows an *ARMA* (*p, p*) process.

10. (***Sum of two auto-regressive processes***) Show that a sum of $AR(p_1)$ and $AR(p_2)$ processes results in an $ARMA(p_1 + p_2, max\{p_1, p_2\})$ process.

7

Glimpses of Multivariate Statistics

THE ADVENT OF the internet and sensor technology has enabled humankind to collect, store, and share data in bulk. In turn, access to a variety of data has amplified a different kind of problem, which is to devise an appropriate strategy to derive meaning from data. Indeed, extracting information from data has acquired the highest priority among tasks performed by engineers and scientists alike. State-of-the-art machine learning algorithms are used to process and analyze data in order to leverage maximum gains in developing new technology and creating a new body of knowledge.

Further, the data-rich tech-universe has inherent complexity in addition to the vastness in terms of numbers. This complexity arises from the fact that often this data is embedded in a higher-dimensional space. For example, the data acquired by a camera hosted on a robot is in the form of multiple grayscale images (frames); each data-frame is constituted of a sequence of numbers that represents the intensity of grayness of each pixel.[1] If each image has a resolution 100×100 (pixel count), then this image data is embedded in a 10000 dimensional space. Additionally, if the camera records 100 frames per second for one minute, then we have 6000 data points in a 10000 dimensional space. This is just an illustrative example of how a high-dimensional large data set may be generated. Quite evidently, not all the 10000 dimensions host most of the information. One of the most important techniques that we will learn in this chapter will allow us to extract a lower-dimensional representation of the data set that will retain sufficient information for the robot to navigate and perform its tasks.

Another important statistical technique, which finds wide applications, provides us with a model framework to sift pure signal(s) from noisy data. Both these techniques rely heavily on concepts from linear algebra and statistical theories. Before getting started with this chapter, readers may want to refer to the appendix section on matrices and linear algebra to refresh relevant concepts. This chapter will begin with a brief discussion of elementary ideas from multivariate statistics and, subsequently, present the two aforementioned techniques, namely principal component analysis (PCA) and independent component analysis (ICA), as centerpiece ideas of this chapter.

Figure 7.1 *When you have eliminated the impossible, whatever remains, however improbable, must be the truth.* – Sherlock Holmes in *The Sign of Four.*

[1] This intensity is often a number between 0 and 255, ranging from black to white progressively.

7.1 Chapter objectives

The chapter objectives are listed as follows.

1. Students will learn to formulate the probability distribution of random vectors.

2. Students will learn to compute the expected value of random vectors and covariance matrices.

3. Students will learn the definition and apply the concept of entropy for signal extraction from noisy data.

4. Students will learn to use the ICA to extract pure signals from signal mixtures.

5. Students will learn to apply the PCA to compress and encrypt higher-dimensional data.

7.2 Chapter project: Minimalism of a butterfly

7.2.1 *Prologue: The minimalist sketch of a butterfly – to a machine and the human eye*

Consider the following picture of a butterfly. The human eye and brain have evolved over millions of years and acquired the cognitive and optical capabilities to recognize this beautiful natural creature as a butterfly. This capacity of ours is so sophisticated that we can often recognize a butterfly even from a low-resolution picture of this insect. This suggests that the most relevant information in a high-dimensional signal or data lives in a much lower-dimensional manifold. In this chapter project, we will use a mathematical technology, known as the *principal component analysis*, to obtain a minimal dimensional representation of a butterfly that is yet recognizable by humans as well as by machines.

Figure 7.2 The intricate natural beauty of a butterfly. See color plate on page 298.

7.3   Random vectors and probability distributions

The concepts discussed in this section are a continuation of the discussion in section 2.4.1 and presented more systematically with respect to the framework of multivariate statistical models. In this context, we may consider a sequence of continuous random variables[2] generated at time stamp τ as $\mathbf{X} = \{X_1(\tau), X_2(\tau), ..., X_n(\tau)\} = \{X_1, X_2, ..., X_n\}$. $\mathbf{X}$ is a random vector that comprises all the n-realizations of the random variable recorded at time τ. The value of the ith realization of the random variable $X_i(\tau) \equiv X_i = x_i$ was denoted as $X_i = x_\tau^{(i)}$ in the previous chapter. However, we may ignore rewriting the time variable τ every time as it may be considered implicit.

The cumulative distribution of a random vector is defined as follows.

$$F_{\mathbf{X}}(\mathbf{x}) \equiv P(\mathbf{X} \le \mathbf{x}) := P(X_1 \le x_1, X_2 \le x_2, ..., X_n \le x_n). \tag{7.1}$$

The probability density function (p.d.f.) is defined below.

$$f_{\mathbf{X}}(\mathbf{x}) := \frac{\partial^n F_{\mathbf{X}}(\mathbf{x})}{\partial x_1 ... \partial x_n}. \tag{7.2}$$

Consequently, $F_{\mathbf{X}}(\mathbf{x}) = \int_{-\infty}^{\mathbf{x}} f_{\mathbf{X}}(\mathbf{x}')\,\mathrm{d}\,\mathbf{x}' = \int_{-\infty}^{x_1} \cdots \int_{-\infty}^{x_n} f_{\mathbf{X}}(\mathbf{x}')\,dx_1' \cdots dx_n'.$

The conditional probability function of $\mathbf{X}$ conditioned on the event $\mathbb{B}$ is defined as follows.

$$F_{\mathbf{X}|\mathbb{B}} := P\left(\mathbf{X} \le \mathbf{x}|\mathbb{B}\right) = \frac{P\left(\mathbf{X} \le \mathbf{x}, \mathbb{B}\right)}{P\left(\mathbb{B}\right)}, \tag{7.3}$$

where $P\left(\mathbb{B}\right) \ne 0$. Likewise, if $\mathbb{B}_i, i = 1, 2, .., n$, form a set of partitioning events, then $F_{\mathbf{X}}\left(\mathbf{x}\right) = \sum_{i=1}^{n} F_{\mathbf{X}|\mathbb{B}_i}\left(\mathbf{x}|\mathbb{B}_i\right), P\left(\mathbb{B}_i\right)$ and a similar formula holds true for the p.d.f. $f_{\mathbf{X}}(\mathbf{x})$.

Now if we have two random vectors $\mathbf{X}$ and $\mathbf{Y}$ of dimensions n and m, respectively, then the joint p.d.f. is

$$f_{\mathbf{XY}}\left(\mathbf{x}, \mathbf{y}\right) = \frac{\partial^{(n+m)} F_{\mathbf{XY}}\left(\mathbf{x}, \mathbf{y}\right)}{\partial x_1 .. \partial x_n \partial y_1 .. \partial y_m}, \tag{7.4}$$

where the joint distribution is $F_{\mathbf{XY}}(\mathbf{x}, \mathbf{y}) = P(\mathbf{X} \le \mathbf{x}, \mathbf{Y} \le \mathbf{y})$.

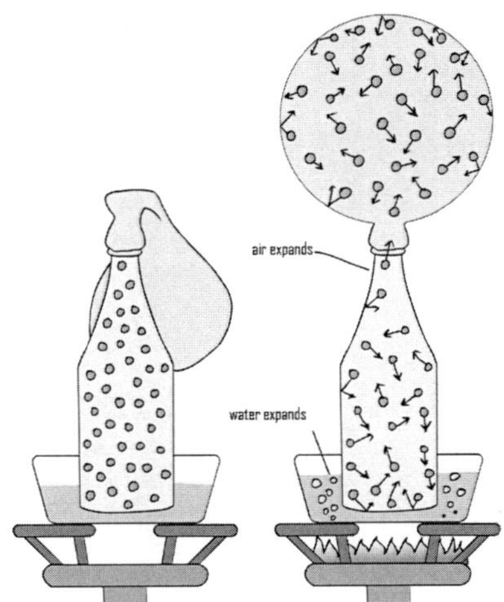

Figure 7.3 Movement of gas molecules while inflating a balloon (*courtesy: Britannica*, https://www.britannica.com/summary/gas-state-of-matter).

Example 7.1   Probability of finding a random gas molecule in a spherical balloon

Consider $\mathbf{X} = (X_1, X_2, X_3)$ to be the coordinates of a gas molecule inside a spherical balloon of radius 1 unit and centered at the origin. If the gas molecule is equally likely to be found anywhere inside the balloon, compute the probability that the molecule will be located within a sub-sphere of radius 2/3.

Solution: The volume of the sphere is $\frac{4\pi}{3}$. Therefore, the joint p.d.f. of $\mathbf{X}$ is $f_{\mathbf{X}}(\mathbf{x}) = \frac{3}{4\pi}$, when $x_1^2 + x_2^2 + x_3^2 < 1$, and 0, otherwise. Let $\mathbb{E}$ be the event the molecule is within the sub-sphere of radius 2/3 designated by the region $\mathfrak{R}$. Then, we have the following result:

$$P(\mathbb{E}) := \int_{\mathfrak{R}} f_{\mathbf{X}}(x_1,\, x_2,\, x_3)\, dx_1 dx_2 dx_3 = \frac{3}{4\pi} \int_{r=0}^{2/3} \int_{\phi=0}^{2\pi} \int_{\theta=0}^{2\pi} r^2 \sin\phi\, dr d\phi d\theta = \frac{8}{27}. \quad (7.5)$$

Now, if we have n functionally independent set of functions $g_i(\mathbf{x})$, $i = 1, 2, \dots, n$, that are used to transform the random vector $\mathbf{X}$ to a new random vector $\mathbf{Y} = \{Y_1, Y_2, \dots, Y_n\}$, then we may be interested in finding the joint p.d.f. of $\mathbf{Y}$, i.e., $f_{\mathbf{Y}}(\mathbf{y})$. The functional transformations are listed as below.

$$
\begin{aligned}
Y_1 &= g_1(X_1, X_2, \dots, X_n), \\
Y_2 &= g_2(X_1, X_2, \dots, X_n), \\
&\;\;\cdot \quad\cdot\quad\cdot \\
&\;\;\cdot \quad\cdot\quad\cdot \\
Y_n &= g_n(X_1, X_2, \dots, X_n).
\end{aligned}
\quad (7.6)
$$

We will assume that the functional maps g_i are invertible and may allow us to compute $\mathbf{X}$ from $\mathbf{Y}$ as follows.

$$
\begin{aligned}
X_1 &= h_1(Y_1, Y_2, \dots, Y_n), \\
X_2 &= h_2(Y_1, Y_2, \dots, Y_n), \\
&\;\;\cdot \quad\cdot\quad\cdot \\
&\;\;\cdot \quad\cdot\quad\cdot \\
X_n &= h_n(Y_1, Y_2, \dots, Y_n).
\end{aligned}
\quad (7.7)
$$

Consequently, we can compute the joint p.d.f. of $\mathbf{Y}$ as

$$f_{\mathbf{Y}}(\mathbf{y}) = \left\| \mathbb{J} \right\|^{-1} f_{\mathbf{x}}(\mathbf{x}), \quad (7.8)$$

where $\left\| \mathbb{J} \right\|$ is the absolute value of the determinant of the *Jacobian* matrix.[3] Thus, we have $\left\| \mathbb{J} \right\| := \left| det(\mathbb{J}) \right| = abs \begin{vmatrix} \dfrac{\partial g_1}{\partial x_1} & \cdots & \dfrac{\partial g_1}{\partial x_n} \\ \cdot & & \cdot \\ \cdot & & \cdot \\ \dfrac{\partial g_n}{\partial x_1} & \cdots & \dfrac{\partial g_n}{\partial x_n} \end{vmatrix}$. Equation (7.8) is a consequence of coordinate transformation from $\mathbf{x} = (x_1, x_2, \cdots, x_n)$ coordinates to $\mathbf{y} = (y_1, y_2, \cdots, y_n)$ coordinates and can be easily verified from any standard textbook on multivariate calculus. Further, assume that the n-component system $g_i(\mathbf{x}) - y_i$, $i = 1, \dots, n$, has r roots; then the joint p.d.f. of the random vector $\mathbf{Y}$ is given by the following expression:

[3] For the curious readers, here is a fun math article on the Jacobian matrix – Robert Smith, The joys of the Jacobian, *Chalkdust*, October 6, 2015, https://chalkdustmagazine.com/features/the-joys-of-the-jacobian/ , accessed April 18, 2024.

$$f_{\mathbf{Y}}(\mathbf{y}) = \sum_{i=1}^{r} \left\| \mathbb{J}_i \right\|^{-1} f_{\mathbf{X}}(\mathbf{x}_i).$$

(7.9)

It may be worthwhile to note here that $f_{\mathbf{X}}(\mathbf{x}) = f_{X_1}(x_1) f_{X_2}(x_2) \cdots f_{X_n}(x_n)$ if the n components of $\mathbf{X}$ are independent.

Example 7.2   Distribution of transformation of a random vector

Consider the following functionally independent set:

$$\begin{aligned}
y_1 &= g_1(\mathbf{x}) = x_1^2 - x_2^2, \\
y_2 &= g_2(\mathbf{x}) = x_1^2 + x_2^2, \\
y_3 &= g_3(\mathbf{x}) = x_3.
\end{aligned}$$

(7.10)

Consider only real roots of the above system and compute $f_{\mathbf{Y}}(\mathbf{y})$ if

$$f_{\mathbf{X}}(\mathbf{x}) = \frac{1}{(2\pi)^{3/2}} e^{-\frac{1}{2}(x_1^2 + x_2^2 + x_3^2)}.$$

Solution: At the outset, we list here the set of four real roots of $g_i(\mathbf{x}) - y_i$, $i = 1, 2, 3$.

$$\begin{aligned}
x_1^{\pm} &= \pm\sqrt{\frac{y_1 + y_2}{2}}, \\
x_2^{\pm} &= \pm\sqrt{\frac{y_2 - y_1}{2}}, \\
x_3 &= y_3.
\end{aligned}$$

(7.11)

These roots (solutions) are labeled succinctly as $\mathbf{x}_1, \mathbf{x}_2, \mathbf{x}_3, \mathbf{x}_4$, where

$$\begin{aligned}
\mathbf{x}_1 &= (x_1^+, x_2^+, x_3), \\
\mathbf{x}_2 &= (x_1^-, x_2^+, x_3), \\
\mathbf{x}_3 &= (x_1^+, x_2^-, x_3), \\
\mathbf{x}_4 &= (x_1^-, x_2^-, x_3).
\end{aligned}$$

Let us now compute the Jacobian.

$$\begin{aligned}
\mathbb{J} &= \begin{vmatrix}
\dfrac{\partial g_1}{\partial x_1} & \dfrac{\partial g_1}{\partial x_2} & \dfrac{\partial g_1}{\partial x_3} \\[2mm]
\dfrac{\partial g_2}{\partial x_1} & \dfrac{\partial g_2}{\partial x_2} & \dfrac{\partial g_2}{\partial x_3} \\[2mm]
\dfrac{\partial g_3}{\partial x_1} & \dfrac{\partial g_3}{\partial x_2} & \dfrac{\partial g_3}{\partial x_3}
\end{vmatrix} \\[3mm]
&= \begin{vmatrix}
2x_1 & -2x_2 & 0 \\
2x_1 & 2x_2 & 0 \\
0 & 0 & 1
\end{vmatrix} \\[3mm]
&= 8x_1 x_2.
\end{aligned}$$

(7.12)

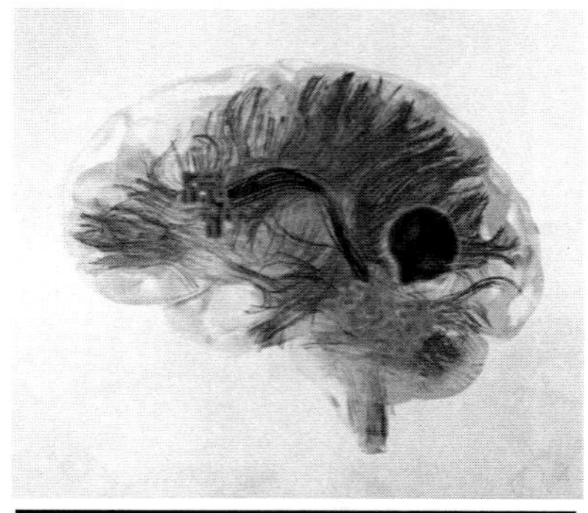

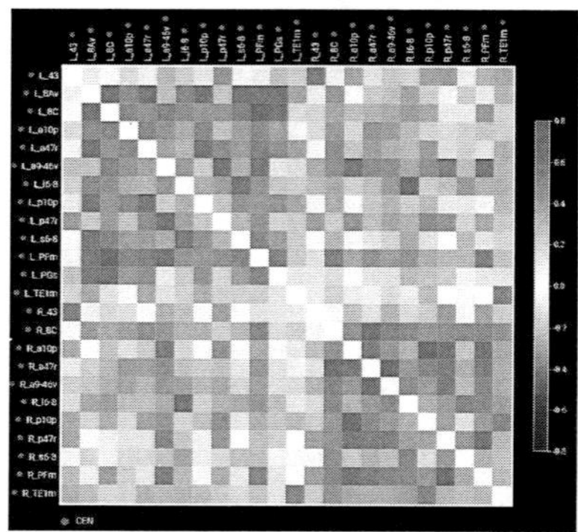

Figure 7.4 *Top*: Brain MRI portrait displaying macroscopic connections between regions in the human brain. *Bottom*: A correlation matrix showing functional connectivity between each brain region in the default mode network (DMN) (*courtesy*: Daniel White, Creative Commons), https://en.wikipedia.org/wiki/Connectomics, accessed April 3, 2024 (license under https://creativecommons.org/licenses/by-sa/4.0/).

The determinants of the Jacobian evaluated (as absolute values) at the roots $\{\mathbf{x}_1,\ \mathbf{x}_2,\ \mathbf{x}_3,\ \mathbf{x}_4\}$ yield $\left\|\mathbb{J}_1\right\| = \left\|\mathbb{J}_2\right\| = \left\|\mathbb{J}_3\right\| = \left\|\mathbb{J}_4\right\| = 4\sqrt{\left(y_2^2 - y_1^2\right)}$. Finally, we use equation (7.9) to compute the joint p.d.f. of $\mathbf{Y}$.

$$f_{\mathbf{Y}}(\mathbf{y}) = \frac{2}{4\sqrt{y_2^2 - y_1^2}} \sum_{i=1}^{4} f_{\mathbf{X}}(\mathbf{x}_i) = \frac{1}{\left(2\pi\right)^{3/2} \sqrt{y_2^2 - y_1^2}} e^{-\frac{1}{2}(y_2^2 + y_3^2)}. \tag{7.13}$$

7.4   Statistical moments of random vectors

In this section, we will discuss various statistical moments of random vectors. These moments play a pivotal role in the statistical analysis of multivariate (multi-dimensional) data. In the subsequent sections of this chapter, we will revisit the utility of these statistical moments in the context of principal and independent component analyses which have wide applications in signal extraction and data compression.

The expected value of a random vector $\mathbf{X} = \left(X_1, X_2, ..., X_n\right)$ is a vector μ whose components are given as

$$\mu_i = \int_{-\infty}^{\infty} x_i f_{X_i}(x_i) dx_i, \tag{7.14}$$

the expression above is in terms of the marginal p.d.f.

$$f_{X_i}(x_i) := \int_{-\infty}^{\infty} \cdots \int_{-\infty}^{\infty} f_{\mathbf{X}}(\mathbf{x}) dx_1 ... dx_{i-1} dx_{i+1} ... dx_n.$$

7.4.1   *Covariance matrix* Γ

The covariance matrix Γ associated with a real random vector $\mathbf{X}$ is defined as follows.

$$\Gamma := E(\mathbf{X} - \mu)(\mathbf{X} - \mu)^T. \tag{7.15}$$

Γ is a symmetric matrix, with entries $\Gamma_{ij} = \Gamma_{ji}$, $\forall i, j = 1, ..., n$, and is constructed of the outer (wedge) product of the vector $(\mathbf{X} - \mu)$ with itself. The diagonal entries of this matrix are the variances of the respective random variables X_i denoted by $\sigma_i^2 \equiv \Gamma_{ii} = \sigma_{X_i}^2$, $\forall i = 1, ..., n$.

7.4.2   *Correlation matrix* $\mathbf{R}$

The correlation matrix of the random vector $\mathbf{X}$ is defined as $\mathbf{R} := E(\mathbf{X}\mathbf{X}^T)$. Using the definition (7.15), we arrive at the following relation:

$$\Gamma = \mathbf{R} - \mu\mu^T. \tag{7.16}$$

7.4.3   *Uncorrelatedness, orthogonality, independence*

Two random vectors $\mathbf{X}$ and $\mathbf{Y}$ of dimensions n are *uncorrelated* if $E(\mathbf{XY}^T) = \mu_{\mathbf{X}}\mu_{\mathbf{Y}}^T$. $\mathbf{X}$ and $\mathbf{Y}$ are *orthogonal* if $E(\mathbf{XY}^T) = \mathbf{0}$ (an $n \times n$ matrix of all zeros). Finally, $\mathbf{X}$ and $\mathbf{Y}$ are *independent* if $f_{\mathbf{XY}}(\mathbf{x}, \mathbf{y}) = f_{\mathbf{X}}(\mathbf{x})f_{\mathbf{Y}}(\mathbf{y})$.

It may be noted that *independence* implies *uncorrelatedness* but the converse is not true. However, for a multivariate Gaussian random variable, the converse is also true, i.e., independence and uncorrelatedness are interchangeable.

7.5 Entropy

Entropy is an important physical concept[4] that we must grasp at this stage of our discussion to fully appreciate one of the two primary statistical machine learning techniques that we will introduce in the next section. Before we formally define entropy, it may be worthwhile to develop some conceptual intuition.

Since high entropy can be associated with higher disorder, we may consider N instances (either temporal or ensemble depending on the context) of a random variable Z, namely $Z_1, Z_2, ..., Z_N$, and ask ourselves how we could measure the extent of uniformity of the distribution of Z. A uniform distribution of the Z values will suggest high disorder because every possible state in the sample space is equally likely to be realized, which will thereby imply greater unpredictability in pinning down which of the possible states might actually be realized. This uncertainty (lack of surety) owing to the absence of order (predictability) corresponds to a *maximal entropy* state. We may also associate this state of higher unpredictability with a state of surprise (or shock). In this sense, entropy measures the average amount of surprise linked to the occurrence of a given event.

Let us illustrate this point with the help of a simple example. Consider the toss of a coin. At this point, we do not make any comment on whether it is a fair coin or a biased one. If it is a biased coin with heads (tails) on both sides, then there is no element of surprise in the outcome of such a coin as it will always show up heads (tails) upon tossing. This is what we may state as perfect order in the context of this experiment and hence corresponds to minimal entropy. On the other hand, if it is a fair coin and the frequency of occurrence of a head or a tail is equal, then the element of surprise associated with any given outcome is the highest. This situation corresponds to maximal entropy. In order to adequately capture this in a suitable definition of entropy, let us try a mathematical description whereby $Z^i = 1$ when the outcome of the ith toss is a head and $Z^i = 0$ for a tail. Further, let us define $p := P(Z^i = 1)$ and hence $(1 - p) = P(Z^i = 0)$. The case $p \to 0$ and $p \to 1$ must correspond to minimal entropy, i.e., $H(p) \to 0$ if H is a measure of entropy. Two candidate functions that behave in this fashion are $-p \log p$ and $-(1-p)\log(1-p)$. The sum (or average) of these two functions also behaves in a similar function, i.e.,

$$H(p) = -p \log p - (1-p)\log(1-p) \to 0 \text{ as } p \to 0 \text{ (or as } p \to 1\text{). By setting } \frac{dH}{dp} = 0\text{, we}$$

find a critical point $p = 1/2$ which may be easily checked to be a global maximum of the function $H(p)$. Thus, we have recovered the case of maximal entropy when $p = 1/2$ as we intuited earlier. So this model definition of $H(p)$ appears to adhere to our intuitive understanding of entropy. All that remains is to make a convincing case for this formulation of $H(p)$ as a formal definition of entropy. We attempt to tackle this below.

Let us consider that the symbol head (labelled H) appears on both sides of a coin. In that case, the probability of observing a head upon tossing, $p = P(H) = 1$. This must not be a surprise for obvious reasons. This may be captured by applying the logarithm function $\log p = 0$. Likewise, the probability of observing a tail by tossing the same coin must be *shocking* and hence $P(T) = 1 - p = 0$, whence $-\log(1-p) = -\infty$ (the possibility of such a shocking occurrence must be *negated*[5]

[4] The genesis of this idea of entropy and its application can be traced back to the field of thermodynamics. Later the development of statistical mechanics and information theory relied heavily on this centerpiece conceptual framework of *relative disorder*.

Figure 7.5 The element of surprise when antigravity consumes your morning coffee and along with it all your Christmas dreams resembles a state of high mental entropy.

[5] Read as *ruled out*.

with infinite conviction). Similarly, for a coin with a biased tail (T inscribed on both sides), $p = P(H) = 0$ is a near impossibility, which must invoke a measure of negative infinity ($-\log p = -\infty$) and $P(T) = 1 - p = 1$ is *not* shocking and hence $-\log(1 - p) = 0$. Therefore, the entropy for this coin tossing system can be defined as

$$H(p) = -p \log p - (1-p)\log(1-p) = -\sum_{i=H,T} p_i \log p_i. \tag{7.17}$$

7.5.1 Definition of entropy

Entropy of a discrete stochastic system   Let p_i, $i = 1, 2, ..., n$, be the probability of the ith discrete event $Z_i = z_i$. Then the entropy of the stochastic system is defined as

$$H(Z; p_i) = -\sum_{i=1}^{n} p_i \log p_i. \tag{7.18}$$

Entropy of a continuous stochastic system   Let $p_Z(z)$ be the p.d.f. of a continuous random variable Z that defines a continuous stochastic process. Then the entropy of the stochastic system is defined as

$$H(Z; p_Z) = -\int_{z=-\infty}^{\infty} p_Z(z)\log p_Z(z)\,dz \equiv -E\big(\log p_Z(z)\big). \tag{7.19}$$

Entropy of a sampling process   Consider a finite number of randomly generated samples of a stochastic process: $Z_1, Z_2, ..., Z_N$. Further, assume that $p_Z(Z_i = z_i) \equiv p_Z(z_i)$ is the probability of the outcome $Z_i = z_i$.[6] The entropy of this sampled stochastic process is defined as

$$H(Z; p_Z) = -\frac{1}{N}\sum_{i=1}^{N} \log p_Z(z_i). \tag{7.20}$$

Entropy of a multi-variate stochastic process   Consider a multivariate (n-dimensional) stochastic process $\mathbf{Z}_1, \mathbf{Z}_2, ..., \mathbf{Z}_M$, where $\mathbf{Z}_i = (Z_{i,1}, Z_{i,2}, ..., Z_{i,n})$. The entropy of this process is defined as

$$H(\mathbf{Z}; p_Z) = -\frac{1}{M}\sum_{i=1}^{M} \log p_{Z_i}(\mathbf{z}_i). \tag{7.21}$$

If the n-marginals of the p.d.f. $p_Z(\mathbf{z})$ are independent,[7] then the multivariate entropy is prescribed by

$$\begin{aligned}
H(\mathbf{Z}; p_Z) &= -E\big(\log p_Z(\mathbf{z})\big) \\
&= -E\left(\log \prod_{i=1}^{n} p_Z(z_i)\right) \\
&= -E\left(\sum_{i=1}^{n} \log p_Z(z_i)\right) \\
&= -\sum_{i=1}^{n} E\big(\log p_Z(z_i)\big) \\
&= -\sum_{i=1}^{n} H(Z; p_Z) \\
&= -nH(Z; p_Z).
\end{aligned} \tag{7.22}$$

[6] Here we have assumed that all the Z_is are characterized by the same distribution function $p_Z(z)$.

[7] $p_Z(\mathbf{z}) = p_{Z_1, Z_2, ..., Z_n}(z_1, z_2, ..., z_n) = p_{Z_1}(z_1)\,p_{Z_2}(z_2)\cdots p_{Z_n}(z_n)$ in the case of independent random variables $Z_1, Z_2, ..., Z_n$.

This result (7.22) will be centrally useful for implementing the ICA technique for signal purification that we will discuss in the next section.

7.6 Independent component analysis

Independent component analysis (ICA) is a statistical technique to extract pure signals from a mixture of signals. In order to explain the mechanics of this technique, let us consider two signal mixtures x_1 and x_2 as follows.

$$\begin{aligned} x_1 &= as_1 + bs_2, \\ x_2 &= cs_1 + ds_2. \end{aligned} \qquad (7.23)$$

Here, s_1 and s_2 are the pure signals. x_1 and x_2 are two different mixtures of s_1 and s_2. As an example, we can think of two vocalists who are located a few meters apart and are singing concurrently. There are two different microphones located a couple of meters from the vocalists. The signals received (and amplified) by each microphone (x_1 and x_2) are a superposition of the individual vocal signals of each singer (s_1 and s_2). We have a situation whereby we have access to the microphone outputs (x_1 and x_2) and our objective is to extract the vocal signatures of each singer s_1 and s_2.

The system (7.23) can be rewritten as

$$\mathbf{x} = \begin{pmatrix} x_1 \\ x_2 \end{pmatrix} = \begin{pmatrix} a & b \\ c & d \end{pmatrix} \begin{pmatrix} s_1 \\ s_2 \end{pmatrix} = A\mathbf{s}. \qquad (7.24)$$

Mathematically, extracting the pure signals s_1 and s_2 should involve an inversion of the system (7.24), i.e., computing the inverse of the matrix $\begin{pmatrix} a & b \\ c & d \end{pmatrix}$ whence the pure signals may be expressed as follows.

$$\mathbf{s} = \begin{pmatrix} s_1 \\ s_2 \end{pmatrix} = \begin{pmatrix} \alpha & \beta \\ \gamma & \delta \end{pmatrix} \begin{pmatrix} x_1 \\ x_2 \end{pmatrix} = A^{-1}\mathbf{x} = W\mathbf{x}. \qquad (7.25)$$

$W = A^{-1}$ is known as the *unmixing matrix*. However, the issue is that we do not know the entries of the matrix A. All we have are the signal mixtures x_1 and x_2. Hence, computing A^{-1} to extract the pure signals is not a viable option. This suggests that the entries of the unmixing matrix must be computed in an iterative fashion based on a suitable stopping criterion determined by the interrelationship between the corresponding candidate signals $\tilde{s}_1$ and $\tilde{s}_2$. As may be evident from the aforementioned example of the vocalists, the pure signals are often of independent origin. Hence, a very effective and suitable stopping criterion for this iterative approach may be to use a measure of independence between the extracted signal iterates. So, the following simple procedure may be attempted.

```
Iterative approach to signal extraction
```

- Propose and initialize the unmixing matrix $\tilde{W}_i$.

- Compute the corresponding source signals $\tilde{\mathbf{s}} = \tilde{W}_i\mathbf{x}$.

- Repeat until the entries of $\tilde{\mathbf{s}}$ are *maximally* independent.

Now, it is important to find a suitable metric for estimating independence between the extracted signals. One of the possible ways is to calculate entropy[8] of the extracted signals $\tilde{s}$ corresponding to the proposed unmixing matrix. Maximizing entropy of the extracted signals enables us to obtain a set of maximally independent set of source signals. The unmixing matrix that maximizes the entropy of the extracted signals also maximizes the amount of shared entropy or mutual information between the pure (extracted) signals and the set of signal mixtures. A uniformly distributed set of bounded random values in an appropriate space (discussed later) corresponds to a state of maximum entropy as was explained in detail in the previous section on entropy (see section 7.5).

We will demonstrate in this section how signal mixtures may be prepared (or how they originate). We will then show how different unmixing matrices may be used to extract the pure-form signals from these mixtures with the help of a certain class of model distribution functions. We will infer the characteristics of the mutual entropy of the extracted signals based on the output profile generated by this model function when applied to the extracted signals. As discussed earlier, this will determine the extent of the independence of the extracted signals and underscore the operational crux of ICA.

Process flow of signal mixing and extraction by ICA

$$\mathbf{s} \xrightarrow[\text{mixing}]{\mathbb{A}} \mathbf{x} \xrightarrow[\text{unmixing}]{\mathbb{W}} \tilde{\mathbf{s}} \xrightarrow[\text{infomax}]{\text{model c.d.f. } g} \mathbf{S} \text{ (extracted pure signal with maximum entropy).}$$

Here, the vectors $\mathbf{s}, \mathbf{x}, \tilde{\mathbf{s}}$, and $\mathbf{S}$ are of dimensions $n \times 1$, the mixing and unmixing matrices $\mathbb{A}$ and $\mathbb{W}$ are of dimensions $n \times n$, and g is the model cumulative distribution function.

7.6.1 *Illustrative example of signal purification and extraction using ICA*

In order to demonstrate the operational aspect of each step in the signal extraction process by ICA, we will consider a set of illustrative examples in this section. For this we will consider two sample signals, namely a pre-recorded audio signal sourced from the MathWorks database, `handel.mat`, and an additive white Gaussian noise signal generated by the Matlab in-built command `awgn`. This situation is akin to the portrait of a vocalist singing in front of a rapturous crowd (see Figure 7.6).

We will consider a mixing process prescribed by the matrix $\mathbb{A} = \begin{pmatrix} 0.15 & 0.85 \\ 0.1 & 0.9 \end{pmatrix}$. This may be likened to a superimposition of acoustic signatures generated by the aforementioned sources and, consequently, made available at two different microphones located not far from each other. Clearly, the mixed signals x_1 and x_2, as shown in Figure 7.7, completely mask the signature of the original audio signal s_1 and at best bear a resemblance with the white noise signal. The ICA technique will be used to extract the signature of s_1 from $\mathbf{x}$. If the mixing process is due to the signal capture at the microphones as described above, then one may be able to estimate the entries of the mixing matrix $\mathbb{A}$ by accounting for the relative location of the microphones and the acoustic sources. Let us suppose that we are able to accurately estimate $\mathbb{A}$ and thereby $\mathbb{W} = \mathbb{A}^{-1}$; then as is shown in Figure 7.7, we can successfully

Figure 7.6 How can we sift the melody of the singing voice from the cacophony of the crowd? See color plate on page 298.

extract the signals by applying a model distribution function $g(x) = \dfrac{1}{2}\dfrac{d}{dx}\log\cosh(x)^9$ to the unmixed signal $\tilde{\mathbf{s}}$. Because our choice of the unmixing matrix $\mathbb{W}$ was correct, the extracted signals $\mathbf{S} = \begin{pmatrix} S_1 \\ S_2 \end{pmatrix}$ have maximal mutual entropy and are hence maximally independent. This fact becomes amply clear if we look at the scatter plots of the signal pairs. Consider the distribution of the mixed signals x_1 and x_2 in Figure 7.8. Clearly, these two signals are correlated. Because they are a mixture of signals, their joint distribution will follow the Gaussian law on account of the central limit theorem in case of a mixture of a large number of signals. The highly correlated signal mixture does *not* form a mutually independent set. However, the extracted signals derived by the unmixing process $\mathbb{W}$ form a maximally independent set as is inferred from their entropy profile constructed by the model function g. This fact is visually obvious from the distribution of the extracted signals shown on the right in Figure 7.8.

Let us juxtapose the choice of the unmixing matrix $\mathbb{W} = \begin{pmatrix} 18 & -17 \\ -2 & 3 \end{pmatrix}$ that is constructed by inverting a suitable guess for the mixing matrix $\mathbb{A} = \begin{pmatrix} 0.15 & 0.85 \\ 0.1 & 0.9 \end{pmatrix}$ by a different unmixing matrix such as $\mathbb{W} = \begin{pmatrix} 18.5 & -13.8211 \\ -3.6 & 3.3 \end{pmatrix}$ for the same set of signal mixtures introduced earlier in this section. The profile of the extracted signals is shown in Figure 7.9. It is clear from this profile that the extracted signals $\tilde{\mathbf{s}}$ do not fully resemble the original pure signals. In fact, the technique entirely fails to extract s_2, the additive Gaussian white noise signal – neither of the extracted signals ($\tilde{s}_1$ or $\tilde{s}_2$) resembles s_2. One can plot the distribution on the S_1 versus S_2 plane and check that the distribution of the values is far from uniform. We will leave this as an exercise for the readers to validate using a simple computer routine. This underscores the sensitivity and importance of the unmixing matrix $\mathbb{W}$ on the accuracy of ICA.

*The choice of the model function as the derivative of the function is widely supported through many examples and applications in diverse areas of machine learning and statistical learning models. Other trial functions may be used, such as the exponential function, and the square root function, as is standard practice in the reconstruction of ICA models. Interested readers may refer to the following archived article: https://arxiv.org/abs/2208.04564, accessed April 8, 2024.

What is perhaps central to an effective ICA technique, besides a suitable choice of the model function g, is an iterative search for an unmixing matrix $\mathbb{W}$ that results in a set of maximally independent extracted signals. This search is as much an art as it is a computational exercise, especially for large datasets. However, if our objective is to identify the most elemental fabric of the original signals, then it is not necessary to be able to find the most precise $\mathbb{W}$ that will give us the extracted signals with infinite accuracy. The ICA technique is fairly robust to a family of well approximated unmixing matrices, i.e., if $\mathbb{W}_0$ is the unknown perfect unmixing matrix, then the technique will likely yield the desired result for a large set of matrices whose entries are in the ε-neighborhood of the entries of W_0. We present one such example with $\mathbb{W} = \begin{pmatrix} 18.7761 & -18.0842 \\ -2.1069 & 3.1294 \end{pmatrix}$ and find that we are able to extract the original signals fairly accurately (see Figure 7.10).

Figure 7.7 Stages of signal mixing and purification by ICA.

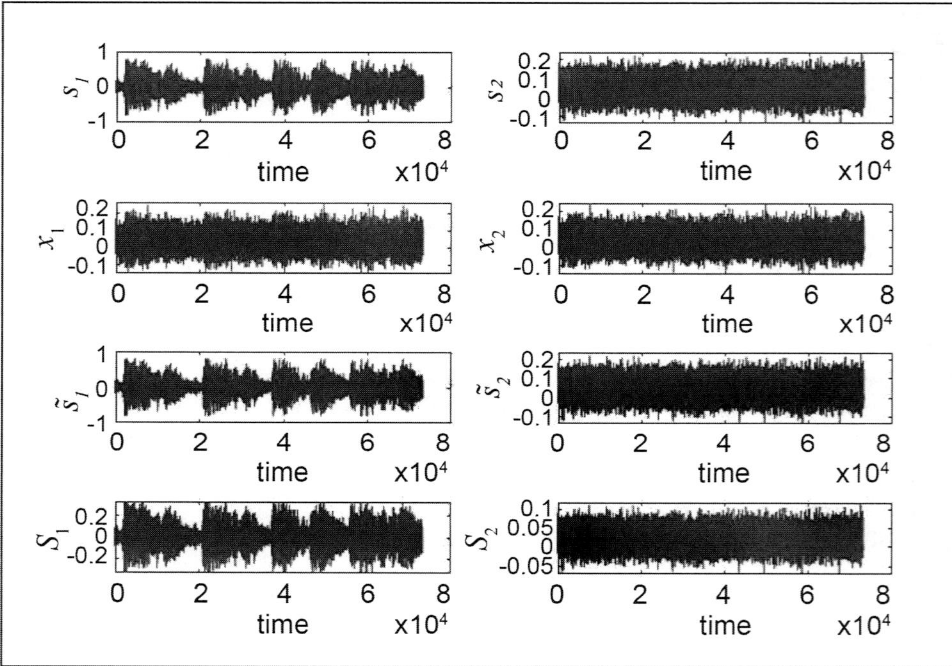

Figure 7.8 *Left*: Signal mixtures are highly correlated and do not form a maximally independent set. *Right*: Uniform distribution of extracted signals S_1 versus S_2 corresponds to a state of maximum entropy and maximal independence between S_1 and S_2.

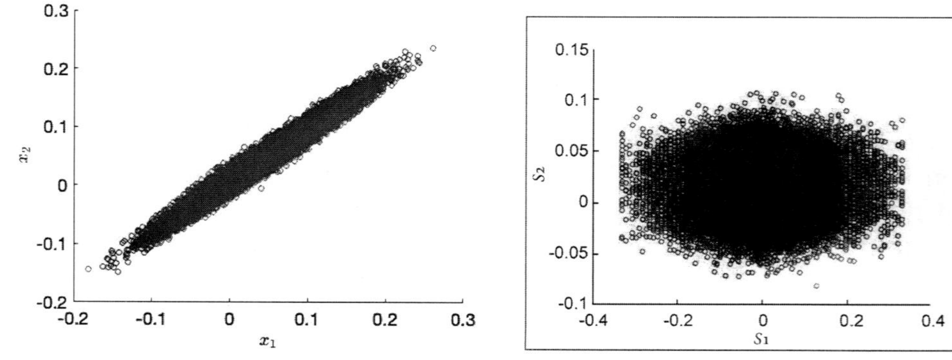

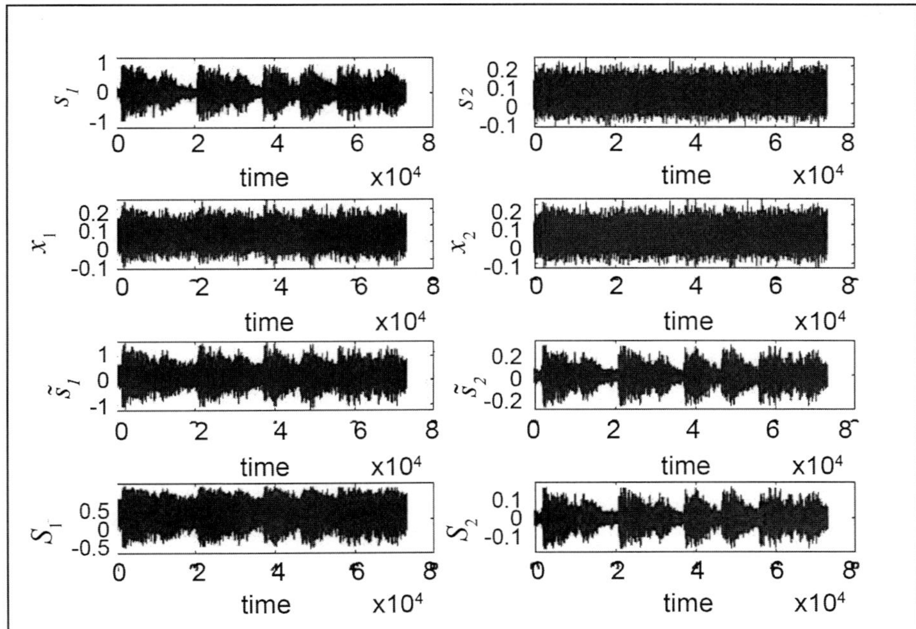

Figure 7.9 Stages of ICA using an unmixing matrix $\mathbb{W}$ that does not result in purification of the signal mixture.

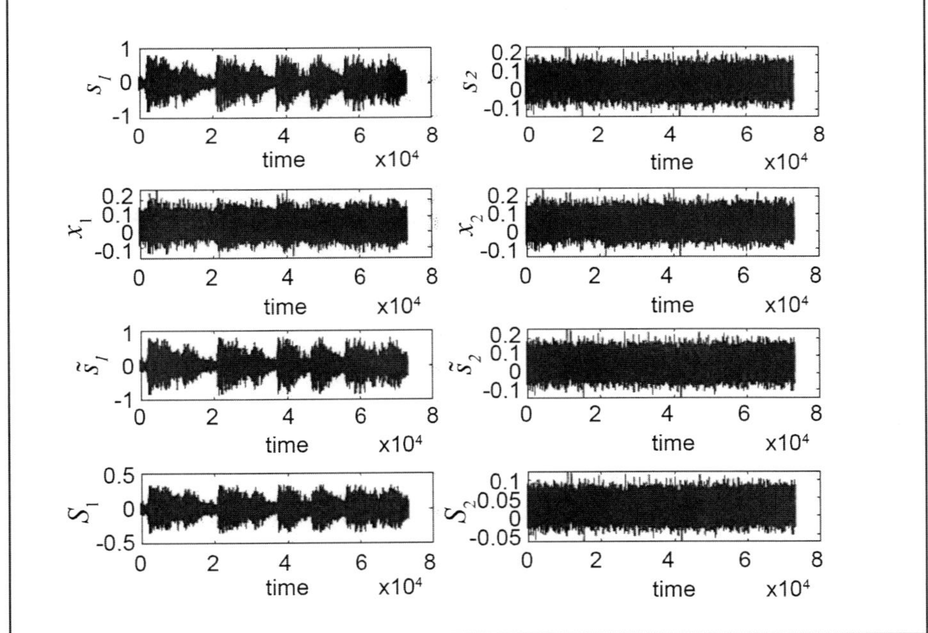

Figure 7.10 Robustness of ICA to an approximately accurate unmixing matrix $\mathbb{W}$.

Figure 7.11 Have you tried our noise cancellation headphones?

All the above examples were simulated using the following Matlab routine:

```
load handel.mat
s1 = y'; % original pure signal-1
s2 = awgn(0.05.*ones(1,length(s1)),1,'measured'); % additive white Gaussian
% noise original pure signal-2
```

```matlab
A = [0.15 0.85; 0.1 0.9]; % mixing matrix (generally not known, but we may
% have some idea based on the problem at hand)

s = [s1 ;s2]; % the original signal (in vector form)
x = A*s; % signal mixture (x is known)

%eps = [0.01 0.03; 0.02 0.01];
%A = A + eps.*A;
W = inv(A); % W is the un-mixing matrix (we guess W iteratively)
s_extract = W*x;  % extracted signal using proposed W at each step

syms z % symbolic variable

g = 0.5*log(cosh(z)); % model pdf used in ICA to analyze distribution
                                % of extracted signal in order to check for max
                                % entropy
dg = diff(g); % model cdf ....

S1 = eval(subs(dg,z,s_extract(1,:))); % reconstructed signal-1 using model cdf
S2 = eval(subs(dg,z,s_extract(2,:))); % re-constructed signal-2 using model cdf

figure('DefaultAxesFontSize',20);

subplot(4,2,1); plot(s1); xlabel('time'); ylabel ('$s_1$','interpreter','latex');
subplot(4,2,2); plot(s2); xlabel('time'); ylabel ('$s_2$','interpreter','latex');
subplot(4,2,3); plot(x(1,:)); xlabel('time'); ylabel ('$x_1$','interpreter','latex');
subplot(4,2,4); plot(x(2,:)); xlabel('time'); ylabel ('$x_2$','interpreter','latex');
subplot(4,2,5); plot(s_extract(1,:)); xlabel('time'); ylabel('$\tilde{s}_1$',...
      'interpreter','latex');
subplot(4,2,6); plot(s_extract(2,:)); xlabel('time'); ylabel('$\tilde{s}_2$',...
      'interpreter','latex');
subplot(4,2,7); plot(S1); xlabel('time'); ylabel ('$S_1$','interpreter','latex');
subplot(4,2,8); plot(S2); xlabel('time'); ylabel ('$S_2$','interpreter','latex');

figure, scatter(s1,s2); xlabel('$s_1$','interpreter', 'latex');
ylabel('$s_2$','interpreter','latex');
set( findall(gcf, '-property', 'fontsize'), 'fontsize', 20);
figure, scatter(x(1,:),x(2,:)); xlabel('$x_1$', 'interpreter','latex');
ylabel('$x_2$','interpreter','latex');
set( findall(gcf, '-property', 'fontsize'), 'fontsize', 20);
figure, scatter(s_extract(1,:),s_extract(2,:)); xlabel('$\tilde{s}_1$',...
      'interpreter','latex');
ylabel('$\tilde{s}_2$','interpreter','latex');
set( findall(gcf, '-property', 'fontsize'), 'fontsize', 20);
figure, scatter(S1,S2); xlabel('$S_1$','interpreter', 'latex');
ylabel('$S_2$','interpreter','latex');
set( findall(gcf, '-property', 'fontsize'),'fontsize', 20);
```

7.7 Chapter project: Minimalism of a butterfly

7.7.1 *Interlude: De-noising images and restoring identity of the original*

Images can be corrupted by noise due to a variety of factors such as failures in device and cell memory, errors in analog to digital conversion, and transmission errors. In this project module, we will learn to use the ICA technique to sift the signal from the noise. In doing so, we will ask the readers to benchmark the efficacy of this technique to other de-noising methods and against different levels of noise. At the outset, let us demonstrate the outlook of a corrupted image: in this case the image is corrupted by what is known as *salt and pepper noise*.[10] Image corruption may generally happen within the capturing device. Here, we have simulated the outcome of such noise corruption synthetically by a simple set of Matlab commands as follows. The noisy images with a varying extent of corruption are shown Figure 7.12.

```
[fly,map]  =  imread('butterfly2.jpg');
percentage_noise = 0.25;
J = imnoise(fly,'salt & pepper',percentage_noise);
figure,  imshow(J)
```

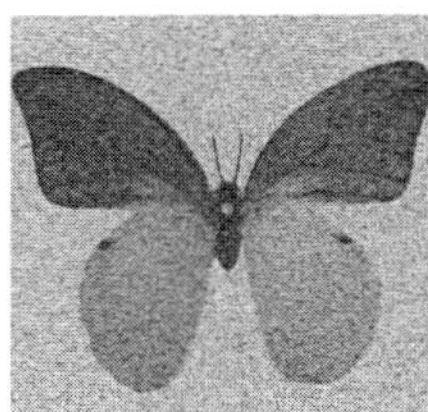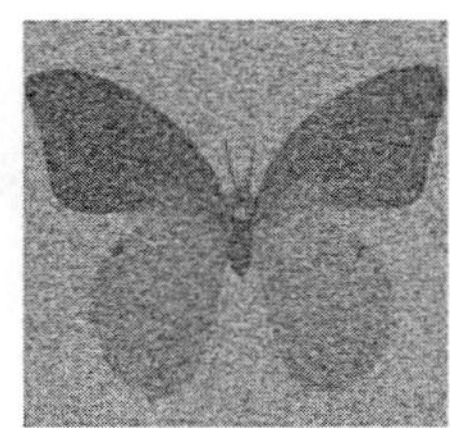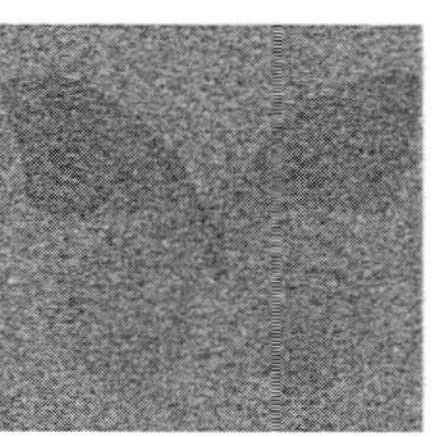

Figure 7.12 Image corrupted by salt and pepper noise to a varying extent.

Attempt the following questions to purify the corrupted images for each of four levels of noise: 25%, 50%, 75%, 90%. The levels of noise can be applied by setting the variable `percentage_noise` to the desired number as shown above.

1. Use the reconstruction ICA framework provided by Matlab (see the documentation page on `rica`) or any other publicly available implementation of ICA[11] to extract the original image signal of the butterfly from the noisy images.

2. Perform the same task of de-noising the images using median filtering and adaptive filtering techniques. You may want to use `medfilt2` and `wiener2` Matlab functions to do these tasks. Compare the performance of ICA with the filtering approaches.

7.8 Principal component analysis: Introductory remarks

We have so far discussed an effective technique to purify signal mixtures in order to ascertain the identity of pure signals masked by noise or by interference from other signals. In a very different context, we may be interested in reducing the dimensional complexity of data. In other words, we may be interested in retaining only those dimensions or attributes of data that lend it the most unique

[10] Charles Boncelet, "Image Noise Models," in *Communications, Networking and Multimedia: Handbook of Image and Video Processing* (Second Edition), ed. Al Bovik (Academic Press, 2005), pp. 397–409, https://doi.org/10.1016/B978-012119792-6/50087-5.

[11] http://research.ics.aalto.fi/ica/fastica/, accessed April 8, 2024.

identity. Consequently, eliminating the dimensions or attributes of data that do not contribute enough to its identity enables us to retain a lower dimensional representation of the data in a more efficient manner (in terms of storage). A few natural questions that emerge are the following: *What are the most important attributes or dimensions of a large and rich data set? And more importantly, how can we find a lower dimensional representation of the data?* These are some of the pressing questions that have implications in many areas of pattern recognition and machine learning. We will attempt to answer them in this section.

In order to appreciate the opening remarks of this section better, let us reconsider the mixed signal $\mathbf{x} = (x_1\ x_2)^T$ from the Figure 7.7. The distribution profile of this mixture is more revealing and is reproduced here in Figure 7.13.

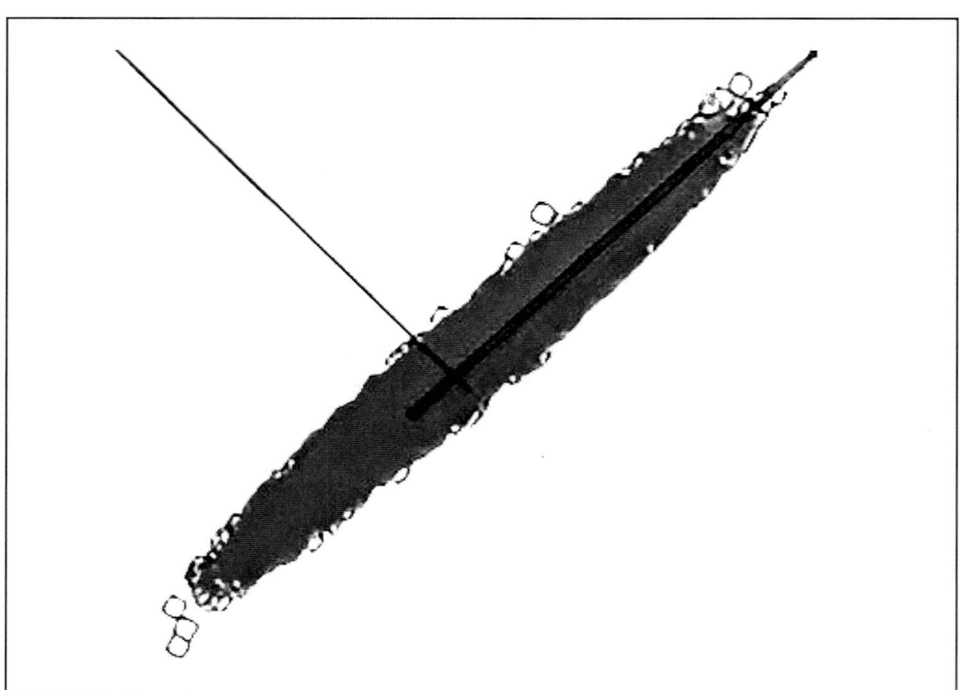

Figure 7.13 Distribution profile of a two-dimensional signal.

Recall from the entries of the mixing matrix $\mathbb{A} = \begin{pmatrix} 0.15 & 0.85 \\ 0.1 & 0.9 \end{pmatrix}$ that the two microphones must have been located reasonably close to each other in our example. If that is the case, then which of the two may be dispensable without compromising much on the identity of the output signal (in this case $\mathbf{x}$)? Mathematically, we can pose the same question as follows. *Which among x_1 and x_2 retains the character of $\mathbf{x}$ reasonably well?*[12] One of the common threads woven into our discussion on ICA in the previous section is the idea of entropy or information. Our objective is to retain maximum information about the data while eliminating some of the redundant dimensions. Richer information coincides with the manifold of higher entropy or higher variability. One may be better placed to grasp this idea by thinking of absence of variability along a certain dimension as a *boring* dimension. So, in essence, we must look for a manifold of higher variability or higher variance.

[12] In some applications, this question may be a false start, but we will guard ourselves from digressing much from the crux of the agenda of this section.

In our example of Figure 7.13, this manifold (dimension) is shown by the thick solid line along the direction of greater variability of data. This is counterposed against the orthogonal direction shown by the thin solid line along which there is minimal spread of the data. Identifying such rich and variable dimension(s) in an appropriate mathematical space is the central objective of the PCA technique. This will be explained in greater mathematical detail in the subsequent section.

7.9 Theoretical underpinning of principal component analysis

Principal component analysis (PCA) is a dimension reduction technique. The reduction in the representation of the dataset to a few important dimensions is carried forth by a suitable linear transformation. Technically, this involves a change of basis in the representation of the given data matrix. PCA de-correlates the original data by finding the dimensions along which the variance is maximized. The mathematical procedure then utilizes these dimensions to find the new basis.

Figure 7.14 Do we retain our identity in a lower-dimensional world?

7.9.1   *Data and covariance matrices, principal components*

Consider an $m \times n$ data matrix X where the n columns are the samples (observations) and the m rows are the variables (dimensions) such that for each k of the m dimensions, the n data points constitute a (row) vector $\tilde{\mathbf{x}}_k$ with mean zero. That is, $\tilde{\mathbf{x}}_k \leftarrow \left(\tilde{\mathbf{x}}_k - \sum_{i=1}^{n} \frac{\tilde{x}_{k,i}}{n} \right)$. Formally,

$$X = \begin{pmatrix} | & | & & & | \\ \mathbf{x}_1 & \mathbf{x}_2 & \cdot & \cdot & \cdot & \mathbf{x}_n \\ | & | & & & | \end{pmatrix} = \begin{pmatrix} \tilde{\mathbf{x}}_1 \\ \tilde{\mathbf{x}}_2 \\ \cdot \\ \cdot \\ \cdot \\ \tilde{\mathbf{x}}_m \end{pmatrix}$$

represents the mean subtracted data where each $\mathbf{x}_k$ represents an m dimensional data point (vector) and there are n of them corresponding to n samples, and each of $\tilde{\mathbf{x}}_k = \left(\tilde{x}_{k,1} \ \tilde{x}_{k,2} \ \cdots \tilde{x}_{k,n} \right)$ represents the n sample observations (data points) of the kth dimension. The data is then transformed into a new representation Y (also an $m \times n$ matrix) by a change of basis matrix P of dimensions $m \times m$ as follows:

$$Y = PX = \begin{pmatrix} \mathbf{p}_1 \cdot \mathbf{x}_1 & \mathbf{p}_1 \cdot \mathbf{x}_2 & \cdot & \cdot & \cdot & \mathbf{p}_1 \cdot \mathbf{x}_n \\ \mathbf{p}_2 \cdot \mathbf{x}_1 & \mathbf{p}_2 \cdot \mathbf{x}_2 & \cdot & \cdot & \cdot & \mathbf{p}_2 \cdot \mathbf{x}_n \\ \cdot & \cdot & \cdot & \cdot & \cdot \\ \cdot & \cdot & \cdot & \cdot & \cdot \\ \cdot & \cdot & \cdot & \cdot & \cdot \\ \mathbf{p}_m \cdot \mathbf{x}_1 & \mathbf{p}_m \cdot \mathbf{x}_2 & \cdot & \cdot & \cdot & \mathbf{p}_m \cdot \mathbf{x}_n \end{pmatrix}. \tag{7.26}$$

Here $P = \begin{pmatrix} ---\mathbf{p_1}--- \\ ---\mathbf{p_2}--- \\ \cdot \\ \cdot \\ \cdot \\ ---\mathbf{p_m}--- \end{pmatrix}$, where rows of P, represented here as $---\mathbf{p_k}---$,

form the new basis. In fact, the rows of P turn out to be the *principal components* (PCs). In the next section, we will demonstrate how to construct this matrix, P. The covariance matrix of the mean subtracted data is

$$C_X = \frac{1}{n-1}XX^T = \frac{1}{n-1}\begin{pmatrix} \tilde{\mathbf{x}}_1\tilde{\mathbf{x}}_1^T & \tilde{\mathbf{x}}_1\tilde{\mathbf{x}}_2^T & \cdot & \cdot & \tilde{\mathbf{x}}_1\tilde{\mathbf{x}}_m^T \\ \tilde{\mathbf{x}}_2\tilde{\mathbf{x}}_1^T & \tilde{\mathbf{x}}_2\tilde{\mathbf{x}}_2^T & \cdot & \cdot & \tilde{\mathbf{x}}_2\tilde{\mathbf{x}}_m^T \\ \cdot & \cdot & \cdot & \cdot & \cdot \\ \cdot & \cdot & \cdot & \cdot & \cdot \\ \cdot & \cdot & \cdot & \cdot & \cdot \\ \tilde{\mathbf{x}}_m\tilde{\mathbf{x}}_1^T & \tilde{\mathbf{x}}_m\tilde{\mathbf{x}}_2^T & \cdot & \cdot & \tilde{\mathbf{x}}_m\tilde{\mathbf{x}}_m^T \end{pmatrix}. \tag{7.27}$$

By construction, C_X is a symmetric positive definite matrix of dimensions $m \times m$. The off-diagonal entries of a covariance matrix reveal how well correlated two distinct variables are (in this case, how well correlated two distinct dimensions are). If two dimensions are significantly correlated, then perhaps retaining both these dimensions is not necessary as they will provide redundant information. This entails that the optimal representation of the data should be such that the off-diagonal terms of the covariance matrix in that representational space must be zero. Further, this also underscores the importance of retaining those dimensions that lend themselves to exhibit the *greatest variety* of *independent and uncorrelated* information inherent within the data (*maximal variance* presents richer information). Therefore, the diagonal entries of the covariance matrix in the transformed space should be as large as possible, especially for the important dimensions that are to be retained in the reduced (transformed) representation of the data.

7.9.2   *Linear transformation of the data*

The objective here is to find a transformation P which transforms the data into an optimal representational space using $Y = PX$. This is done by constructing the covariance matrix of the transformed data C_Y by

1. maximizing the diagonal entries of C_Y (maximizing variance of every dimension) and

2. minimizing the off-diagonal entries of C_Y (minimizing covariance between dimensions).

This means our objective is to find the P that makes C_Y diagonal. Also, since the rows of P must form the new basis, P is an orthonormal matrix ($PP^T = I$).

By definition,

$$C_Y = \frac{1}{n-1}YY^T = \frac{1}{n-1}(PX)(PX)^T = \frac{1}{n-1}(PX)(X^TP^T) = \frac{1}{n-1}P(XX^T)P^T = PSP^T,$$

where $S = \frac{1}{n-1}XX^T$ is an $m \times m$ symmetric positive definite matrix. Since every square symmetric matrix is diagonalizable, we have $S = EDE^T = \frac{1}{n-1}XX^T$, where E is an $m \times m$ orthonormal matrix whose columns are the eigenvectors of S and D is a diagonal matrix with the respective eigenvalues along the diagonal. Now, if we choose the rows of P to be the eigenvectors of S, we have $P = E^T$. From above, we find that $C_Y = PSP^T = E^T(EDE^T)E = D$ because $E^TE = PP^T = I$ as stated above. In summary, *if we choose $P = E^T$ (choose the PCs or rows of P as the eigenvectors of XX^T), then we recover C_Y in the desired diagonal form.*

7.9.3   *Singular value decomposition of the covariance matrix*

In linear algebra, there is a well-known methodology to compute the eigenvalues and eigenvectors of Z^TZ by using the singular value decomposition (SVD) of $Z = U\Sigma V^T$, where $Z := \frac{X^T}{\sqrt{n-1}}$, Σ is diagonal with the singular values of Z, and U and V are orthonormal matrices. The non-zero singular values of Z are the square roots of the non-zero eigenvalues of $Z^TZ = C_X$. So, $Z^TZ = \left(U\Sigma V^T\right)^T\left(U\Sigma V^T\right) = V\left(\Sigma^T\Sigma\right)V^T$. Hence, Z^TZ is similar to $\Sigma^T\Sigma$ and it has the same eigenvalues as that of $\Sigma^T\Sigma$. Therefore, *the non-zero singular values of Z, upon rearrangement in descending order along the diagonal of Σ and upon squaring, give us the eigenvalues of Z^TZ (or equivalently, the eigenvalues of $S = \frac{1}{n-1}XX^T$), also in descending order. Further, the columns of V, rearranged in accordance with the respective singular values of Z, are the same as the columns of E (rearranged) and constitute the rows of P which are the PCs. Therefore,* $\boxed{P = E^T = V^T}$.

7.9.4   *Data recovery and data compression*

There must now be a straightforward way to recover the original data from the transformed space (perhaps with some loss accounted for). Since $Y = PX = V^TX$, $\boxed{X = VY}$ returns the original data X from the transformed data Y.

Suppose that before projecting the data using the relation $Y = V^TX$, we were to truncate the matrix V so that we kept only the first $r < m$ columns. We would thus have a matrix of dimensions $n \times r$. The projection $\widehat{Y} = \widehat{V}^T X$ is the new reduced (transformed) data with dimensions $r \times n$. Suppose that we then wished to transform this data back to the original basis by computing $\widehat{X} = \widehat{V}\widehat{Y}$; we would recover the dimensions of X because $\widehat{X}$ also has dimensions $m \times n$ even though $\widehat{X} \neq X$.

Figure 7.15 How small can we pack them all?

The matrices X and $\widehat{X}$ are of the same dimensions, but they are not the same matrix, since we truncated the matrix of PCs V in order to obtain $\widehat{X}$. It is therefore reasonable to conclude that the matrix $\widehat{X}$ has, in some sense, *less information* in it than the matrix X. Of course, in terms of memory allocation on a computer, this is certainly not the case since both matrices have the same dimensions and would therefore be alloted the same amount of memory. However, the matrix $\widehat{X}$ can be computed as the product of two smaller matrices ($\widehat{V}$ and $\widehat{Y}$). This, together with the fact that the *essential* information in the matrix is captured by the first few PCs suggests a possible method for image compression.

Let us suppose we have an image of dimensions 512×512 and we want to reduce it to dimensions 512×40. So, $\widehat{V}$ and $\widehat{Y}$ have dimensions 512×40 and 40×512, respectively. In addition to this, we have another vector of size 512×1 that stores the means from the mean-subtraction step. Therefore, we have reduced the number of columns required from 512 to $40 + 40 + 1 = 81$. This gives us a compression ratio of 512 : 81 or approximately 6.33 : 1. In the chapter project, you will be asked to investigate different compression ratios and a qualitative assessment of the amount of information retained.

7.10 Chapter project: Minimalism of a butterfly

7.10.1 *Epilogue: Deconstructing the being and nothingness of the minimalist butterfly*

So we ask the question once again. What is the being and nothingness of the minimalist butterfly? By the former we mean: what constitutes the most elemental features of *being* a butterfly to the eyes of a curious onlooker? And by the latter we refer to the nothingness of all the redundant dimensions of *being* a butterfly to the curious onlooker.

We will achieve our objectives of this section by attempting to answer the following questions:

1. Use the image in Figure 7.2 and apply the image compression algorithm given below to compress the given image by considering the top 40 PCs. Plot the top 40 eigenvalues as a bar graph. What is the compression ratio?

2. Repeat the above experiment by using only top 5 PCs. What is the compression ratio?

3. What can you conclude qualitatively from the above two experiments?

4. What happens if you replace $[USV] = \mathbf{svd}(Z)$ in your code by $[USV] = \mathbf{svd}(C_X)$? Explain your observations.

Figure 7.16 Deconstruction of a minimal representational space.

<u>**Pseudocode of image compression by PCA**</u>

```
INPUT: image, numPCs.
```

- store **image** as a matrix   % use Matlab command **imread**
- find X, mean subtracted data matrix   % use Matlab commands **mean** and **repmat**
- construct $Z := \frac{1}{\sqrt{n-1}} X^T$ % dimension of X is $m \times n$
- construct covariance matrix $C_X = Z^T Z$
- compute SVD of Z: $[USV] = \mathbf{svd}(Z)$ % S stores singular values of Z
- compute eigenvalues of $Z^T Z$ by squaring singular values of Z
- store top **numPCs** eigenvectors: $\widehat{V} = V(:, 1: \mathbf{numPCs})$
- store transformed reduced data in $\widehat{Y} = \widehat{V}^T X$
- compute compression ratio: $\mathbf{ratio} = \frac{n}{2(\mathbf{numPCs})+1}$
- convert image data in original space: $\widehat{X} = \widehat{V}\widehat{Y}$
- add the subtracted means to $\widehat{X}$ to construct **compressed image**

```
OUTPUT: compressed image, ratio.
```

Note: Lines to the right of % symbol are comments, not executable lines of code.

7.11 Selected bibliography

Blom, Gunnar. *Probability and Statistics: Theory and Applications* (1st edition). Springer-Verlag, 1989.

Deisenroth, Marc Peter, A. Aldo Faisal, and Cheng Soon Ong. *Mathematics For Machine Learning* (1st edition). Cambridge University Press, 2020.

Feller, William. *An Introduction to Probability Theory and Its Applications.* Vol. 2 (2nd edition). John Wiley & Sons Inc., 1971.

Stark, Henry and John W. Woods. *Probability and Random Processes with Applications to Signal Processing* (2nd impression). Pearson, 2007.

Stone, James V. *Independent Component Analysis: A Tutorial Introduction* (1st edition). The MIT Press, 2004.

7.12 Exercise problems

1. (*Play of independence and correlation between random variables*) Consider two random variables X_1 and X_2 with joint p.d.f. $f_{X_1 X_2}(x_1, x_2) = x_1 + x_2,\ 0 < x_1 \le 1,\ 0 < x_2 \le 1,$ and 0 everywhere else. Show that X_1 and X_2 are *not* independent but that they are essentially uncorrelated.[13]

2. (*Constructing independent random vectors*) A zero mean random vector

   $\mathbf{X} = \begin{pmatrix} X_1 \\ X_2 \end{pmatrix}$ that obeys the Gaussian distribution has a covariance matrix

   $\Gamma(\mathbf{X}) = \begin{pmatrix} 1 & -0.33 \\ -0.33 & 1 \end{pmatrix}$. Find a transformation $\mathbf{Y} = \mathbb{D}\mathbf{X}$ such that the new random vector $\mathbf{Y}$ is a Gaussian random vector with independent components and of unit variance.

3. (*De-correlating random vectors*) Consider a random vector $\mathbf{X} = \begin{pmatrix} X_1 \\ X_2 \\ X_3 \end{pmatrix}$ whose covariance is given by $\Gamma(\mathbf{X}) = \begin{pmatrix} 1 & -0.5 & 0.5 \\ -0.5 & 1 & 0 \\ 0.5 & 0 & 1 \end{pmatrix}$. You may consider $X_i,\ i = 1,2,3,$ as the random states of an electrical circuit element (e.g., the output state of an amplifier). Design a non-zero transformer (a circuit that is constructed by adding and/or multiplying the X_is) that transforms $\mathbf{X}$ to a new random vector $\mathbf{Y}$ whose components are uncorrelated. Further, devise a strategy by which the three components of $\mathbf{Y}$ may be transformed such that $\Gamma_{ii}(\mathbf{Y})$ are the same for all $i = 1,2,3$.

4. Explain why the matrix $\begin{pmatrix} 4 & 0 & 0 \\ 0 & 6 & 0 \\ 0 & 0 & -2 \end{pmatrix}$ cannot be a covariance matrix of a real random vector.

[13] You may use the following measure of correlation: $\rho = \frac{\Gamma_{12}}{\sigma_1 \sigma_2}$, where Γ_{ij} are the entries of the covariance matrix Γ.

5. (***Marginal distribution of a random vector***) Consider the joint p.d.f. of an n-dimensional random vector $\mathbf{X}$ as below.

$$f_{\mathbf{X}}(\mathbf{x}) = \frac{(2\pi)^{-n/2}}{\sigma_1 \cdots \sigma_n} e^{-\frac{1}{2}\sum_{i=1}^{n}\left(\frac{x_i}{\sigma_i}\right)^2}.$$

The above is defined for all real values of x_i, $i = 1, 2, \ldots, n$. Show that all the marginal distributions follow the Gaussian law.

6. (***Rank of a correlation matrix***) Explain why for any random vector $\mathbf{X}$, the matrix $\mathbf{X}\mathbf{X}^{\mathrm{T}}$ always has unit rank but the correlation matrix $\mathbf{R} := \mathrm{E}(\mathbf{X}\mathbf{X}^{\mathrm{T}})$ may be full rank.

7. Let $\mathbf{X}$ be a d-dimensional random variable having a Gaussian distribution $N(\mu, \Sigma)$. Consider an m-dimensional random variable given by $\mathbf{Y} = A\mathbf{X} + \mathrm{b}$, where A is an $m \times d$ dimensional matrix. Show that $\mathbf{Y}$ also has a Gaussian distribution. Find the expressions for the mean and covariance of $\mathbf{Y}$. Discuss all cases: $m < d$, $m = d$, $m > d$.

8. Explain the geometrical significance of the eigenvectors of the covariance matrix for a given data set.

9. (***PCA by pencil and paper***) In this exercise, consider a two-dimensional data $\{(x,y)\} = \{(1,-1), (0,1), (-1,0)\}$. Using pencil and paper, construct the data matrix and the covariance matrix. Perform the PCA and identify the PCs. In each step of your calculation, show on the two-dimensional Euclidean plane the relevant eigenvectors and the PCs. Based on your calculation, identify which of the two dimensions has richer information. Also construct the one-dimensional representation of the data set.

10. (***Sketching data from covariance matrices***) Draw a data distribution qualitatively consistent with the following covariance matrices.

(i) $\Gamma = \begin{pmatrix} 1 & -0.5 \\ -0.5 & 1 \end{pmatrix}$, (ii) $\Gamma = \begin{pmatrix} 1 & 0 \\ 0 & 0.5 \end{pmatrix}$, and (iii) $\Gamma = \begin{pmatrix} 1 & 1 \\ 1 & 1 \end{pmatrix}$.

Appendix $\mathbf{I}$

A Brief Tour of Matrices

A1.1 Basic matrix operations

A1.1.1 *Definition (matrix)*

A matrix is an ordered rectangular array of numbers (or functions). The numbers (or functions) are called the elements or the entries of the matrix.

A1.1.2 *Dimension/size of a matrix*

A matrix having m rows and n columns is a matrix of dimension $m \times n$.

A1.1.3 *Notation of a matrix*

A general $m \times n$ matrix has the following rectangular arrangement.

$$A = \begin{bmatrix} a_{11} & a_{12} & a_{13} & \cdots & a_{1j} & \cdots & a_{1n} \\ a_{21} & a_{22} & a_{23} & \cdots & a_{2j} & \cdots & a_{2n} \\ \vdots & \vdots & \vdots & & \vdots & & \vdots \\ a_{i1} & a_{i2} & a_{i3} & \cdots & a_{ij} & \cdots & a_{in} \\ \vdots & \vdots & \vdots & & \vdots & & \vdots \\ a_{m1} & a_{m2} & a_{m3} & \cdots & a_{mj} & \cdots & a_{mn} \end{bmatrix}$$

or

$$A = \left(a_{ij} \right)_{m \times n} \quad 1 \leq i \leq m, \ 1 \leq j \leq n, \ i, j \in \mathbb{N}.$$

A1.1.4 *Addition of matrices*

Two matrices A and B can be added if they are of the same dimension. The matrices A and B are denoted as

$$A = \left(a_{ij}\right)_{m \times n} \text{ and } B = \left(b_{ij}\right)_{m \times n} \quad 1 \le i \le m,\ 1 \le j \le n,\ i, j \in \mathbb{N};$$

then $\exists\ C = A + B$ of dimension $m \times n$ such that $C = (a_{ij} + b_{ij})_{m \times n} = (c_{ij})_{m \times n}.$

Example A1.2

$$A = \begin{bmatrix} 1 & 2 \\ 3 & 4 \end{bmatrix},\ B = \begin{bmatrix} 1 \\ 2 \end{bmatrix},\ C = \begin{bmatrix} 1 & 0 \\ 0 & 1 \end{bmatrix}.$$

Here, $A + B$ does not exist since A and B are of different dimensions.

Although $A + C = \begin{bmatrix} 1+1 & 2+0 \\ 3+0 & 4+1 \end{bmatrix} = \begin{bmatrix} 2 & 2 \\ 3 & 5 \end{bmatrix}.$

A1.1.5 *Product of two matrices*

The product AB of matrices A and B is defined if and only if the number of columns of A is equal to the number of rows of B.

Consider $A = \left(a_{ij}\right)_{m \times n}$ and $B = (b_{ij})_{n \times p};$
then

$$AB = C = (c_{ij})_{m \times p},$$

where C is of dimension $m \times p$ and

$$c_{ij} = \sum_{j=1}^{n} a_{ij} b_{jk}. \tag{A1.1}$$

Mathematical representation of matrix multiplication

$$C = AB = \begin{bmatrix} a_{11} & a_{12} & a_{13} & \cdots & a_{1j} & \cdots & a_{1n} \\ a_{21} & a_{22} & a_{23} & \cdots & a_{2j} & \cdots & a_{2n} \\ \vdots & \vdots & \vdots & & \vdots & & \vdots \\ a_{i1} & a_{i2} & a_{i3} & \cdots & a_{ij} & \cdots & a_{in} \\ \vdots & \vdots & \vdots & & \vdots & & \vdots \\ a_{m1} & a_{m2} & a_{m3} & \cdots & a_{mj} & \cdots & a_{mn} \end{bmatrix}_{m \times n} \begin{bmatrix} b_{11} & b_{12} & b_{13} & \cdots & b_{1j} & \cdots & b_{1p} \\ b_{21} & b_{22} & b_{23} & \cdots & b_{2j} & \cdots & b_{2p} \\ \vdots & \vdots & \vdots & & \vdots & & \vdots \\ b_{i1} & b_{i2} & b_{i3} & \cdots & b_{ij} & \cdots & b_{ip} \\ \vdots & \vdots & \vdots & & \vdots & & \vdots \\ b_{n1} & b_{n2} & b_{n3} & \cdots & b_{nj} & \cdots & b_{np} \end{bmatrix}_{n \times p},$$

where (say) the element c_{22} (of AB) $= a_{21}b_{12} + a_{22}b_{22} + a_{23}b_{32} + \ldots + a_{2n}b_{n2}$, and so on.

Example A1.3

Consider $A = \begin{bmatrix} 1 & 2 \\ 3 & 4 \end{bmatrix}_{2\times 2}$ and $B = \begin{bmatrix} 1 & 2 \end{bmatrix}_{1\times 2}$.

Here, AB does not exist.

Whereas when $A = \begin{bmatrix} 1 & 2 \\ 3 & 0 \\ 4 & 5 \end{bmatrix}_{3\times 2}$ and $B = \begin{bmatrix} 1 & 0 & 1 \\ 0 & 1 & 0 \end{bmatrix}_{2\times 3}$,

then

$$AB = \begin{bmatrix} 1 & 2 & 1 \\ 3 & 0 & 3 \\ 4 & 5 & 4 \end{bmatrix}_{3\times 3}.$$

A1.1.6 _Transpose of a matrix_

Consider $A = (a_{ij})_{m\times n}$. Then the transpose of an $m \times n$ matrix is defined to be a matrix of dimension $n \times m$, denoted by A^{T} such that

$$A^{\mathrm{T}} = \left(a_{ji}\right)_{n\times m} \quad \forall \; i, j$$

Example A1.4

$A = \begin{bmatrix} 1 & 2 & 3 \\ 4 & 5 & 6 \end{bmatrix}_{2\times 3}$. Then $A^{\mathrm{T}} = \begin{bmatrix} 1 & 4 \\ 2 & 5 \\ 3 & 6 \end{bmatrix}_{3\times 2}$,

i.e., the rows of matrix A become the columns of the new matrix A^{T}.

A1.1.7 _Determinant_

A determinant is defined as a scalar quantity associated with a square matrix denoted by $\det(A)$ or $|A|$. The determinant of a 2×2 matrix is defined as

$$\begin{vmatrix} a & b \\ c & d \end{vmatrix} = ad - bc.$$

Similarly, the determinant of a 3×3 matrix is defined as

$$\det \begin{bmatrix} a_{11} & a_{12} & a_{13} \\ a_{21} & a_{22} & a_{23} \\ a_{31} & a_{32} & a_{33} \end{bmatrix} = \begin{vmatrix} a_{11} & a_{12} & a_{13} \\ a_{21} & a_{22} & a_{23} \\ a_{31} & a_{32} & a_{33} \end{vmatrix} = a_{11}\begin{vmatrix} a_{22} & a_{23} \\ a_{32} & a_{33} \end{vmatrix} - a_{12}\begin{vmatrix} a_{21} & a_{23} \\ a_{31} & a_{33} \end{vmatrix} + a_{13}\begin{vmatrix} a_{21} & a_{22} \\ a_{31} & a_{32} \end{vmatrix}$$

$$= a_{11}\left(a_{22}a_{33} - a_{23}a_{32}\right) - a_{12}\left(a_{21}a_{33} - a_{23}a_{31}\right) + a_{13}\left(a_{21}a_{32} - a_{22}a_{31}\right).$$

Example A1.5

i. $A = \begin{bmatrix} 0 & 2 \\ 5 & 0 \end{bmatrix} \Rightarrow \det(A) = -10.$

ii. $A = \begin{bmatrix} 1 & 2 & 3 \\ 2 & 3 & 1 \\ 1 & 2 & 2 \end{bmatrix}$ then $\det(A) = 1\begin{vmatrix} 3 & 1 \\ 2 & 2 \end{vmatrix} - 2\begin{vmatrix} 2 & 1 \\ 1 & 2 \end{vmatrix} + 3\begin{vmatrix} 2 & 3 \\ 1 & 2 \end{vmatrix} = 1(6-2) - 2(4-1) +$

$3(4-3) = 4 - 6 + 3 = 1.$

A1.1.8 *Minor*

Let A be a matrix of dimension $n \times n$. The minor of element a_{ij} is equal to the determinant of a sub-matrix of order $(n-1) \times (n-1)$ which is obtained by leaving the ith row and the jth column of A. It is denoted by M_{ij}.

Example A1.6

For $A = \begin{bmatrix} a_{11} & a_{12} & a_{13} \\ a_{21} & a_{22} & a_{23} \\ a_{31} & a_{32} & a_{33} \end{bmatrix}$, the minor of a_{11} is equal to $M_{11} = \begin{vmatrix} a_{22} & a_{23} \\ a_{32} & a_{33} \end{vmatrix}$ (leaving out

1st row and 1st column of A).

A1.1.9 *Co-factor matrix*

Let $A = \left(a_{ij}\right)_{n \times n}$ be a matrix. The $(i, j)^{th}$ entry of the co-factor matrix corresponding to element a_{ij} is equal to $A_{ij} = (-1)^{i+j} M_{ij}$. Here, M_{ij} is the minor of a_{ij}.

A1.1.10 *Adjoint of a matrix*

Let $A = \left(a_{ij}\right)_{n \times n}$ and its co-factor matrix be $C = \left(A_{ij}\right)_{n \times n} = \left((-1)^{i+j} M_{ij}\right)_{n \times n}$.

The transpose of the co-factor matrix of A is known as the adjoint of A, denoted by $adj(A) = \left[A_{ij}\right]^{\mathrm{T}} = C^{\mathrm{T}}$,

i.e., if

$$A = \begin{bmatrix} a_{11} & a_{12} & a_{13} & \cdots & a_{1j} & \cdots & a_{1n} \\ a_{21} & a_{22} & a_{23} & \cdots & a_{2j} & \cdots & a_{2n} \\ \vdots & \vdots & \vdots & & \vdots & & \vdots \\ a_{i1} & a_{i2} & a_{i3} & \cdots & a_{ij} & \cdots & a_{in} \\ \vdots & \vdots & \vdots & & \vdots & & \vdots \\ a_{n1} & a_{n2} & a_{n3} & \cdots & a_{nj} & \cdots & a_{nn} \end{bmatrix}_{n \times n},$$

then

$$adj(A) = \begin{bmatrix} A_{11} & A_{21} & A_{31} & \cdots & A_{j1} & \cdots & A_{n1} \\ A_{12} & A_{22} & A_{32} & \cdots & A_{j2} & \cdots & A_{n2} \\ \vdots & \vdots & \vdots & & \vdots & & \vdots \\ A_{1i} & A_{2i} & A_{3i} & \cdots & A_{ji} & \cdots & A_{ni} \\ \vdots & \vdots & \vdots & & \vdots & & \vdots \\ A_{1n} & A_{2n} & A_{3n} & \cdots & A_{jn} & \cdots & A_{nn} \end{bmatrix}_{n \times n},$$

where A_{ij} is the $(i, j)^{th}$ entry of the co-factor matrix.

Example A1.7

Find the $adj(A)$ for $A = \begin{bmatrix} 1 & 4 & 5 \\ 0 & 2 & 6 \\ 0 & 0 & 3 \end{bmatrix}$.

Solution: First, we find the co-factor matrix.

$$A_{11} = (-1)^{1+1} \begin{vmatrix} 2 & 6 \\ 0 & 3 \end{vmatrix} = 6, \quad A_{12} = (-1)^{1+2} \begin{vmatrix} 0 & 6 \\ 0 & 3 \end{vmatrix} = 0, \quad A_{13} = (-1)^{1+3} \begin{vmatrix} 0 & 2 \\ 0 & 0 \end{vmatrix} = 0, \tag{A1.2}$$

$$A_{21} = (-1)^{2+1} \begin{vmatrix} 4 & 5 \\ 0 & 3 \end{vmatrix} = -12, \quad A_{22} = (-1)^{2+2} \begin{vmatrix} 1 & 5 \\ 0 & 3 \end{vmatrix} = 3, \quad A_{23} = (-1)^{2+3} \begin{vmatrix} 1 & 4 \\ 0 & 0 \end{vmatrix} = 0, \tag{A1.3}$$

$$A_{31} = (-1)^{3+1} \begin{vmatrix} 4 & 5 \\ 2 & 6 \end{vmatrix} = 24 - 10 = 14, \quad A_{32} = (-1)^{3+2} \begin{vmatrix} 1 & 5 \\ 0 & 6 \end{vmatrix} = -6, \quad A_{33} = (-1)^{3+3} \begin{vmatrix} 1 & 4 \\ 0 & 2 \end{vmatrix} = 2.$$

$$\tag{A1.4}$$

The co-factor matrix of A,

$$C = \begin{bmatrix} 6 & 0 & 0 \\ -12 & 3 & 0 \\ 14 & -6 & 2 \end{bmatrix}.$$

$$adj(A) = C^T = \begin{bmatrix} 6 & -12 & 14 \\ 0 & 3 & -6 \\ 0 & 0 & 2 \end{bmatrix}.$$

A1.1.11 *Inverse of a square matrix*

$A_{n \times n}$ is said to be invertible if $\exists\, B_{n \times n}$ such that $AB = BA = I_n$, where I_n is the identity matrix of dimension $n \times n$. If B exists, it is denoted by A^{-1} and given by

$$A^{-1} = \frac{1}{\det(A)}(adjA) \quad \left(A^{-1} \text{ exist} \Leftrightarrow \det(A) \neq 0\right).$$

Example A1.8

Find the inverse of $A = \begin{bmatrix} a & b \\ c & d \end{bmatrix}$ given that $ad - bc \neq 0$.

Solution: Since $\det(A) = ad - bc \neq 0$, A is an invertible matrix.

$$A_{11} = (-1)^{1+1} d = d, \ A_{12} = -c, \ A_{21} = -b, \ A_{22} = a. \tag{A1.5}$$

So the co-factor matrix is

$$\left[A_{ij} \right] = \begin{bmatrix} d & -c \\ -b & a \end{bmatrix}$$

and

$$adj(A) = \begin{bmatrix} d & -b \\ -c & a \end{bmatrix}.$$

Hence,

$$A^{-1} = \frac{1}{ad - bc} \begin{bmatrix} d & -b \\ -c & a \end{bmatrix}.$$

Example A1.9

Find the inverse of the matrix $A = \begin{bmatrix} 5 & 0 & 7 \\ 2 & 1 & 3 \\ 0 & 4 & 6 \end{bmatrix}$.

Solution: $\det(A) = 5(6 - 12) + 7(8) = -30 + 56 = 26 \neq 0$.
Hence, A is invertible.

$$\text{Co-factor matrix} \left(\left[A_{ij} \right] \right) = \begin{bmatrix} -6 & -12 & 8 \\ 28 & 30 & -20 \\ -7 & -1 & 5 \end{bmatrix}$$

$$adj(A) = \text{Transpose of the co-factor matrix} = \begin{bmatrix} -6 & 28 & -7 \\ -12 & 30 & -1 \\ 8 & -20 & 5 \end{bmatrix}.$$

Hence,

$$A^{-1} = \frac{1}{\det(A)} \left(adj(A) \right) = \frac{1}{26} \begin{bmatrix} -6 & 28 & -7 \\ -12 & 30 & -1 \\ 8 & -20 & 5 \end{bmatrix}.$$

A1.2 Inner product space

A1.2.1 *Definition*

An inner product on a vector space V over $\mathbb{R}$ is a map $\langle,\rangle : V \times V \to \mathbb{R}$ that satisfies the following properties. For $\mathbf{x}, \mathbf{y}, \mathbf{z} \in V$ and $\alpha \in \mathbb{R}$,

   i. $\langle \mathbf{x}, \mathbf{x} \rangle \geq 0$ and $\langle \mathbf{x}, \mathbf{x} \rangle = 0$ if and only if $\mathbf{x} = 0$,

   ii. $\langle \mathbf{x}, \mathbf{y} \rangle = \langle \mathbf{y}, \mathbf{x} \rangle$,

   iii. $\langle \mathbf{x} + \mathbf{z}, \mathbf{y} \rangle = \langle \mathbf{x}, \mathbf{y} \rangle + \langle \mathbf{z}, \mathbf{y} \rangle$ and $\langle \mathbf{x}, \mathbf{y} + \mathbf{z} \rangle = \langle \mathbf{x}, \mathbf{y} \rangle + \langle \mathbf{x}, \mathbf{z} \rangle$, and

   iv. $\langle \alpha \mathbf{x}, \mathbf{y} \rangle = \alpha \langle \mathbf{x}, \mathbf{y} \rangle$.

A vector space V with a given inner product structure is called an inner product space. A complete inner product space is a Hilbert space which finds wide applications in many areas of mathematics, physics, and computer science.

> ### *Example A1.10*
>
> $$\mathbf{x} = \begin{pmatrix} x_1 \\ x_2 \end{pmatrix} \in \mathbb{R}^2 \text{ and } y = \begin{pmatrix} y_1 \\ y_2 \end{pmatrix} \in \mathbb{R}^2.$$
>
> Then the inner product (dot product) of x and y, denoted by $\langle \mathbf{x}, \mathbf{y} \rangle$, is defined as
>
> $$\langle \mathbf{x}, \mathbf{y} \rangle = x_1 y_1 + x_2 y_2.$$
>
> One can easily check that properties (i)–(iv) in definition (2.1) are satisfied.

A1.3 Outer product space

A1.3.1 *Definition*

Let $\mathbf{u} = \begin{bmatrix} u_1 \\ u_2 \\ \vdots \\ u_m \end{bmatrix}_{m \times 1}$ and $\mathbf{v} = \begin{bmatrix} v_1 \\ v_2 \\ \vdots \\ v_n \end{bmatrix}_{n \times 1}$ be two vectors of sizes m × 1 and n × 1, respectively.

Then the outer product of $\mathbf{u}$ and $\mathbf{v}$, denoted by $\mathbf{u} \otimes \mathbf{v}$, is defined as the m × n matrix $\mathbf{A}$ obtained by element-wise multiplication.

$$\mathbf{u} \otimes \mathbf{v} = \mathbf{A} = \begin{bmatrix} u_1 v_1 & u_1 v_2 & \cdots & u_1 v_n \\ u_2 v_1 & u_2 v_2 & \cdots & u_2 v_n \\ \vdots & \vdots & \ddots & \vdots \\ u_m v_1 & u_m v_2 & \cdots & u_m v_n \end{bmatrix} = \mathbf{u}\mathbf{v}^T.$$

A1.4 Orthogonality

A1.4.1 *Definition (orthogonal vectors)*

Let V be an inner product space. Then the two vectors $\mathbf{u}$ and $\mathbf{v}$ in V are said to be orthogonal if $\langle \mathbf{u}, \mathbf{v} \rangle = 0$ $(\mathbf{u} \perp \mathbf{v})$.

Example A1.11

Let V be a finite dimensional inner product space over $\mathbb{R}$.

i. $\mathbf{u} = \begin{bmatrix} 1 \\ 2 \\ 4 \\ 1 \end{bmatrix}, \mathbf{v} = \begin{bmatrix} -1 \\ -2 \\ 1 \\ 1 \end{bmatrix}$, then $\langle \mathbf{u}, \mathbf{v} \rangle = -1 - 4 + 4 + 1 = 0$. Hence, $\mathbf{u}$ and $\mathbf{v}$ are orthogonal vectors.

ii. The standard basis vectors $\left\{ \begin{bmatrix} 1 \\ 0 \\ 0 \\ \vdots \\ 0 \end{bmatrix}, \begin{bmatrix} 0 \\ 1 \\ 0 \\ \vdots \\ 0 \end{bmatrix}, \ldots, \begin{bmatrix} 0 \\ 0 \\ 0 \\ \vdots \\ 1 \end{bmatrix} \right\}$ in $\mathbb{R}^n$ are mutually orthogonal.

A1.5 Projection of a vector

A1.5.1 *Scalar projection*

A scalar projection of $\mathbf{u}$ onto $\mathbf{v}$ is the length of the shadow that $\mathbf{u}$ casts on $\mathbf{v}$.

$$\text{Scalar projection of } \mathbf{u} \text{ on } \mathbf{v} = \frac{\langle \mathbf{u}, \mathbf{v} \rangle}{\| \mathbf{v} \|}.$$

A1.5.2 *Vector projection*

A vector projection of $\mathbf{u}$ onto $\mathbf{v}$ is the vector in the same direction as $\mathbf{v}$ whose length is the scalar projection of $\mathbf{u}$ on $\mathbf{v}$.

$$proj_v \mathbf{u} = \left(\frac{\langle \mathbf{u}, \mathbf{v} \rangle}{\| \mathbf{v} \|} \right) \frac{\mathbf{v}}{\| \mathbf{v} \|} = \frac{\langle \mathbf{u}, \mathbf{v} \rangle}{\langle \mathbf{v}, \mathbf{v} \rangle} \mathbf{v}.$$

Example A1.12

Find the scalar and vector projection of $\mathbf{u} = \begin{pmatrix} 1 \\ 2 \end{pmatrix}$ on $\mathbf{v} = \begin{pmatrix} 3 \\ 4 \end{pmatrix}$.

Solution:

$$\langle \mathbf{u}, \mathbf{v} \rangle = 3 + 8 = 11 \text{ and } \| \mathbf{v} \| = \sqrt{3^2 + 4^2} = 5.$$

Hence,

$$\text{scalar projection of } \mathbf{u} \text{ on } \mathbf{v} = \frac{\langle \mathbf{u}, \mathbf{v} \rangle}{\| \mathbf{v} \|} = \frac{11}{5},$$

and

$$proj_v \mathbf{u} = \left(\frac{\langle \mathbf{u}, \mathbf{v} \rangle}{\| \mathbf{v} \|} \right) \frac{\mathbf{v}}{\| \mathbf{v} \|} = \left(\frac{11}{5} \right) \begin{pmatrix} 3/5 \\ 4/5 \end{pmatrix} = \begin{pmatrix} 33/25 \\ 44/25 \end{pmatrix}.$$

A1.6 Field

A1.6.1 *Definition (field)*

A field is a set $\mathbb{F}$ of numbers with the property that if $a, b \in \mathbb{F}$, then $a + b$, $a - b$, ab, and $\dfrac{a}{b}$ are also in $\mathbb{F}$ (assuming that $b \neq 0$ in the expression $\dfrac{a}{b}$).

Example A1.13

$\mathbb{Q}$, $\mathbb{R}$, and $\mathbb{C}$ are fields but $\mathbb{Z}$ and $\mathbb{N}$ are not fields.

A1.7 Vector spaces

A1.7.1 *Definition (vector space)*

A vector space $\mathcal{V}$ consists of a set $\mathbb{V}$ of vectors, a field $\mathbb{F}$ of scalars, and two operations:

1. Vector addition: if $\mathbf{v}, \mathbf{w} \in \mathbb{V}$, then $\mathbf{v} + \mathbf{w} \in \mathbb{V}$.

2. Scalar multiplication: $c \in \mathbb{F}$ and $\mathbf{v} \in \mathbb{V}$, then $c\mathbf{v} \in \mathbb{V}$.

These scalars and vectors satisfy the following rules:

1. Associativity of addition: $(\mathbf{v} + \mathbf{u}) + \mathbf{w} = \mathbf{v} + (\mathbf{u} + \mathbf{w})$, $\forall \mathbf{u}, \mathbf{v}, \mathbf{w} \in \mathbb{V}$.

2. Associativity of multiplication: $(ab)\mathbf{u} = a(b\mathbf{u})$, for any $a, b \in \mathbb{F}, \mathbf{u} \in \mathbb{V}$.

3. Distributive property: $(a+b)\mathbf{u} = a\mathbf{u} + b\mathbf{u}$ and $a(\mathbf{u} + \mathbf{v}) = a\mathbf{u} + a\mathbf{v}$, $\forall a, b \in \mathbb{F}, \forall \mathbf{u}, \mathbf{v} \in \mathbb{V}$.

4. Unitarity: $1\mathbf{u} = \mathbf{u}$, $\forall \mathbf{u} \in \mathbb{V}$.

5. Existence of zero: $\exists\, 0 \in \mathbb{V}$ such that $\mathbf{u} + 0 = \mathbf{u}$, $\forall \mathbf{u} \in \mathbb{V}$.

6. Negation: For every $\mathbf{u} \in \mathbb{V}, \exists (-\mathbf{u}) \in \mathbb{V}$ such that $\mathbf{u} + (-\mathbf{u}) = 0 \in \mathbb{V}$.

Example A1.14

i. $\mathbb{V}$ is the set of $n \times 1$ column matrices (vectors), $\mathbb{F}$ is the field of reals $\mathbb{R}$, and the laws of vector addition and scalar multiplication are defined as:

$$\begin{pmatrix} x_1 \\ x_2 \\ \cdot \\ \cdot \\ \cdot \\ x_n \end{pmatrix} + \begin{pmatrix} y_1 \\ y_2 \\ \cdot \\ \cdot \\ \cdot \\ y_n \end{pmatrix} = \begin{pmatrix} x_1 + y_1 \\ x_2 + y_2 \\ \cdot \\ \cdot \\ \cdot \\ x_n + y_n \end{pmatrix} \text{ and } c\begin{pmatrix} x_1 \\ x_2 \\ \cdot \\ \cdot \\ \cdot \\ x_n \end{pmatrix} = \begin{pmatrix} cx_1 \\ cx_2 \\ \cdot \\ \cdot \\ \cdot \\ cx_n \end{pmatrix}.$$

ii. $\mathbb{V}$ is the set of all continuous functions, $f : \mathbb{R} \to \mathbb{R}$, let the field of scalars be $\mathbb{R}$, and let the operations be usually defined.

iii. $\mathbb{V}$ is the set of all continuous functions, $f : \mathbb{R} \to \mathbb{R}$, that satisfy the equation $f'' = -f$.

A1.8 Linear independence of vectors

A1.8.1 *Definition (linearly dependent vectors)*

Let V be a vector space and $\mathcal{X} \subset V$ be a non-empty subset. Then $\mathcal{X}$ is linearly dependent if there are distinct vectors $\mathbf{v}_1, \mathbf{v}_2, \ldots, \mathbf{v}_k \in \mathcal{X}$, and scalars $c_1, c_2, \ldots, c_k$ (not all of them zero), such that $c_1\mathbf{v}_1 + c_2\mathbf{v}_2 + \ldots + c_k\mathbf{v}_k = 0$.

This is equivalent to saying that at least one of the vectors $\mathbf{v}_i$ can be expressed as a linear combination of the others.

$$\mathbf{v}_i = -\sum_{j \neq i}\left(\frac{c_j}{c_i}\right)\mathbf{v}_j.$$

Example A1.15

$\left\{\begin{pmatrix}1\\1\end{pmatrix}, \begin{pmatrix}2\\2\end{pmatrix}\right\}$ is a linearly dependent subset of $\mathbb{R}^2$.

A1.8.2 *Definition (linearly independent vectors)*

A subset which is not linearly dependent is said to be linearly independent. Thus, a set of distinct vectors $\{\mathbf{v}_1, \mathbf{v}_2, \ldots, \mathbf{v}_k\}$ is linearly independent if and only if an equation of the form $c_1\mathbf{v}_1 + c_2\mathbf{v}_2 + \cdots + c_k\mathbf{v}_k = 0$ is true only when $c_1 = c_2 = \ldots = c_k = 0$.

Example A1.16

$\left\{\begin{pmatrix}1\\0\end{pmatrix}, \begin{pmatrix}0\\1\end{pmatrix}\right\}$ is a linearly independent subset of $\mathbb{R}^2$.

A1.9 Geometrical interpretation of linear dependence

Let $\mathbf{v}_1, \mathbf{v}_2, \mathbf{v}_3$ be the vectors in a 3D-Euclidean space $\mathbb{R}^3$ with a common origin. If these vectors form a linearly dependent set, then one of them, say $\mathbf{v}_1$, can be expressed as a linear combination of the other two: $\mathbf{v}_1 = a\mathbf{v}_2 + b\mathbf{v}_3$. This implies, by the parallelogram law, that the three vectors are co-planar.

In fact, *linearly dependent vectors with a common origin are always co-planar and vice-versa.*

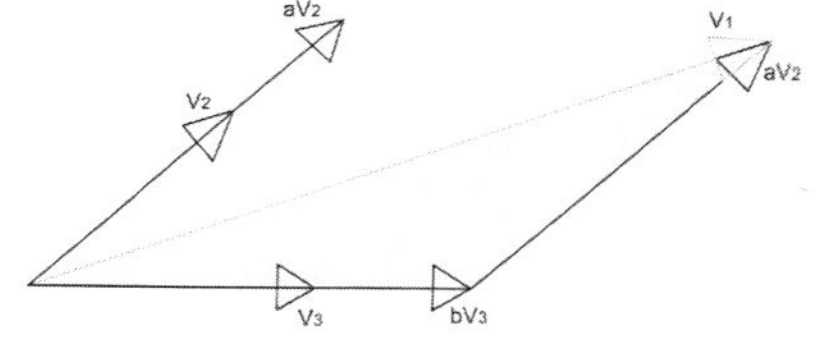

Figure A1.1 Linear dependence of vectors is equivalent to co-planar geometry.

A1.10 Basis of a vector space

A1.10.1 *Definition (basis)*

Let $\mathbb{X}$ be a non-empty subset of a vector space V. Then $\mathbb{X}$ is called a basis of V if each of the following is true:

    i. $\mathbb{X}$ is linearly independent,

    ii. $\mathbb{X}$ generates V (i.e., $\mathbb{X}$ spans V).

Example A1.17

i. Basis of $\mathbb{R}^n$: $\mathbf{e}_1 = \begin{pmatrix} 1 \\ 0 \\ \vdots \\ 0 \end{pmatrix}$, $\mathbf{e}_2 = \begin{pmatrix} 0 \\ 1 \\ \vdots \\ 0 \end{pmatrix}$, $\cdots$, $\mathbf{e}_n = \begin{pmatrix} 0 \\ 0 \\ \vdots \\ 1 \end{pmatrix}$ form a basis of $\mathbb{R}^n$ because they

(*a*) are linearly independent (by inspection) and (*b*) span $\mathbb{R}^n$ owing to the

fact that $c_1\mathbf{e}_1 + c_1\mathbf{e}_2 + \cdots c_n\mathbf{e}_n = \begin{pmatrix} c_1 \\ c_2 \\ \vdots \\ c_n \end{pmatrix}$ generates any vector in $\mathbb{R}^n$ depending

on the values of c_i, $\forall i = 1, 2, \ldots, n$.

ii. Let $\mathbb{P}_n$ be a vector space of all polynomial functions of degree n or less. The basis of $\mathbb{P}_n$ is $\{1, x, x^2, \ldots, x^n\}$, the set of monomials.

iii. Let $\mathbb{M}_{m\times n}(\mathbb{F})$ denote the set of $m \times n$ matrices with entries in $\mathbb{F}$. Then $\mathbb{M}_{m\times n}(\mathbb{F})$ is a vector space over $\mathbb{F}$. Vector addition is just matrix addition and scalar multiplication is defined in the obvious way (by multiplying each entry of the matrix by the same scalar). The zero vector is just the zero matrix. One possible choice of basis is the matrices with a single entry equal to 1 and all other entries as 0.

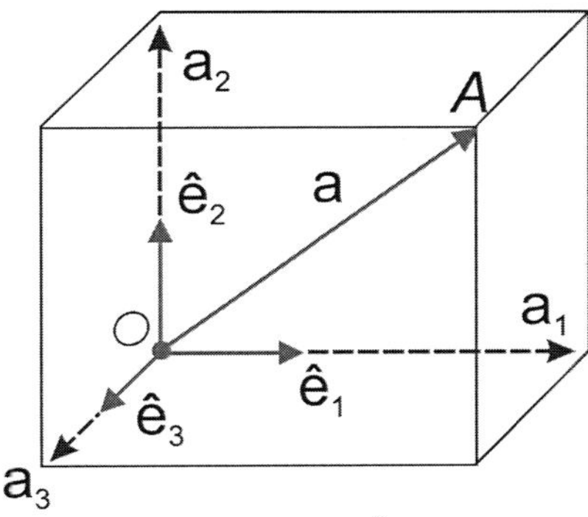

Figure A1.2 Basis of $\mathbb{R}^3$.

A1.10.2 *Properties of bases*

i. Must every vector space have a basis?
Ans: Every non-zero, finitely generated vector space has a basis.

ii. Does a vector space have a unique basis?

Ans: Usually a vector space will have many bases. For example, the vector space $\mathbb{R}^2$ has the basis $\left\{ \begin{pmatrix} 1 \\ -1 \end{pmatrix}, \begin{pmatrix} 2 \\ -1 \end{pmatrix} \right\}$ as well as the standard basis $\left\{ \begin{pmatrix} 1 \\ 0 \end{pmatrix}, \begin{pmatrix} 0 \\ 1 \end{pmatrix} \right\}$.

iii. What is the dimension of a vector space?
Ans: $\dim(\mathcal{V}) = $ no. of elements (vectors) in the basis (basis set).

A1.11 Vector space of $m \times n$ matrices over the reals, $\mathbb{M}_{m\times n}(\mathbb{R})$

A1.11.1 *Matrix algebra*

Please refer to the section A1.1 of this appendix. We summarize some of the elementary matrix operations below:

$$A + B = B + A \text{ (commutative law of addition)}$$

$$(A + B) + C = A + (B + C) \text{ (associative law of addition)}$$

$$A + 0 = A$$

$$(AB)C = A(BC) \text{ (associative law of multiplication)}$$

$$AI = A = IA$$

$$A(B + C) = AB + AC \text{ (distributive law)}$$

$$(A + B)C = AC + BC \text{ (distributive law)}$$
$$A - B = A + (-1)B$$
$$(cd)A = c(dA)$$
$$c(AB) = (cA)B = A(cB)$$
$$c(A + B) = cA + cB$$
$$(c + d)A = cA + dA$$
$$(A + B)^\mathrm{T} = A^\mathrm{T} + B^\mathrm{T}$$
$$(AB)^\mathrm{T} = B^\mathrm{T}A^\mathrm{T}$$

Here, $A, B, C \in \mathbb{M}_{m \times n}(\mathbb{R})$, the matrix entries $a_{ij} \in \mathbb{R}$, $c, d \in \mathbb{R}$.

A1.11.2

The rules of matrix algebra guarantee that $\mathbb{M}_{m \times n}(\mathbb{R})$ is a vector space.

Example A1.18

Consider the set of two linear equations:

$$2x - y = 0$$
$$-x + 2y = 3. \tag{A1.6}$$

We begin by sketching out the respective straight lines $2x - y = 0$ and $-x + 2y = 3$. The solution of this system is the intersection point of these two straight lines, $x = 1$, $y = 2$. If we express this system of linear equations in matrix-vector notation, then we have $\boldsymbol{A}\mathbf{x} = \mathbf{b}$, where $A = \begin{pmatrix} 2 & -1 \\ -1 & 2 \end{pmatrix}$ is the coefficient matrix, $\mathbf{x} = \begin{pmatrix} x \\ y \end{pmatrix}$, and $\mathbf{b} = \begin{pmatrix} 0 \\ 3 \end{pmatrix}$. The system (A1.6) corresponds to the row picture. The solution strategy presented above lends a geometrical interpretation of this row picture.

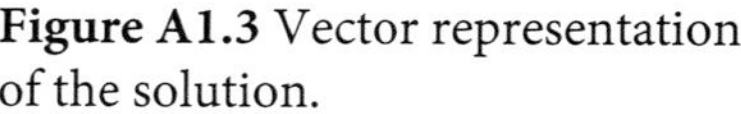

Figure A1.3 Vector representation of the solution.

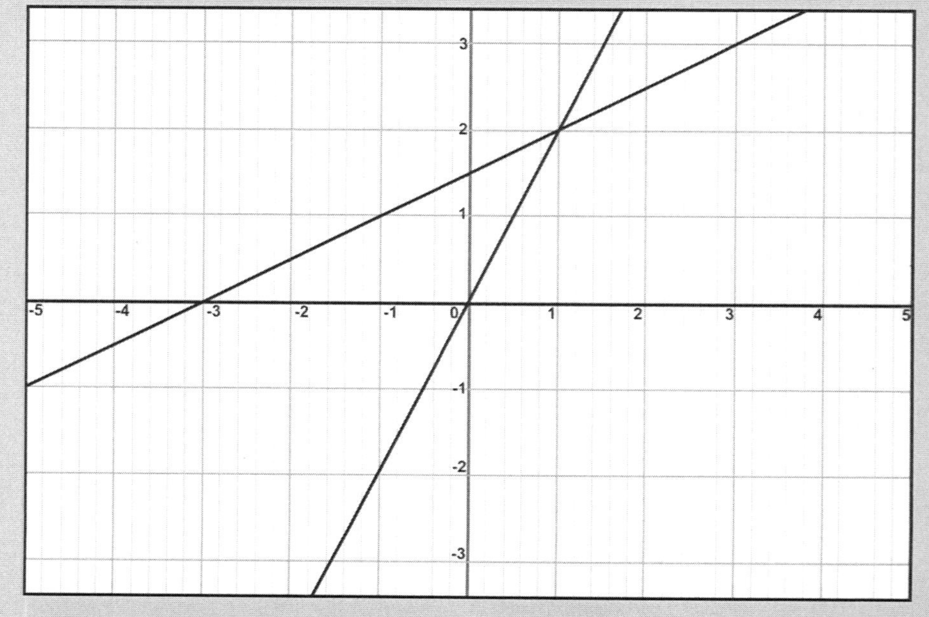

There is an alternative (and sometimes more useful) geometrical picture, the column picture, which lends a different interpretation of the situation at hand. By careful inspection, we notice that the above system of equations can be rewritten as follows:

$$x\begin{pmatrix} 2 \\ -1 \end{pmatrix} + y\begin{pmatrix} -1 \\ 2 \end{pmatrix} = \begin{pmatrix} 0 \\ 3 \end{pmatrix}.$$

Here the system of equations is expressed in terms of a linear combination of the column vectors of A. If we treat the columns as vectors in 2D Euclidean space, and consider the correct solutions (say, we somehow know that $x = 1$ and $y = 2$), then we have the picture in Figure A1.4. The result of adding the two vectors on the LHS is identically equal to the vector $\mathbf{b}$ on the RHS.

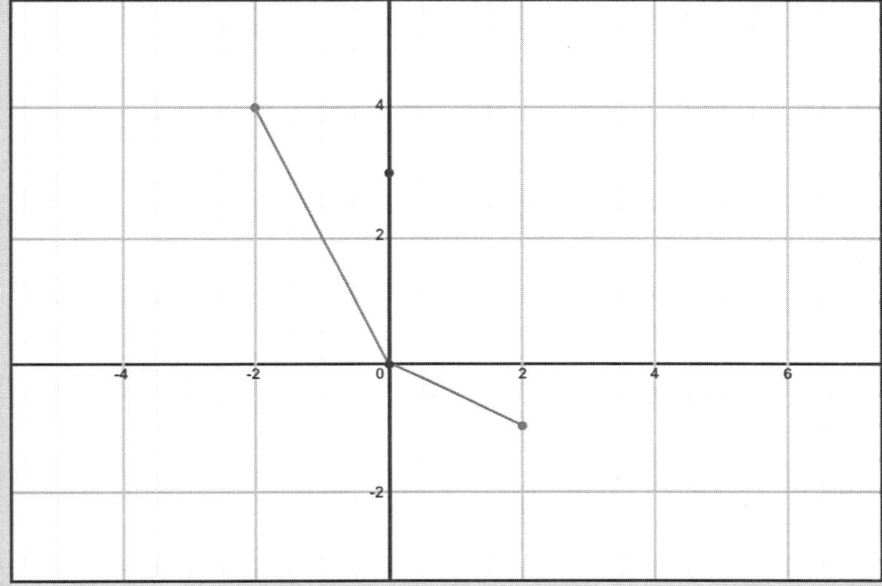

Figure A1.4 Another way to represent the solution.

Of course, the question arises whether we can always solve this system as follows: $\mathbf{x} = A^{-1}\mathbf{b}$ for any given 2D vector $\mathbf{b}$. This will be possible only if A is invertible. In that case, the columns of A will span the entire 2D Euclidean plane (and $\mathbf{x} = A^{-1}\mathbf{b}$ will be the solution for any 2D vector $\mathbf{b}$). This should lead us to ask when A is invertible. That is, when will the columns of A span the entire 2D Euclidean plane? The answer should be obvious by inspecting the Figure A1.4 2D graph: whenever the columns of A are linearly independent.

A1.12 Row echelon form, reduced row echelon form, and rank of a matrix

A1.12.1 *Definition (reduced row echelon form)*

A matrix is said to be in reduced row echelon form (rref) if it satisfies all the following conditions:

i.   If a row has non-zero entries, then the first non-zero entry is 1, known as the leading 1 or the **pivot** element of that row.

ii. If a column has a pivot element, then all the other entries in that column are 0.

iii. If a row contains a pivot element, then each row above it contains a leading 1 further to the left.

Note

1. The third condition implies that rows of zeros, if any, appear at the bottom of the matrix.

2. If a matrix satisfies only the conditions i and iii, then the matrix is in the row echelon form (ref).

Examples of ref $\quad \begin{bmatrix} 1 & 1 \\ 0 & 1 \end{bmatrix}, \begin{bmatrix} 1 & 5 & 7 \\ 0 & 1 & 8 \\ 0 & 0 & 0 \end{bmatrix}, \begin{bmatrix} 1 & 2 & 3 & 4 \\ 0 & 1 & 6 & 5 \\ 0 & 0 & 0 & 1 \end{bmatrix}.$

Examples of rref $\quad \begin{bmatrix} 1 & 0 \\ 0 & 1 \end{bmatrix}, \begin{bmatrix} 1 & 0 & -33 \\ 0 & 1 & 8 \\ 0 & 0 & 0 \end{bmatrix}, \begin{bmatrix} 1 & 0 & -9 & 0 \\ 0 & 1 & 6 & 0 \\ 0 & 0 & 0 & 1 \end{bmatrix}.$

A1.12.2 *How to convert the matrix into rref or ref*

Using the following elementary row operations, we convert a matrix into rref or ref.

1. Divide a row by a non-zero scalar.

2. Add a multiple of a row with another row.

3. Swap two rows.

Example A1.19

Convert the matrix A into ref and rref.

$$A = \begin{bmatrix} 2 & 6 & 16 \\ 1 & 3 & 8 \\ 1 & 4 & 10 \end{bmatrix}$$

Solution:

$$A = \begin{bmatrix} 2 & 6 & 16 \\ 1 & 3 & 8 \\ 1 & 4 & 10 \end{bmatrix} \tag{A1.7}$$

$$\sim \begin{bmatrix} 1 & 3 & 8 \\ 1 & 3 & 8 \\ 1 & 4 & 10 \end{bmatrix} \qquad R_1 \mapsto \frac{R_1}{2} \tag{A1.8}$$

$$\sim \begin{bmatrix} 1 & 3 & 8 \\ 0 & 0 & 0 \\ 0 & 1 & 2 \end{bmatrix} \qquad R_2 \mapsto R_2 - R_1, R_3 \mapsto R_3 - R_1 \tag{A1.9}$$

$$\sim \begin{bmatrix} 1 & 3 & 8 \\ 0 & 1 & 2 \\ 0 & 0 & 0 \end{bmatrix} \qquad R_2 \leftrightarrow R_3 \qquad \text{(A1.10)}$$

This is the ref of the matrix A.

Further using $R_1 \rightarrow R_1 - 3R_2$, we get the $rref(A) = \begin{bmatrix} 1 & 0 & 2 \\ 0 & 1 & 2 \\ 0 & 0 & 0 \end{bmatrix}$. $\qquad$ (A1.11)

A1.12.3 *Definition (rank)*

The rank of a matrix A is the number of pivot elements in the rref of the matrix A.

For the matrix $A = \begin{bmatrix} 2 & 6 & 16 \\ 1 & 3 & 8 \\ 1 & 4 & 10 \end{bmatrix}$, the rref is $\begin{bmatrix} 1 & 0 & 2 \\ 0 & 1 & 2 \\ 0 & 0 & 0 \end{bmatrix}$,

hence, the $rank(A) = 2$.

Note: If the rank of the matrix is equal to the number of the columns, then the matrix is full rank.

In the previous example the matrix A is not full rank.

A1.13 Linear transformation

A1.13.1 *Definition (linear transformation)*

A function $T : \mathbb{R}^n \rightarrow \mathbb{R}^m$ is called a linear transformation if $\exists A \in \mathrm{M}_{m \times n}(\mathbb{R})$ such that $T(\mathbf{x}) = \mathbf{Ax},\ \forall \mathbf{x} \in \mathbb{R}^n$.

Example A1.20

The rotation matrix $\begin{pmatrix} \cos\theta & -\sin\theta \\ \sin\theta & \cos\theta \end{pmatrix}$ is a linear transformation that rotates a vector in $\mathbb{R}^2$ by θ.

A1.13.2 *Find the matrix of linear transformation* $T : \mathbb{R}^n \rightarrow \mathbb{R}^m$

Let $\{e_1, e_2, \ldots, e_n\}$ be the standard basis of $\mathbb{R}^n$. Then the matrix of the linear transformation is given by

$$A = \begin{pmatrix} | & | & & | \\ T(\mathbf{e}_1) & T(\mathbf{e}_2) & \cdots & T(\mathbf{e}_n) \\ | & | & & | \end{pmatrix}.$$

Note: A square matrix is invertible if and only if its linear transformation is invertible.

Theorem A1.1 *An $n \times n$ matrix A is invertible* $\Leftrightarrow \operatorname{rref}(A) = I_n$. *In this case,* rank(A) = n.

A1.13.3 *Finding inverse of a matrix*

$A \in \mathbb{M}_{n\times n}(\mathbb{R})$. In order to find A^{-1}, form the augmented matrix $\tilde{A} = \left(A \mid I_n\right)$ and compute rref($\tilde{A}$).

i. If $\operatorname{rref}\left(\tilde{A}\right)$ is of the form $\left(I_n \mid B\right)$, then $A^{-1} = B$.

ii. If $\operatorname{rref}\left(\tilde{A}\right)$ is of any other form, then A is not invertible.

A1.13.4 *Definition (image or range of a matrix/linear transformation)*

Im(A) = Im(T) is the span of the column vectors of A.

Example A1.21

Find a basis of the image of $A = \begin{pmatrix} 1 & 2 & 2 & -5 & 6 \\ -1 & -2 & -1 & 1 & -1 \\ 4 & 8 & 5 & -8 & 9 \\ 3 & 6 & 1 & 5 & -7 \end{pmatrix} = \begin{pmatrix} | & | & | & | & | \\ \mathbf{a}_1 & \mathbf{a}_2 & \mathbf{a}_3 & \mathbf{a}_4 & \mathbf{a}_5 \\ | & | & | & | & | \end{pmatrix}$

and determine the dim(Im(A)).

Solution: To find the basis of Im(A), we need to identify the redundant columns of A from amongst all the column vectors of A. By inspection of A, it will be hard to tell which of the columns of A are redundant (linearly dependent on the others). So we will transform A to B = rref(A).

$$B = \operatorname{rref}(A) = \begin{pmatrix} 1 & 2 & 0 & 3 & -4 \\ 0 & 0 & 1 & -4 & 5 \\ 0 & 0 & 0 & 0 & 0 \\ 0 & 0 & 0 & 0 & 0 \end{pmatrix} = \begin{pmatrix} | & | & | & | & | \\ \mathbf{b}_1 & \mathbf{b}_2 & \mathbf{b}_3 & \mathbf{b}_4 & \mathbf{b}_5 \\ | & | & | & | & | \end{pmatrix}.$$

The redundant columns of B correspond to the redundant columns of A. The redundant columns of B are also easy to spot. They are the columns that do not contain the pivots, namely, $\mathbf{b}_2 = 2\mathbf{b}_1$, $\mathbf{b}_4 = 3\mathbf{b}_1 - 4\mathbf{b}_3$, and $\mathbf{b}_5 = -4\mathbf{b}_1 + 5\mathbf{b}_3$. Thus, the redundant columns of A are $\mathbf{a}_2 = 2\mathbf{a}_1$, $\mathbf{a}_4 = 3\mathbf{a}_1 - 4\mathbf{a}_3$, and $\mathbf{a}_5 = -4\mathbf{a}_1 + 5\mathbf{a}_3$. And the non-redundant columns of A are $\mathbf{a}_1$ and $\mathbf{a}_3$; they form a basis of the image of A.

Therefore, a basis of the image of A is $\left\{ \begin{pmatrix} 1 \\ -1 \\ 4 \\ 3 \end{pmatrix}, \begin{pmatrix} 2 \\ -1 \\ 5 \\ 1 \end{pmatrix} \right\}$ and *dim (Im(A))* = 2.

A1.13.5 *Definition (kernel of A (or equivalently the null space of A, Null(A)))*

The set of all $\mathbf{x} \in \mathbb{R}^n$ such that $T(\mathbf{x}) = A\mathbf{x} = 0$.

Example A1.22

Find a basis of the kernel of $A = \begin{pmatrix} 1 & 2 & 2 & -5 & 6 \\ -1 & -2 & -1 & 1 & -1 \\ 4 & 8 & 5 & -8 & 9 \\ 3 & 6 & 1 & 5 & -7 \end{pmatrix}$ and determine the dim $(Null(A))$.

Solution: Most importantly, $\ker(A) = \ker(rref(A)) = \ker(B)$. So we might as well solve for $\mathbf{x} = \begin{pmatrix} x_1 \\ x_2 \\ x_3 \\ x_4 \\ x_5 \end{pmatrix}$ such that $B\mathbf{x} = 0$. This is done by considering the augmented matrix $\tilde{B} = \left(B \mid 0 \right)$ from which we have the following:

$$x_1 + 2x_2 + 0x_3 + 3x_4 - 4x_5 = 0$$
$$0x_1 + 0x_2 + x_3 - 4x_4 + 5x_5 = 0$$

or equivalently

$$x_1 = -2x_2 - 3x_4 + 4x_5$$
$$x_3 = 4x_4 - 5x_5$$

which gives us $x_2 = \alpha$, $x_4 = \beta$, $x_5 = \gamma$ are set arbitrarily. Therefore,

$$\mathbf{x} = \begin{pmatrix} -2\alpha - 3\beta + 4\gamma \\ \alpha \\ 4\beta - 5\gamma \\ \beta \\ \gamma \end{pmatrix} = \begin{pmatrix} -2\alpha & -3\beta & +4\gamma \\ \alpha & & \\ & 4\beta & -5\gamma \\ & \beta & \\ & & \gamma \end{pmatrix} = \alpha \begin{pmatrix} -2 \\ 1 \\ 0 \\ 0 \\ 0 \end{pmatrix} + \beta \begin{pmatrix} -3 \\ 0 \\ 4 \\ 1 \\ 0 \end{pmatrix} + \gamma \begin{pmatrix} 4 \\ 0 \\ -5 \\ 0 \\ 1 \end{pmatrix}.$$

The $\ker(A)$ is spanned by these basis vectors $\left\{ \begin{pmatrix} -2 \\ 1 \\ 0 \\ 0 \\ 0 \end{pmatrix}, \begin{pmatrix} -3 \\ 0 \\ 4 \\ 1 \\ 0 \end{pmatrix}, \begin{pmatrix} 4 \\ 0 \\ -5 \\ 0 \\ 1 \end{pmatrix} \right\}$ and the $\dim(Null(A)) = 3$.

Theorem A1.2 $A \in \mathbb{M}_{m \times n}(\mathbb{R})$. Then $\ker(A) = \{0\} \Leftrightarrow \operatorname{rank}(A) = n$.

Theorem A1.3   *Theorem (rank-nullity theorem): For any $m \times n$ matrix A, the following is known as the fundamental theorem of linear algebra:*

$$\dim(\text{Null}(A)) + \dim(\text{Im}(A)) = n,$$

or equivalently

$$(\text{nullity of } A) + (\text{rank of } A) = n.$$

A1.14 Norm of a vector space

A1.14.1 *Vector norms*

Let $\mathbf{x} \in \mathbb{R}^n$, i.e., $\mathbf{x} = \begin{pmatrix} x_1 \\ x_2 \\ \vdots \\ x_n \end{pmatrix}$, with $x_i \in \mathbb{R}$. A vector norm on $\mathbb{R}^n$ is a function $\|\cdot\| : \mathbb{R}^n \to \mathbb{R}$

that satisfies the following properties:

1. $\|\mathbf{x}\| \geq 0$ for all $\mathbf{x} \in \mathbb{R}^n$,

2. $\|\mathbf{x}\| = 0 \Leftrightarrow \mathbf{x} = 0$,

3. $\|\alpha \mathbf{x}\| = |\alpha| \, \|\mathbf{x}\|$ for all $\alpha \in \mathbb{R}$ and for all $\mathbf{x} \in \mathbb{R}^n$,

4. $\|\mathbf{x} + \mathbf{y}\| \leq \|\mathbf{x}\| + \|\mathbf{y}\|$ for all $\mathbf{x}, \mathbf{y} \in \mathbb{R}^n$.

There are many types of norms on $\mathbb{R}^n$, e.g.,

i. $\|\mathbf{x}\|_2 = \sqrt{x_1^2 + x_2^2 + \cdots + x_n^2}$, $\hspace{2cm}$ (A1.12)

ii. $\|\mathbf{x}\|_\infty = \max_{1 \leq i \leq n} |x_i|$. $\hspace{2cm}$ (A1.13)

All norms are equivalent.

A1.14.2 *Cauchy–Schwarz inequality*

$$\left| \langle \mathbf{x}, \mathbf{y} \rangle \right| \equiv \left| \mathbf{x}^T \mathbf{y} \right| = \left| \sum_{1 \leq i \leq n} x_i y_i \right| \leq \|\mathbf{x}\|_2 \, \|\mathbf{y}\|_2.$$

Theorem A1.4 *For each* $\mathbf{x} \in \mathbb{R}^n$, $\|\mathbf{x}\|_\infty \leq \|\mathbf{x}\|_2 \leq \sqrt{n} \, \|\mathbf{x}\|_\infty$.

A1.14.3 *Matrix norms*

Let $A \in M_{n \times n}(\mathbb{R})$. A matrix norm is a function $\|\cdot\| : M_{n \times n} \to \mathbb{R}$ that satisfies the following properties:

1. $\|A\| \geq 0$.

2. $\|A\| = 0 \Leftrightarrow A$ is a "0" matrix with all entries 0.

3. $\|\alpha A\| = |\alpha| \, \|A\|$ for all $\alpha \in \mathbb{R}$ and for all $A \in M_{n \times n}(\mathbb{R})$.

4. $\|A + B\| \leq \|A\| + \|B\|$ for all $A, B \in M_{n \times n}(\mathbb{R})$.

5. $\|AB\| \leq \|A\| \, \|B\|$ for all $A, B \in M_{n \times n}(\mathbb{R})$.

Theorem A1.5 (Induced Matrix Norm) *If $\|\cdot\|$ is a vector norm on $\mathbb{R}^n$, then*
$$\|A\| = \max_{\|\mathbf{x}\|=1} \|A\mathbf{x}\| \text{ is a matrix norm.}$$

Alternatively, $\displaystyle \|A\| = \max_{\mathbf{x} \neq 0} \left\| A\left(\frac{\mathbf{x}}{\|\mathbf{x}\|}\right) \right\| = \max_{\mathbf{x} \neq 0} \frac{\|A\mathbf{x}\|}{\|\mathbf{x}\|}.$

Theorem A1.6 *For $A = \left(a_{ij}\right) \in M_{n \times n}\left(\mathbb{R}\right)$, we denote $\displaystyle \|A\|_1 = \max_{1 \leq j \leq n} \sum_{1 \leq i \leq n} \left|a_{ij}\right|$. Then $\|A\|_1$ is a matrix norm.*

Theorem A1.7 *For $A = \left(a_{ij}\right) \in M_{n \times n}\left(\mathbb{R}\right)$, we denote $\displaystyle \|A\|_\infty = \max_{1 \leq i \leq n} \sum_{1 \leq j \leq n} \left|a_{ij}\right|$. Then $\|A\|_\infty$ is a matrix norm.*

Example A1.23

Let $A = \begin{pmatrix} 1 & 2 & -1 \\ 0 & 3 & -1 \\ 5 & -1 & 1 \end{pmatrix}$. Then $\displaystyle \sum_{1 \leq j \leq 3} \left|a_{1j}\right| = |1| + |2| + |-1| = 4$, $\displaystyle \sum_{1 \leq j \leq 3} \left|a_{2j}\right| = |0| + |3| + |-1| = 4$,

$\displaystyle \sum_{1 \leq j \leq 3} \left|a_{3j}\right| = |5| + |-1| + |1| = 7$. Hence, $\|A\|_\infty = \max\{4, 4, 7\} = 7$.

A1.14.4 *Spectral radius of a matrix*

Let $A \in M_{n \times n}\left(\mathbb{R}\right)$. Define $\rho\left(A\right) := \max_{1 \leq i \leq n} \left|\lambda_i\right|$, where $\lambda_1, \ldots, \lambda_n$ are eigenvalues[1] of A in $\mathbb{C}$. The following theorem shows how spectral radius is closely related to the norm of the matrix.

[1] Eigenvalues are discussed in section A1.17.

Theorem A1.8 *Let $A \in M_{n \times n}\left(\mathbb{R}\right)$. Then*

  i. $\|A\|_2 = \sqrt{\rho\left(A^T A\right)},$ (A1.14)

  ii. $\rho\left(A\right) \leq \|A\|, \quad where \|\cdot\| \text{ is a matrix norm.}$ (A1.15)

A1.15 Orthogonal basis and Gram–Schmidt orthogonalization

A1.15.1 *Definition (orthogonal vectors)*

Two vectors $\mathbf{u}_1$ and $\mathbf{u}_2$ are orthogonal if and only if $\langle \mathbf{u}_1, \mathbf{u}_2 \rangle = 0$.

A1.15.2 *Definition (orthonormal vectors)*

The vectors $\mathbf{u}_1, \mathbf{u}_2, \cdots, \mathbf{u}_m \in \mathbb{R}^n$ are orthonormal if and only if $\langle \mathbf{u}_i, \mathbf{u}_j \rangle = \delta_{ij}$.

Example A1.24

The vectors $\mathbf{e}_1, \mathbf{e}_2, \ldots, \mathbf{e}_n \in \mathbb{R}^n$ are orthonormal.

A1.15.3 *Properties of orthonormal vectors*

1. Orthonormal vectors are (automatically) linearly independent.

2. Orthonormal vectors $\mathbf{u}_1, \mathbf{u}_2, \cdots, \mathbf{u}_n \in \mathbb{R}^n$ form a basis in $\mathbb{R}^n$.

A1.15.4 *Orthogonal projection and orthogonal complement*

[2] The elements of V are of the form $\begin{pmatrix} x \\ y \\ 0 \end{pmatrix}$, where $x, y \in \mathbb{R}$

1. Let $\mathbf{x} \in \mathbb{R}^n$ and a subspace V^2 of $\mathbb{R}^n$. Then we can write $\mathbf{x} = \mathbf{x}^\| + \mathbf{x}^\perp$, where $\mathbf{x}^\| \in V$ and $\mathbf{x}^\perp \in V^\perp$ and this representation is unique.

2. Here $V^\perp = \left\{ \mathbf{x} \in \mathbb{R}^n : \langle \mathbf{v}, \mathbf{x} \rangle = 0, \ \forall \mathbf{v} \in V \right\}$. The transformation $T(\mathbf{x}) = proj_v(\mathbf{x}) = \mathbf{x}^\|$ from $\mathbb{R}^n$ to V is linear and $V^\perp = \ker(\mathrm{T})$.

The shaded area denoted by V in Figure A1.5 is an infinite plane through the origin. The following are notationally equivalent:

$$\vec{x} \equiv \mathbf{x}, \vec{\mathbf{x}}^\| \equiv \mathbf{x}^\|, \text{ and } \vec{\mathbf{x}}^\perp \equiv \mathbf{x}^\perp.$$

How do we compute $\mathbf{x}^\|$? Consider an orthonormal basis of V: $\mathbf{u}_1, \mathbf{u}_2, \dots \mathbf{u}_m \in V$, which is a subspace of $\mathbb{R}^n$. Then

$$\mathbf{x}^\| = \langle \mathbf{u}_1, \mathbf{x} \rangle \mathbf{u}_1 + \cdots + \langle \mathbf{u}_m, \mathbf{x} \rangle \mathbf{u}_m; \ \forall \mathbf{x} \in \mathbb{R}^n.$$

Consequently, consider an orthonormal basis of $\mathbb{R}^n$: $\mathbf{u}_1, \mathbf{u}_2, \dots \mathbf{u}_n$. Then for any $\in$

$$\mathbf{x} = \langle \mathbf{u}_1, \mathbf{x} \rangle \mathbf{u}_1 + \cdots + \langle \mathbf{u}_n, \mathbf{x} \rangle \mathbf{u}_n.$$

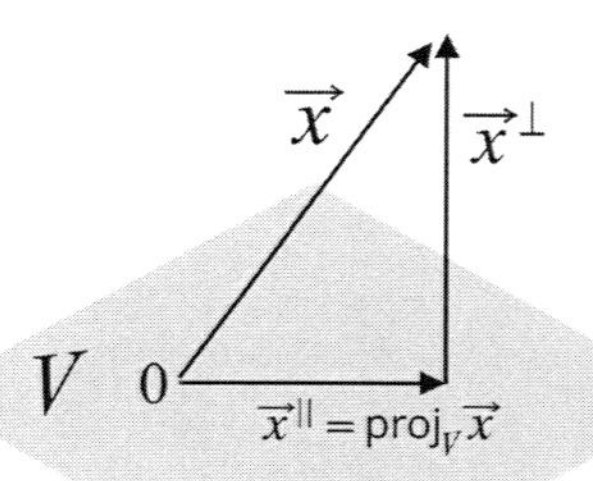

Figure A1.5 Vector projection and orthogonal complement.

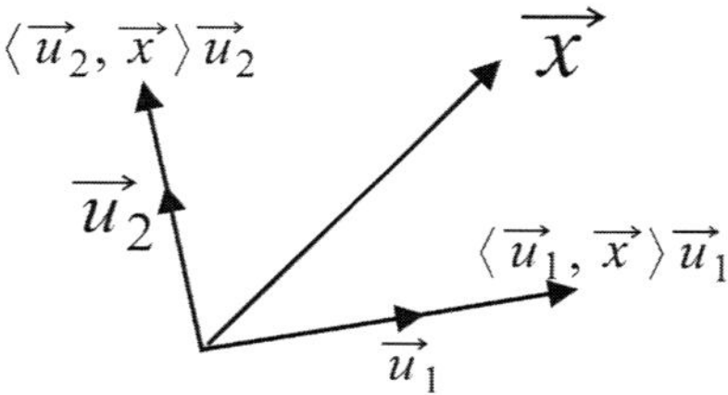

Figure A1.6 Linear combination of orthonormal basis.

A1.15.5 *Properties of orthogonal complement*

Consider a subspace $V \in \mathbb{R}^n$.

1. $V^\perp$ is a subspace of $\mathbb{R}^n$.

2. $V \cap V^\perp = \{0\}$.

3. $\dim(V) + \dim(V^\perp) = n$.

4. $\left(V^\perp \right)^\perp = V$.

Example A1.25

Consider the subspace $V = \mathrm{Im}(A)$ of $\mathbb{R}^4$, where $A = \begin{pmatrix} 1 & 1 \\ 1 & -1 \\ 1 & -1 \\ 1 & 1 \end{pmatrix}$. Find $\mathbf{x}^\|$ for $\mathbf{x} = \begin{pmatrix} 1 \\ 3 \\ 1 \\ 7 \end{pmatrix}$.

Solution: Recall that the column space of A is $\mathrm{Im}(A)$. It can be easily checked that the column vectors of A are orthogonal by taking their scalar product. Thus, we can construct an orthonormal basis of Im(A). The basis vectors are: $\mathbf{u}_1 = \begin{pmatrix} 1/2 \\ 1/2 \\ 1/2 \\ 1/2 \end{pmatrix}$ and $\mathbf{u}_2 = \begin{pmatrix} 1/2 \\ -1/2 \\ -1/2 \\ 1/2 \end{pmatrix}$.

Then $\mathbf{x}^{\parallel} = \langle \mathbf{u}_1, \mathbf{x} \rangle \mathbf{u}_1 + \langle \mathbf{u}_2, \mathbf{x} \rangle \mathbf{u}_2 = 6\mathbf{u}_1 + 2\mathbf{u}_2 = \begin{pmatrix} 3 \\ 3 \\ 3 \\ 3 \end{pmatrix} + \begin{pmatrix} 1 \\ -1 \\ -1 \\ 1 \end{pmatrix} = \begin{pmatrix} 4 \\ 2 \\ 2 \\ 4 \end{pmatrix}$. In order to check

that this answer is indeed correct, verify that $\left(\mathbf{x} - \mathbf{x}^{\parallel} \right) \perp \mathbf{u}_1, \mathbf{u}_2$.

A1.15.6 *Why are orthonormal basis vectors useful?*

i.   We know that if we have some basis $\mathbf{v}_1, \mathbf{v}_2, \dots, \mathbf{v}_n$ of an n-dimensional vector space W, then any vector $\mathbf{x} \in W$ can be written as $\mathbf{x} = \alpha_1 \mathbf{v}_1 + \alpha_2 \mathbf{v}_2 + \cdots + \alpha_n \mathbf{v}_n$ (as a linear combination of the basis vectors) but there is no first-principle or convenient way of finding the unique coefficients $\alpha_1, \alpha_2, \dots, \alpha_n$ except by explicit guesswork calculations. Now, instead, if we have an orthonormal basis set $\mathbf{u}_1, \mathbf{u}_2, \dots, \mathbf{u}_n$, then any vector can be written as a linear combination of this orthonormal basis set as follows:

$\mathbf{x} = \beta_1 \mathbf{u}_1 + \beta_2 \mathbf{u}_2 + \cdots + \beta_n \mathbf{u}_n$, where the coefficients can now be uniquely determined as $\beta_i = \langle \mathbf{u}_i, \mathbf{x} \rangle, \forall i = 1, 2, \dots, n$.

ii. Orthogonality guarantees linear independence.

A1.15.7 *Why are orthogonal transformations useful?*

1.   Orthogonal transformations are metric preserving transformations, i.e., if $T : \mathbb{R}^n \to \mathbb{R}^n$ is orthogonal, then $\| T(\mathbf{x}) \| = \| \mathbf{x} \|$, $\forall \mathbf{x} \in \mathbb{R}^n$.

2.   Orthogonal transformations are angle preserving transformations for pairs of vectors. If $\mathbf{u} \perp \mathbf{w}$, then $T(\mathbf{u}) \perp T(\mathbf{w})$.

A1.16 Orthonormalization, Gram–Schmidt process, and QR factorization

A1.16.1 *How to construct orthonormal basis vectors?*

Let us illustrate this procedure by an example of a vector space V which is a plane. Given that V has bases $\mathbf{v}_1$ and $\mathbf{v}_2$ which are not orthogonal, how do we construct an orthonormal basis?

Step 1 The first basis vector is easy to construct: $\mathbf{u}_1 = \frac{1}{\|\mathbf{v}_1\|} \mathbf{v}_1$.

Step 2 Construct $\mathbf{v}_2^{\parallel} = \mathrm{proj}_L (\mathbf{v}_2) = \langle \mathbf{u}_1, \mathbf{v}_2 \rangle \mathbf{u}_1$. See Figure A1.7.

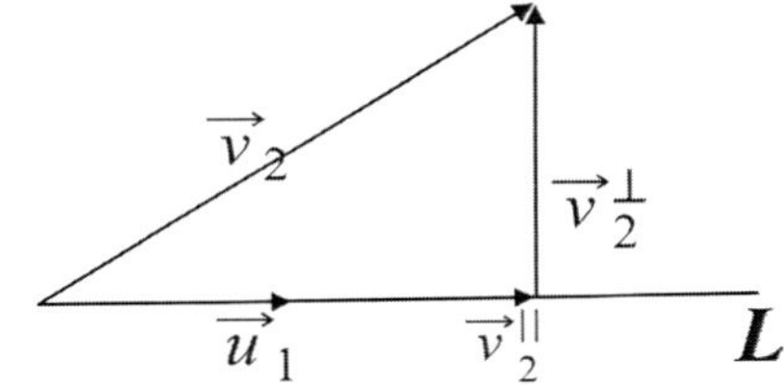

Figure A1.7 Gram–Schmidt orthonormalization.

Step 3 Construct $\mathbf{v}_2^{\perp} = \mathbf{v}_2 - \mathbf{v}_2^{\parallel}$.

Step 4 Construct $\mathbf{u}_2 = \dfrac{1}{\left\|\mathbf{v}_2^{\perp}\right\|}\mathbf{v}_2^{\perp}$.

Example A1.26

Find an orthonormal basis $\mathbf{u}_1$, $\mathbf{u}_2$ of the subspace $V = \mathrm{span}\left(\begin{pmatrix} 1 \\ 1 \\ 1 \\ 1 \end{pmatrix}, \begin{pmatrix} 1 \\ 9 \\ 9 \\ 1 \end{pmatrix}\right)$ of $\mathbb{R}^4$ with

basis $\mathbf{v}_1 = \begin{pmatrix} 1 \\ 1 \\ 1 \\ 1 \end{pmatrix}$, $\mathbf{v}_2 = \begin{pmatrix} 1 \\ 9 \\ 9 \\ 1 \end{pmatrix}$.

Solution: Using the procedure mentioned above, we arrive at

$$\mathbf{u}_1 = \begin{pmatrix} 1/2 \\ 1/2 \\ 1/2 \\ 1/2 \end{pmatrix}, \quad \mathbf{u}_2 = \begin{pmatrix} -1/2 \\ 1/2 \\ 1/2 \\ -1/2 \end{pmatrix}.$$

A1.16.2 *Gram–Schmidt orthonormalisation process*

Consider the basis $\mathbf{v}_1, \ldots, \mathbf{v}_m$ of a subspace V of $\mathbb{R}^n$. For $j = 2, \ldots, m$, we resolve the vector $\mathbf{v}_j$ into its components parallel and perpendicular to the span of the preceding vectors $\mathbf{v}_1, \ldots, \mathbf{v}_{j-1}$.

$$\mathbf{v}_j = \mathbf{v}_j^{\parallel} + \mathbf{v}_j^{\perp}, \ \text{ w.r.t. } \mathrm{span}\left(\mathbf{v}_1, \ldots, \mathbf{v}_{j-1}\right).$$

Then $\mathbf{u}_1 = \dfrac{1}{\left\|\mathbf{v}_1\right\|}\mathbf{v}_1, \mathbf{u}_2 = \dfrac{1}{\left\|\mathbf{v}_2^{\perp}\right\|}\mathbf{v}_2^{\perp}, \ldots, \mathbf{u}_j = \dfrac{1}{\left\|\mathbf{v}_j^{\perp}\right\|}\mathbf{v}_j^{\perp}, \ldots, \mathbf{u}_m = \dfrac{1}{\left\|\mathbf{v}_m^{\perp}\right\|}\mathbf{v}_m^{\perp}$ is an orthonormal basis of V.

Here $\mathbf{v}_j^{\perp} = \mathbf{v}_j - \mathbf{v}_j^{\parallel} = \mathbf{v}_j - \langle \mathbf{u}_1, \mathbf{v}_j \rangle \mathbf{u}_1 - \cdots - \langle \mathbf{u}_{j-1}, \mathbf{v}_j \rangle \mathbf{u}_{j-1}$.

A1.16.3 *QR factorization*

The Gram–Schmidt process represents a change of bases from the old basis $\mathbf{v}_1, \ldots, \mathbf{v}_m$ to a set of new orthonormal bases $\mathbf{u}_1, \ldots, \mathbf{u}_m$ of V. The QR factorization involves a change of basis matrix R such that

$$\begin{pmatrix} | & & & & & | \\ \mathbf{v}_1 & \cdot & \cdot & \cdot & & \mathbf{v}_m \\ | & & & & & | \end{pmatrix} = \begin{pmatrix} | & & & & & | \\ \mathbf{u}_1 & \cdot & \cdot & \cdot & & \mathbf{u}_m \\ | & & & & & | \end{pmatrix} R,$$

i.e., $M = QR$,

where R is an upper triangular matrix with entries:

$r_{11} = \left\|\mathbf{v}_1\right\|, r_{jj} = \left\|\mathbf{v}_j^{\perp}\right\|$ (for $j = 2, \ldots, m$), and $r_{ij} = \langle \mathbf{u}_i, \mathbf{v}_j \rangle$ (for $i < j$).

Example A1.27

Find the QR factorization of the matrix $M = \begin{pmatrix} 2 & 2 \\ 1 & 7 \\ -2 & -8 \end{pmatrix}$.

Solution: $Q = \dfrac{1}{3} \begin{pmatrix} 2 & -2 \\ 1 & 2 \\ -2 & -1 \end{pmatrix}$ and $R = \begin{pmatrix} 3 & 9 \\ 0 & 6 \end{pmatrix}$.

A1.17 Eigenvalues and eigenvectors

A1.17.1 *Eigenvalues and eigenvectors*

Let $A \in M_{n\times n}(\mathbb{F})$, where $\mathbb{F} = \mathbb{R}$ or $\mathbb{F} = \mathbb{C}$. A nonzero vector $\mathbf{x} \in \mathbb{F}^n$ is an eigenvector of A if $A\mathbf{x} = \lambda\mathbf{x}$ for some $\lambda \in \mathbb{F}$. λ is said to be an eigenvalue of A corresponding to the eigenvector $\mathbf{x}$.

Example A1.28

For $A = \begin{pmatrix} 2 & -1 \\ 2 & 4 \end{pmatrix}$, find the eigenvalue and eigenvectors.

Solution: $A - \lambda I = \begin{pmatrix} 2-\lambda & -1 \\ 2 & 4-\lambda \end{pmatrix}$.

$$\det(A - \lambda I) = \det\begin{pmatrix} 2-\lambda & -1 \\ 2 & 4-\lambda \end{pmatrix} = (2-\lambda)(4-\lambda) + 2 = \lambda^2 - 6\lambda + 10.$$

The above polynomial in λ is called the characteristic polynomial for the matrix A. Thus,

$$\det(A - \lambda I) = 0 \Rightarrow \lambda = 3 \pm i.$$

Let us call $\lambda_1 = 3 + i$ and $\lambda_2 = 3 - i$.

To find the eigenvector w.r.t. λ_1, solve $A\mathbf{x} = \lambda_1\mathbf{x}$. After solving, we obtain $(1 + i)x_1 + x_2 = 0$, i.e., $x_2 = -(1 + i)x_1$. We can take x_1 to be any nonzero scalar from $\mathbb{R}$, say k, so as to write $\begin{pmatrix} x_1 \\ x_2 \end{pmatrix} = \begin{pmatrix} k \\ -(1+i)k \end{pmatrix} = k\begin{pmatrix} 1 \\ -1-i \end{pmatrix}$. Hence, any nonzero multiple of the vector $\begin{pmatrix} 1 \\ -1-i \end{pmatrix}$ is an eigenvector of the matrix A w.r.t the eigenvalue λ_1. Similarly, one can find the eigenvector corresponding to the eigenvalue λ_2.

When a matrix is of higher dimensions, we will use the **power method**[3] to find the eigenvalues and the corresponding eigenvectors.

[3] For power method, refer to chapter 9 of Richard L. Burden and J. Douglas Faires, *Numerical Analysis,* 9th ed. (Cengage Learning, 2002).

Theorem A1.9   *For finite dimensional vector spaces U,V over $\mathbb{F}$, a linear transformation $T:U \to V$ is invertible if and only if T is one to one and onto.*

A1.18 Diagonalizable matrices

A1.18.1 *Definition (similarity transformation)*

$A \in M_{n \times n}(\mathbb{F})$ is diagonalizable over $\mathbb{F}$ if there exists an invertible matrix S over $\mathbb{F}$ such that $A = SDS^{-1}$, or, equivalently, $D = S^{-1}AS$.

Note that the eigenvectors of A and D will be the same and the above relation $D = S^{-1}AS$ is known as the similarity transformation.

A1.18.2 *When is a matrix diagonalizable?*

$A \in M_{n \times n}(\mathbb{F})$ is diagonalizable if and only if A has n linearly independent eigenvectors in $\mathbb{F}^n$.

Note that an $n \times n$ complex matrix that has n distinct eigenvalues is diagonalizable.

Example A1.29

Find a matrix that diagonalizes $A = \begin{pmatrix} 2 & -1 \\ 2 & 4 \end{pmatrix}$.

Solution: We solve $\det(A - \lambda I) = 0$ to obtain $\lambda_1 = 3 + i$ and $\lambda_2 = 3 - i$. From $Ax = \lambda_i x$ for $i = 1, 2$, we obtain

$$\mathbf{x}_1 = \begin{pmatrix} 1 \\ -1-i \end{pmatrix}, \ \mathbf{x}_2 = \begin{pmatrix} 1 \\ -1+i \end{pmatrix}$$

as eigenvectors of A w.r.t. the eigenvalues λ_1 and λ_2, respectively. We note that $S = \begin{pmatrix} 1 & 1 \\ -1-i & -1+i \end{pmatrix}$ diagonalizes A. We can check

$$S^{-1}AS = \begin{pmatrix} \dfrac{-1+i}{2i} & -\dfrac{1}{2i} \\ \dfrac{1+i}{2i} & \dfrac{1}{2i} \end{pmatrix} \begin{pmatrix} 2 & -1 \\ 2 & 4 \end{pmatrix} \begin{pmatrix} 1 & 1 \\ -1-i & -1+i \end{pmatrix} \tag{A1.16}$$

$$= \begin{pmatrix} 3+i & 0 \\ 0 & 3-i \end{pmatrix} \tag{A1.17}$$

$$= \begin{pmatrix} \lambda_1 & 0 \\ 0 & \lambda_2 \end{pmatrix} \tag{A1.18}$$

$$= D \tag{A1.19}$$

The column vectors of S form an eigenbasis for A and the diagonal entries of D are the associated eigenvalues.

Example A1.30

Find the eigenspace of $A = \begin{pmatrix} 1 & 1 & 1 \\ 0 & 0 & 1 \\ 0 & 0 & 1 \end{pmatrix}$.

Solution: The eigenvalues are given by 0 and 1 with algebraic multiplicity 1 and 2, respectively. To find the eigenvectors, consider

$$\mathbf{x}_1 = \ker(A - 1I) \tag{A1.20}$$

$$= \ker \begin{pmatrix} 0 & 1 & 1 \\ 0 & -1 & 1 \\ 0 & 0 & 0 \end{pmatrix} \tag{A1.21}$$

$$= \ker \begin{pmatrix} 0 & 1 & 0 \\ 0 & 0 & 1 \\ 0 & 0 & 0 \end{pmatrix} \tag{A1.22}$$

$$= \operatorname{span}\left\{ \begin{pmatrix} 1 \\ 0 \\ 0 \end{pmatrix} \right\}, \tag{A1.23}$$

where $\begin{pmatrix} 0 & 1 & 0 \\ 0 & 0 & 1 \\ 0 & 0 & 0 \end{pmatrix}$ is the rref of the matrix $\begin{pmatrix} 0 & 1 & 1 \\ 0 & -1 & 1 \\ 0 & 0 & 0 \end{pmatrix}$. The calculation is as follows:

$$\begin{pmatrix} 0 & 1 & 0 \\ 0 & 0 & 1 \\ 0 & 0 & 0 \end{pmatrix} \begin{pmatrix} x_1 \\ x_2 \\ x_3 \end{pmatrix} = \begin{pmatrix} 0 \\ 0 \\ 0 \end{pmatrix} \tag{A1.24}$$

$$\Rightarrow \begin{pmatrix} x_2 \\ x_3 \\ 0 \end{pmatrix} = \begin{pmatrix} 0 \\ 0 \\ 0 \end{pmatrix}. \tag{A1.25}$$

The above calculation shows that $x_2 = x_3 = 0$. Thus, we can take any nonzero value as x_1 to obtain an eigenvector of A w.r.t. the eigenvalue 1. For convenience, we take $x_1 = 1$ to obtain $\begin{pmatrix} 1 \\ 0 \\ 0 \end{pmatrix}$ as an eigenvector. Likewise, $\mathbf{x}_2 = \ker(A - 0I) = \operatorname{span}\left\{ \begin{pmatrix} -1 \\ 1 \\ 0 \end{pmatrix} \right\}$.

Thus, we are able to find only two linearly independent eigenvectors. Hence, we will not have an eigenbasis because there is no S to diagonalize A.

Theorem A.10 *A matrix A is orthogonally diagonalizable* ($D = Q^{-1}AQ = Q^t AQ$) $\Leftrightarrow A$ *is symmetric.*

A1.19 Matrix decomposition

A1.19.1 *LU factorization*

Let A be an $n \times n$ square matrix. Then there exists a decomposition of A of the form

$$A = LU,$$

where L^4 is a lower triangular matrix and U is an upper triangular matrix. We can find L and U using the elementary row operations on A.

[4] Sometimes we find the decomposition of the form $A = PLU$, where P is a permutation matrix.

Example A1.31

Find the LU decomposition of the matrix $A = \begin{bmatrix} 1 & 3 & 8 \\ 2 & 8 & 20 \\ 3 & 10 & 25 \end{bmatrix}$.

Solution:

$$A = \begin{bmatrix} 1 & 3 & 8 \\ 2 & 8 & 20 \\ 3 & 10 & 25 \end{bmatrix} \xrightarrow[R_2 \to R_2 - 2R_1]{R_3 \to R_3 - 3R_1} \begin{bmatrix} 1 & 3 & 8 \\ 0 & 2 & 4 \\ 0 & 1 & 1 \end{bmatrix} \xrightarrow{R_3 \to R_3 - \frac{1}{2}R_2} \begin{bmatrix} 1 & 3 & 8 \\ 0 & 2 & 4 \\ 0 & 0 & -1 \end{bmatrix}$$

hence, $L = \begin{bmatrix} 1 & 0 & 0 \\ 2 & 1 & 0 \\ 3 & \frac{1}{2} & 1 \end{bmatrix}$ and $U = \begin{bmatrix} 1 & 3 & 8 \\ 0 & 2 & 4 \\ 0 & 0 & -1 \end{bmatrix}$. The entries of L can be directly obtained by inspecting the elementary rows operations that were performed on A to obtain U.

$$A = \begin{bmatrix} 1 & 3 & 8 \\ 2 & 8 & 20 \\ 3 & 10 & 25 \end{bmatrix} = \begin{bmatrix} 1 & 0 & 0 \\ 2 & 1 & 0 \\ 3 & \frac{1}{2} & 1 \end{bmatrix} \begin{bmatrix} 1 & 3 & 8 \\ 0 & 2 & 4 \\ 0 & 0 & -1 \end{bmatrix}$$ is one of the LU decompositions of A.

A1.19.2 *Cholesky decomposition*

[5] If all the eigenvalues of the matrix are positive, then the matrix is positive definite.

Let $A = \left(a_{ij}\right)_{n \times n}$ be a symmetric and positive definite matrix,[5] then matrix A can be factorized as

$$A = LL^T,$$

$$\text{where } L = \left(l_{ij}\right)_{n \times n} = \begin{cases} \sqrt{\left(a_{ii} - \sum_{j=1}^{i-1} l_{ij}^2\right)} & \text{if } i = j, \\[2em] \dfrac{\left(a_{ij} - \sum_{k=1}^{j-1} l_{ik} l_{jk}\right)}{l_{jj}} & \text{if } i > j, \\[2em] 0 & \text{if } i < j \end{cases}$$ is a lower triangular matrix. $L = \left(l_{ij}\right)_{n \times n}$

is deduced by comparing the entries of $A = LL^T$.

Example A1.32

Find the Cholesky decomposition of the matrix $\begin{bmatrix} 4 & 2 & 6 \\ 2 & 2 & 5 \\ 6 & 5 & 22 \end{bmatrix}$.

Solution:

$$l_{11} = \sqrt{4} = 2, \quad l_{22} = \sqrt{2-1} = 1, \tag{A1.26}$$

$$l_{21} = \frac{2}{2} = 1, \quad l_{32} = 5 - 3 = 2, \tag{A1.27}$$

$$l_{31} = \frac{6}{2} = 3, \quad l_{33} = \sqrt{22 - 9 - 4} = 3. \tag{A1.28}$$

Hence,

$$L = \begin{bmatrix} 2 & 0 & 0 \\ 1 & 1 & 0 \\ 3 & 2 & 3 \end{bmatrix}$$

and

$$A = LL^T = \begin{bmatrix} 2 & 0 & 0 \\ 1 & 1 & 0 \\ 3 & 2 & 3 \end{bmatrix} \begin{bmatrix} 2 & 1 & 3 \\ 0 & 1 & 2 \\ 0 & 0 & 3 \end{bmatrix}.$$

A1.19.3 *Spectral decomposition*

Let A be a real symmetric $n \times n$ matrix with eigenvalues $\lambda_1, \lambda_2, \ldots, \lambda_n$ and corresponding orthonormal eigenvectors $\mathbf{v}_1, \mathbf{v}_2, \ldots, \mathbf{v}_n$; then

$$A = \begin{pmatrix} \vdots & \vdots & \vdots \\ \mathbf{v}_1 & \mathbf{v}_2 & \mathbf{v}_3 \\ \vdots & \vdots & \vdots \end{pmatrix} \begin{pmatrix} \lambda_1 & & \mathbf{o} \\ & \ddots & \\ \mathbf{o} & & \lambda_n \end{pmatrix} \begin{pmatrix} \cdots & \mathbf{v}_1^T & \cdots \\ \cdots & \mathbf{v}_2^T & \cdots \\ & \vdots & \\ \cdots & \mathbf{v}_n^T & \cdots \end{pmatrix} \tag{A1.29}$$

$$= QDQ^T. \tag{A1.30}$$

Example A1.33

Find the spectral decomposition of the matrix $A = \begin{bmatrix} 5 & 2 \\ 2 & 2 \end{bmatrix}$.

Solution: Eigenvalues of A are 6 and 1 and the corresponding eigenvectors

are $\begin{bmatrix} \dfrac{2}{\sqrt{3}} \\ \dfrac{1}{\sqrt{3}} \end{bmatrix}$ and $\begin{bmatrix} -\dfrac{1}{\sqrt{3}} \\ \dfrac{2}{\sqrt{3}} \end{bmatrix}$.

Hence,

$$A = \begin{bmatrix} 5 & 2 \\ 2 & 2 \end{bmatrix} = \begin{bmatrix} \dfrac{2}{\sqrt{3}} & \dfrac{1}{\sqrt{3}} \\ -\dfrac{1}{\sqrt{3}} & \dfrac{2}{\sqrt{3}} \end{bmatrix} \begin{bmatrix} 6 & 0 \\ 0 & 1 \end{bmatrix} \begin{bmatrix} \dfrac{2}{\sqrt{3}} & -\dfrac{1}{\sqrt{3}} \\ \dfrac{1}{\sqrt{3}} & \dfrac{2}{\sqrt{3}} \end{bmatrix}.$$

A1.19.4 *Schur decomposition*

Let A be an $n \times n$ square matrix with real entries; then A can be expressed as

$$A = QTQ^{-1},$$

where Q is an orthogonal matrix and T is an upper triangular matrix with the eigenvalues of A along its diagonal. This decomposition is not unique.

Example A1.34

Find the Schur decomposition of the matrix $A = \begin{bmatrix} 7 & -2 \\ 12 & -3 \end{bmatrix}$.

Solution: Eigenvalues of the matrix A are 3 and 1.

The eigenvector corresponding to the eigenvalue 3 is $u = \dfrac{1}{\sqrt{5}} \begin{bmatrix} 1 \\ 2 \end{bmatrix}$.

We find an orthonormal basis $\{v\}$ of $span\{u\}^{\perp} = span\left\{ \dfrac{1}{\sqrt{5}} \begin{bmatrix} 1 \\ 2 \end{bmatrix} \right\}^{\perp} = span\left\{ \dfrac{1}{\sqrt{5}} \begin{bmatrix} -2 \\ 1 \end{bmatrix} \right\}$.

Let $Q = \dfrac{1}{\sqrt{5}} \begin{bmatrix} 1 & -2 \\ 2 & 1 \end{bmatrix}$ be an orthogonal matrix and

$$A = QTQ^{T}, \tag{A1.31}$$

$$T = \begin{bmatrix} 3 & -14 \\ 0 & 1 \end{bmatrix}. \tag{A1.32}$$

$A = \begin{bmatrix} 7 & -2 \\ 12 & -3 \end{bmatrix} = \dfrac{1}{\sqrt{5}} \begin{bmatrix} 1 & -2 \\ 2 & 1 \end{bmatrix} \begin{bmatrix} 3 & -14 \\ 0 & 1 \end{bmatrix} \dfrac{1}{\sqrt{5}} \begin{bmatrix} 1 & 2 \\ -2 & 1 \end{bmatrix}$ is Schur decompositon of A.

Appendix **II**

A Breeze through Matlab

A2.1 Mathematical operations

Try the following operations and commands on the command line in the command window of Matlab.

A2.1.1 *Defining an array and arranging in ascending and descending order*

```
>> A = [2   4 -3 43 2 10 log(200)]

A =

    2.0000    4.0000   -3.0000   43.0000    2.0000   10.0000    5.2983

>> sort(A)

ans =

   -3.0000    2.0000    2.0000    4.0000    5.2983   10.0000   43.0000

>> sort(A,'descend')

ans =

   43.0000   10.0000    5.2983    4.0000    2.0000    2.0000   -3.0000
```

A2.1.2 *Defining a matrix and arranging in ascending and descending order*

```
>> B = [2 3 1 4; -4 -2 0 -13; 66 3 0.2 1/3]

B =

    2.0000    3.0000    1.0000    4.0000
   -4.0000   -2.0000         0  -13.0000
   66.0000    3.0000    0.2000    0.3333
```

```
>> sort(B)

ans =

   -4.0000   -2.0000         0  -13.0000
    2.0000    3.0000    0.2000    0.3333
   66.0000    3.0000    1.0000    4.0000

>> sort(B,2)

ans =

    1.0000    2.0000    3.0000    4.0000
  -13.0000   -4.0000   -2.0000         0
    0.2000    0.3333    3.0000   66.0000
```

A2.1.3 *Powers of a matrix and its elements*

```
>> C = [ 1 2 3;  4 2 3;  5 6 2]

C =

     1     2     3
     4     2     3
     5     6     2

>> C^2

ans =

    24    24    15
    27    30    24
    39    34    37

>> C.^2

ans =

     1     4     9
    16     4     9
    25    36     4

>> D = [ 3 3 2;  5 0 1;  -2 3.5 6]

D =

    3.0000    3.0000    2.0000
    5.0000         0    1.0000
   -2.0000    3.5000    6.0000
```

```
>> C*D

ans =

    7.0000   13.5000   22.0000
   16.0000   22.5000   28.0000
   41.0000   22.0000   28.0000

>> C.*D

ans =

    3    6    6
   20    0    3
  -10   21   12
```

A2.1.4 *Finding maximum and minimum entries in an array (or matrix)*

```
>> A = [ 4 2 -7 -0.2 33 12]

A =

    4.0000    2.0000   -7.0000   -0.2000   33.0000   12.C000

>> max(A)

ans =

    33

>> min(A)

ans =

    -7
```

Finding `argmax` *and* `argmin` *in an array*

```
>> find(A == max(A))

ans =

    5

>> find(A == min(A))

ans =

    3
```

```
>> D

D =

      3.0000      3.0000      2.0000
      5.0000           0      1.0000
     -2.0000      3.5000      6.0000

>> max(D)

ans =

      5.0000      3.5000      6.0000

>> max(D,[],2)

ans =

      3
      5
      6
```

Summing (and cumulative summing) rows and columns of a matrix

```
>> A = [ 3 1 -2 5; 0 2 11 -4; 9 0 7 6]

A =

      3      1     -2      5
      0      2     11     -4
      9      0      7      6

>> sum(A)

ans =

     12      3     16      7

>> sum(A,2)

ans =

      7
      9
     22

>> cumsum(A)

ans =
```

```
     3       1      -2       5
     3       3       9       1
    12       3      16       7

>> cumsum(A,2)

ans =

     3       4       2       7
     0       2      13       9
     9       9      16      22
```

Product of entries of a matrix

```
>> prod(A)

ans =

     0       0    -154    -120

>> prod(A,2)

ans =

   -30
     0
     0
```

A2.1.5 *Accessing certain sections of a matrix*

```
>> A = [22 3 -4; 0 1 7; 2 -1 2]

A =

    22       3      -4
     0       1       7
     2      -1       2

>> A(:,2)

ans =

     3
     1
    -1

>> A(3,:)

ans =
```

```
       2     -1      2

>> [-2 -1 -3].*A(1,:)

ans =

   -44    -3    12
```

A2.1.6 *Some elementary mathematical functions*

```
>> exp(2)

ans =

    7.3891

>> sin(pi/4)

ans =

    0.7071

>> log(100)

ans =

    4.6052

>> log10(100)

ans =

    2

>> exp([1 2; 3 4])

ans =

     2.7183     7.3891
    20.0855    54.5982
```

A2.1.7 *Random number generator*

```
>> rand

ans =

    0.3112

>> rand()
```

```
ans =

    0.5285

>> rand(3)

ans =

    0.1656    0.6541    0.4505
    0.6020    0.6892    0.0838
    0.2630    0.7482    0.2290

>> rand(1,2)

ans =

    0.9133    0.1524
>> a=2

a =

    2

>> b=5

b =

    5

>> r = a+(b-a).*rand(5,1)        % random numbers between a and b
                                 % from uniform distribution

r =

    4.4775
    3.6150
    4.9884
    2.2345
    3.3280

>> randi([-3 3],5,1)

ans =

    -3
     3
     1
     0
    -2
```

```
>> randperm(4)

ans =

     3     4     2     1

>> nchoosek(5,2)

ans =

    10
>> perms([1 2 3])

ans =

     3     2     1
     3     1     2
     2     3     1
     2     1     3
     1     3     2
     1     2     3
```

Normally distributed random numbers

```
>> randn(5)

ans =

    1.1275   -1.7502   -0.5336   -0.0348    1.3514
    0.3502   -0.2857   -2.0026   -0.7982   -0.2248
   -0.2991   -0.8314    0.9642    1.0187   -0.5890
    0.0229   -0.9792    0.5201   -0.1332   -0.2938
   -0.2620   -1.1564   -0.0200   -0.7145   -0.8479
```

A2.1.8 *Calculus and functional operations*

Writing Matlab script files

It must be noted that the command window is not the only interface in Matlab to perform mathematical operations. In fact, there is a more optimal and preferred editor where one can write Matlab script files and define functions of their choice. To familiarize the user with the Matlab editor, the following sequences of operations are written as a script file using the editor. In order to execute the code, you simply have to hit the run icon on the editor at the top. The answers and results of the code appear on the command window as well as in the workspace.[1]

[1] The commented sections of the code appear to the right of the % symbol and are non-executable portions of the code. The comments should be self-explanatory.

```
% function handles and function evaluations
f = @(y) exp(y)
f(1)
```

```matlab
syms x              % define x as a symbolic variable
g = 5*exp(x) + sin(x) % symbolic function definition
dg = diff(g)        % symbolic differentiation
subs(dg,x,pi)       % replace x with pi in the expression given by dg


% function handles and function evaluations
A = [1 2; 3 4] % A is a matrix whose eigen system we want to find
fn = @eig
% two alternate ways of finding eigenvector and eigenvalues
[EV ev] = fn(A)
% OR
[V e] = feval(fn,A)
```

A2.2 Graphical representation of data

A2.2.1 *Plotting functions*

The following piece of program gives you an idea of plotting functions graphically, assigning appropriate labels, defining boundaries of graphs, and using desired font size and markers. Additionally, you will also learn how to generate multiple plots within a single graphic frame using the Matlab `subplot` functionality. Tweak the parameters and practise to get a better understanding of the different available options for pictorially depicting data in Matlab.

```matlab
x=[0:0.05:4*pi];
y=sin(x);
z=cos(x);
figure, subplot(3,1,1);
plot(x,y,'r*-',x,z,'bo-');
xlim([0 4*pi]);
ylim([-1.3 1.3]);
xlabel('x');
ylabel('fundamental oscillations');
legend('sin(x)','cos(x)');
title('periodic functions');


subplot(3,1,2);
xx=[0:0.1:4*pi];
t=tan(xx);
stem(xx,t,'mx')
xlim([0 4*pi]);
ylim([-10 10]);
xlabel('x');
ylabel('tan x');
title('periodic singular function');


%% drawing rectangular pulse %%
% This example generates a pulse train using the default rectangular
% pulse of unit width. The repetition frequency is 0.5 Hz,
% the signal length is 60 s, and the sample rate is 1 kHz.
% The gain factor is a sinusoid of frequency 0.05 Hz.
```

```
t = 0:1/1e3:60;
d = [0:2:60;sin(2*pi*0.05*(0:2:60))];
x = @rectpuls;
y = pulstran(t,d,x);
subplot(3,1,3);
plot(t,y,'-bs',...
    'LineWidth',2,...
    'MarkerSize',10,...
    'MarkerEdgeColor','g',...
    'MarkerFaceColor',[0.5,0.5,0.5]);
hold on
xlabel('Time (s)')
ylabel('Waveform')
title('rectangular pulse train');
```

Figure A2.1 How to make subplots in Matlab with labels and legends?

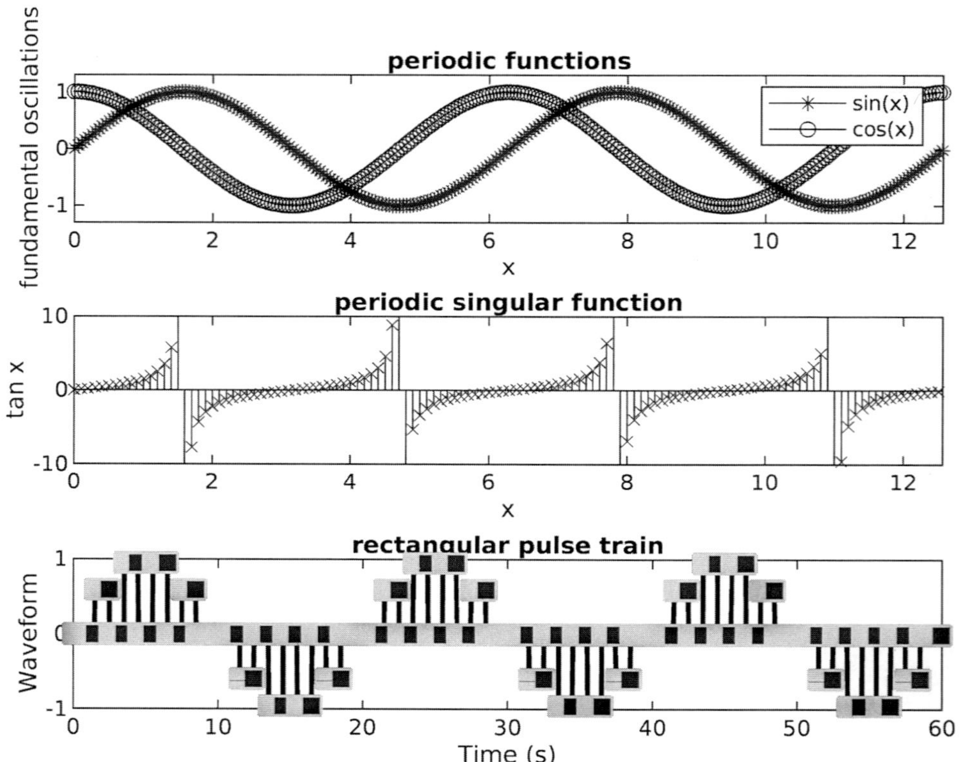

A2.2.2 *Plotting data on a bar graph*

In a Matlab script file, store data as a matrix as shown below. Then call a user-defined function `mydataPlot`, which takes the input matrix as its argument and draws a joint-histogram of the data over time (years). The data must be normalized on a scale of 0–2 and depicted as a bar graph.

The script file may contain the following lines of code.

```
%%%%%%%%%%%%%%% making histograms %%%%%%%%%%%
% Ip stores input data in 3 columns:
% (Year, Rainfall, Temperature) in appropriate units
Ip=[2009 1000 39; 2010 997 34;2011 1152 41; 2012 855 31; ...
    2013 1013 40; 2014 878 30; 2015 1243 43];
mydataPlot(Ip);
```

Writing user-defined functions

`mydataPlot` is a user-defined function that needs to be programmed by writing
a short Matlab function and saving it as `mydataPlot.m`. The function may be
written as follows.

```
function [X] = mydataPlot(Ip)

A = Ip(:,1);
B = Ip(:,2);
C = Ip(:,3);
Bnew=(B(:)-min(B))/(max(B)-min(B));
Cnew=(C(:)-min(C))/(max(C)-min(C));
newData = [Bnew+1 Cnew+1]; % setting it on a normalized scale 0-2
figure, bar(A,newData(:,1:2));
xlabel('Year','fontsize',18);
ylabel('Normalized data on a scale of 0-2','fontsize',18);
legend('Amount of rainfall','temperature','Location','northwest');
title('Visualizing trend & correlation qualitatively using joint-histogram',...
    'fontsize',14);

end % denoting end of function
```

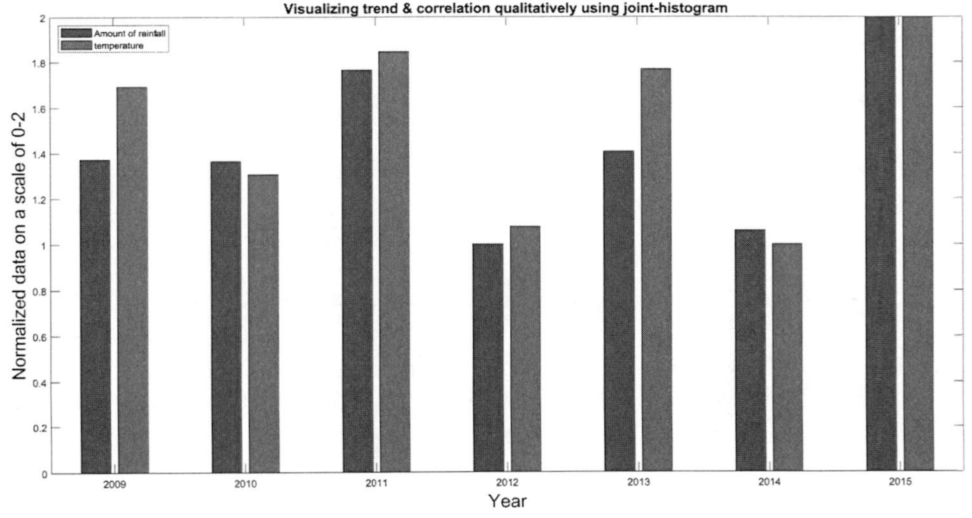

Figure A2.2 How to make bar
graphs in Matlab with labels and
legends?

A2.3 Loops and conditional statements

Matlab provides in-built *loop* functionalities using the following syntax.

```
for
. . . . . . . . .
. . . . . . . .
end
or
while
. . . . . . . .
. . . . . . . .
end
```

A2.3.1 *for loop to generate a matrix with labels in lexicographical order*

```
% use for loops to construct a matrix with entries
% that store the serial location (lexicographical
% order)
% also compute the sum of all entries of the same
% matrix

imax=4; jmax=5;
running_sum=0;
for i=1:imax
    for j=1:jmax
        A(i,j) = (j + (i-1)*jmax);
        running_sum = running_sum + A(i,j);
    end
end
```

A2.3.2 *Conditional statement using if-else for comparing results*

```
if running_sum == sum(sum(A))
    disp('my calculation is correct: Hurrah!');
else
    disp('I made an error in calculation');
end
```

A2.3.3 *for loops for printing prime numbers less than or equal to N*

Exercise: The following is a pseudocode for printing all the prime numbers less than or equal to N, where $N = 30$, for example. Convert the pseudocode to Matlab executable code and display your answer.

Pseudocode for printing prime numbers

```
INPUT: N.
 for all i from 2 to N
    reset prime number flag to default ON
    for all j from 2 to i/2
        if remainder of i ÷ j is not equal to 0
            continue the current for loop
        else
            turn OFF prime number flag
            break from the current for loop
        end if-else condition
    end for loop for j
    if prime number flag is ON
        store prime number i in a dynamic array
    end if condition
 end for loop for i
OUTPUT: display array containing the prime numbers.
```

While implementing the above algorithm, you will learn to use the following Matlab commands: `mod`, `continue`, and `break`. Use `>> doc ⟨function name⟩` or `>> help ⟨function name⟩` on the command line to learn how to use them. Also, the comparative clause *"is not equal to"* is written in Matlab as `~=` .

A2.4 Data structures: loading, organizing, accessing, and writing data

There are many different data structures available in Matlab. Here, we will discuss only some of the important ones that may be relevant to us.

A2.4.1 *Data as graphs of matrices*

A *graph* is a data structure that consists of the following two components.

1. A finite set of *vertices* also known as *nodes*.

2. A finite set of ordered pairs of the form (u, v) known as *edges*. The pair is ordered because (u, v) is not the same as (v, u) in case of a directed graph (di-graph). The pair of the form (u, v) indicates that there is an edge from vertex u to vertex v. The edges may contain weight/value/cost.

Graphs are used to represent many real-life applications. They are used to represent networks. The networks may include paths in a city or a telephone network or circuit network. Graphs are also used in social networks such as LinkedIn and Facebook. For example, in Facebook, each person is represented with a vertex (or node). Each node is a structure and contains information such as id, name, gender, and location of a person.

Adjacency matrix

A graph may be represented as an *adjacency matrix*. An *adjacency matrix* is a 2D array (or matrix) of size $V \times V$, where V is the number of vertices in a graph. Let the 2D array be *adj(i,j)*; a slot *adj(i,j)* = 1 indicates that there is an edge from vertex i to vertex j. The adjacency matrix for an undirected graph is always symmetric. The adjacency matrix is also used to represent weighted graphs. If *adj(i,j)* = w, then there is an edge from vertex i to vertex j with weight w.

Drawing a graph from adjacency matrix

```
% Define a matrix adj.
adj = [0 1 1 0 ; 1 0 0 1 ; 1 0 0 1 ; 0 1 1 0];

% Draw a picture showing the connected nodes.
cla % clear current axis
subplot(1,2,1);
gplot(adj,[0 1;1 1;0 0;1 0],'.-');
% gplot(A,xy) plots the graph (as in 'graph theory') specified by A and xy
text([-0.2, 1.2 -0.2, 1.2],[1.2, 1.2, -.2, -.2],('1234')', ...
    'HorizontalAlignment','center')
axis([-1 2 -1 2],'off')
title('undirected graph','fontsize',14);

% Draw a picture showing the adjacency matrix.
subplot(1,2,2);
xtemp = repmat(1:4,1,4); ytemp = reshape(repmat(1:4,4,1),16,1)';
text(xtemp-.5,ytemp-.5,char('0'+adj(:)),'HorizontalAlignment','center');
line([.25 0 0 .25 NaN 3.75 4 4 3.75],[0 0 4 4 NaN 0 0 4 4])
axis off tight
title('adjacency matrix','fontsize',14);
```

Drawing the adjacency matrix of a given graph

For fun, we will first use the Matlab in-built function `bucky` to draw the graph of a *geodesic dome.*[2]

```
%% using bucky and find the adjacency matrix of a given
% graph
[B,V] = bucky;
G = graph(B);
figure,
p = plot(G);
```

In order to investigate the graph G in more detail, type `>> G.Edges` on the command line.

Now, let us say we were given the graph G (and not the matrix B), and we had to find the corresponding adjacency matrix A; we would do as follows:

```
A = adjacency(G);
H = graph(A(1:10,1:10)); % graph only a section of the
% adjacency matrix
figure,
h = plot(H);
```

[2] https://en.wikipedia.org/wiki/Geodesic_dome, accessed June 24, 2024.

The new plot graphs the selected section of the adjacency matrix A specified by the first 10×10 entries of A. The new adjacency matrix A can be seen from the command line as >> A, and the new truncated graph is shown in Figure A2.5. The information pertaining to the new truncated graph can be found in the following manner.

undirected graph

adjacency matrix

$$\begin{bmatrix} 0 & 1 & 1 & 0 \\ 1 & 0 & 0 & 1 \\ 1 & 0 & 0 & 1 \\ 0 & 1 & 1 & 0 \end{bmatrix}$$

Figure A2.3 A *graph* and its corresponding *adjacency matrix*.

```
>> H.Edges

ans =

   11×2 table

     EndNodes        Weight
     ________        ______

      1      2          1
      1      5          1
      1      6          1
      2      3          1
      3      4          1
      4      5          1
      6      7          1
      6     10          1
      7      8          1
      8      9          1
      9     10          1
```

Figure A2.4 Graph of the geodesic dome using the Matlab function bucky.

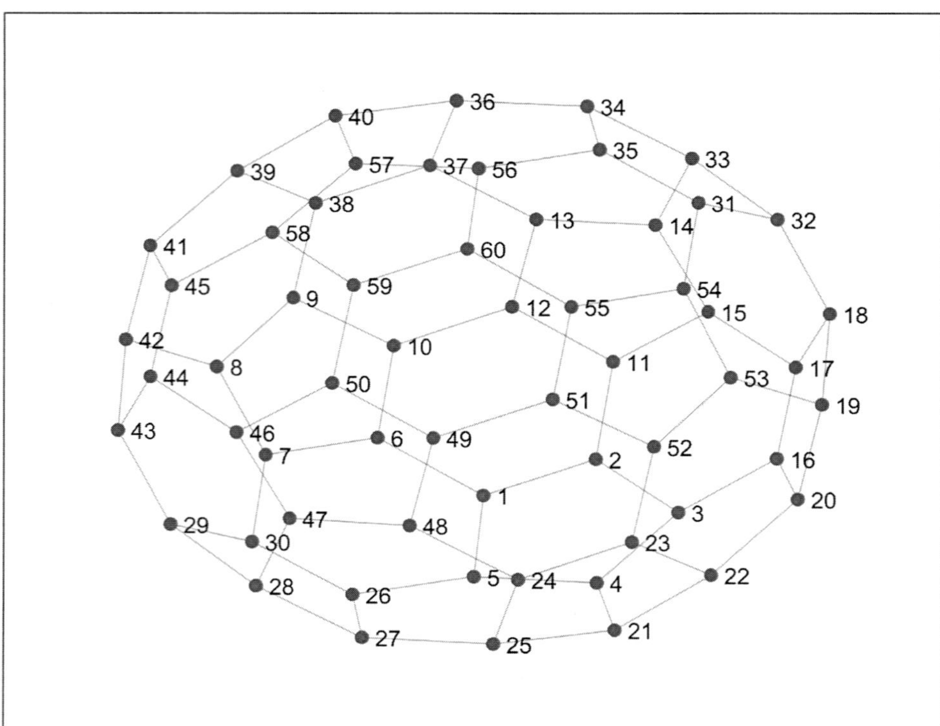

You can also use similar commands to generate the adjacency matrix of the graph in the previous example (the box graph) and compare it with the matrix `adj` you defined.

```
Gnew=graph(adj)
Anew = adjacency(Gnew);
Hnew = graph(Anew);
figure,
hnew = plot(Hnew);
```

Compare Anew and adj to check for consistency.

```
>> Anew

Anew =

    (2,1)        1
    (3,1)        1
    (1,2)        1
    (4,2)        1
    (1,3)        1
    (4,3)        1
    (2,4)        1
    (3,4)        1
```

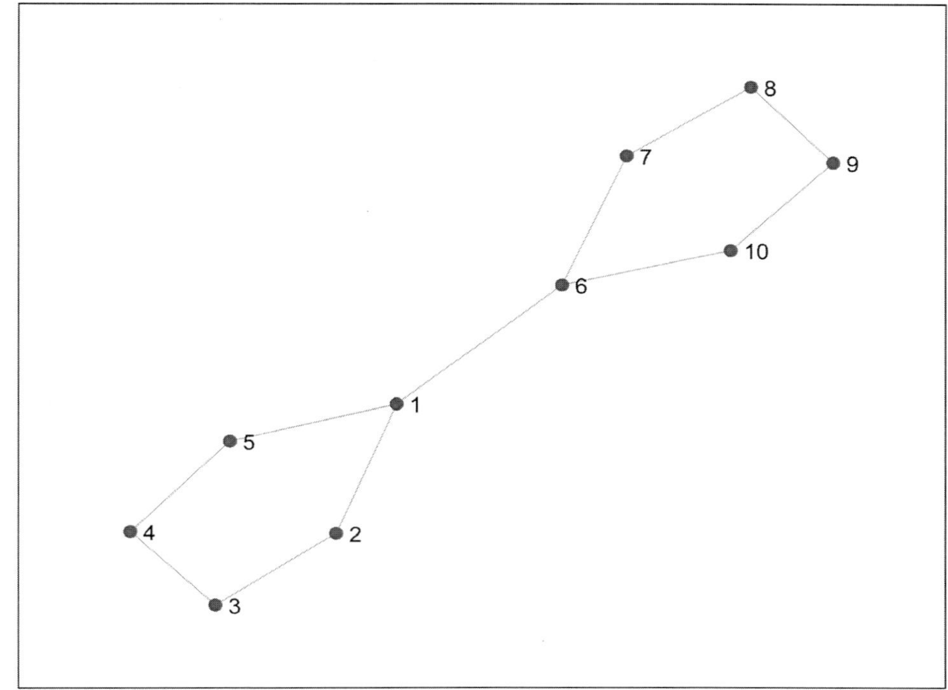

Figure A2.5 Graph of the truncated geodesic dome.

```
>> adj
  adj =

        0       1       1       0
        1       0       0       1
        1       0       0       1
        0       1       1       0
```

A2.4.2 Data as tables

Often we will encounter situations where we may have to import data from an excel file (.csv file). This data in the excel file will be assumed to be stored in tabular (columnar) format under field headings. An example at hand is the file named

```
EnergyConsumptionMP_1996-2018.csv
```

where the data is stored in two columns under the field headings:

```
Year
```

and

```
EnergySupply_MU_
```

This data may be imported in Matlab local workspace as follows:

```
%% reading data from .csv file
T = readtable('EnergyConsumptionMP_1996-2018.csv',
'ReadVariableNames', ...
```

```
        true, 'Format', '%f %f');
xdata = T.Year;
ydata = T.EnergySupply_MU_;
>> xdata

xdata =

        1996
        1998
        2000
        2002
        2004
        2006
        2008
        2010
        2012
        2014
        2016
        2018

>> ydata

ydata =

       27094
       28599
       30624
       32232
       33435
       36073
       35503
       35563
       38799
       42945
       47858
       51976
```

A2.4.3 *Local data structures in Matlab*

Structure arrays

Below is a code to show a single variable (named student) that contains two fields
that are "sub-components" of what it means to be a student. Each field has its own
name and its own type.

```
students(1).name   = 'jim';
students(1).age    = 21;

students(2).name   = 'Jane';
students(2).age    = 33;

students(3).name   = 'Joe';
```

```
students(3).age   = 25;

students(4).name  = 'Janet';
students(4).age   = 24;
```

A query about the first student yields the following information.

```
>> students(1)

ans =

  struct with fields:

    name: 'jim'
     age: 21
```

Cell arrays

Cell arrays contain data in cells that you access by numeric indexing. Common applications of cell arrays include storing separate pieces of text and storing heterogeneous data from spreadsheets. For example, store temperature data for three cities over time in a cell array.

```
temperature(1,:)  =  {'2009-12-31',  [45,  49,  0]};
temperature(2,:)  =  {'2010-04-03',  [54,  68,  21]};
temperature(3,:)  =  {'2010-06-20',  [72,  85,  53]};
temperature(4,:)  =  {'2010-09-15',  [63,  81,  56]};
temperature(5,:)  =  {'2010-12-09',  [38,  54,  18]};

>> temperature

temperature = 5x2 cell array
    {'2009-12-31'}     {1x3  double}
    {'2010-04-03'}     {1x3  double}
    {'2010-06-20'}     {1x3  double}
    {'2010-09-15'}     {1x3  double}
    {'2010-12-09'}     {1x3  double}
```

Plot the temperatures for each city by date.

```
allTemps = cell2mat(temperature(:,2));
dates = datetime(temperature(:,1));

plot(dates,allTemps)
title('Temperature Trends for Different Locations')
xlabel('Date')
ylabel('Degrees (Fahrenheit)')
```

Figure A2.6 Plot of temperature trend.

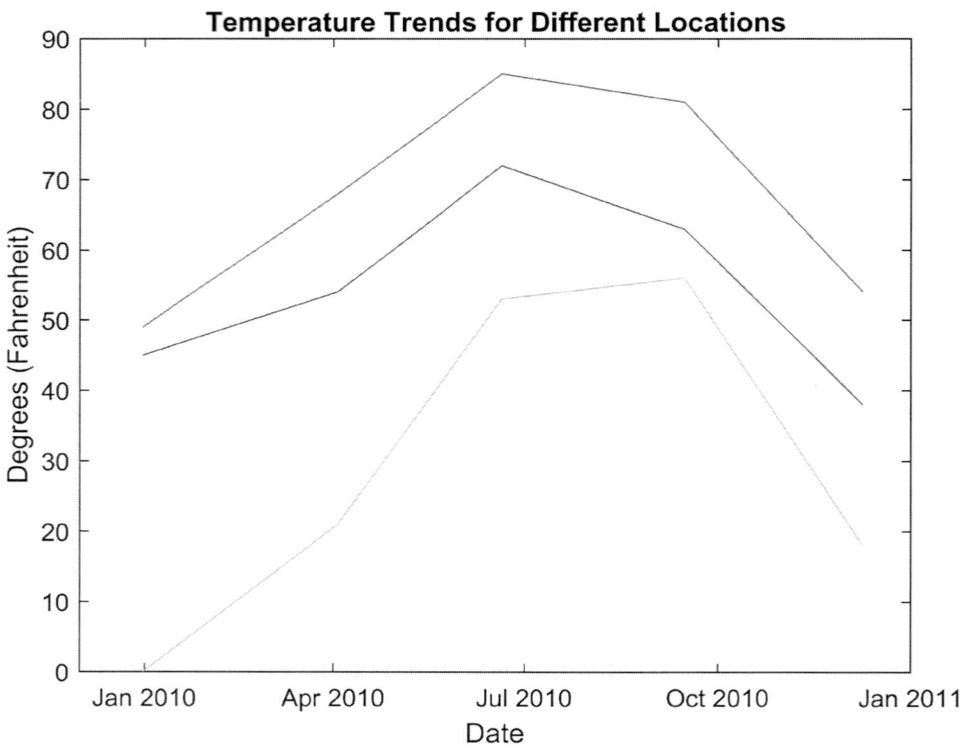

A2.4.4 *Writing to a binary file and reading from a binary file*

Write a nine-element vector to a sample file, `nine.bin`.

```
fileID = fopen('nine.bin','w');
fwrite(fileID,[1:9]);
fclose(fileID);
```

Now, read the contents of the same file and store it in a local variable A.

```
fileID = fopen('nine.bin');
A = fread(fileID)
```

A2.4.5 *Reading a graphic image*

Read a sample image.

```
A = imread('ngc6543a.jpg');
```

`imread` returns a 650-by-600-by-3 array, A. Now, display the image.

```
>> image(A)
```

Use

```
>> doc imread
```

to learn more about the function. Likewise, you may use the Matlab command `imwrite` to write data to an image file.

A2.4.6 *Recording and playing a movie in Matlab*

For recording a set of graphic frames and constructing a movie, you may use `getframe` as shown below immediately after the graphic frame is generated each time in a loop.

```
while n <= 100
    . . . . . . .
    . . . . . . .
    plot(x,y,'r--');
    F(n) = getframe;
    n=n+1;
end
```

In order to make a movie and play it twice, use the following.

```
>> figure, movie(F,2);
```

A2.5 Final remarks

Finally, extensively use `>> doc doc` and `>> help ⟨function name⟩` to explore different functionalities available in Matlab. Also visit the mathworks online help center at https://in.mathworks.com/help/ in order to learn how to use Matlab more effectively.

Index

Color Plates

Figure 2.24 (page 51)

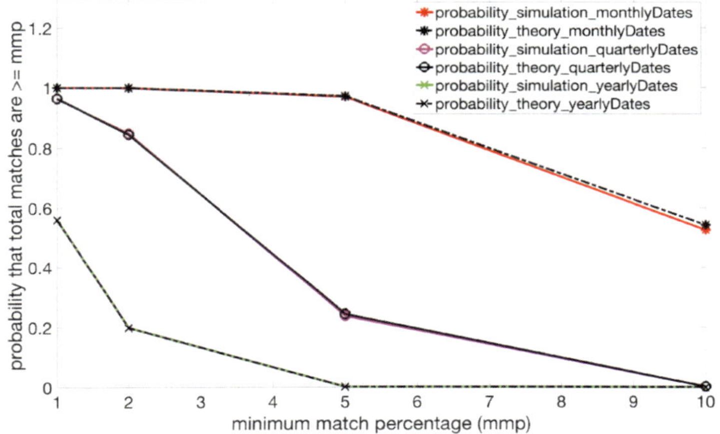

He was too young to have been blighted
by the cold world's corrupt finesse;
his soul still blossomed out, and lighted
at a friend's word, a girl's caress.
In heart's affairs, a sweet beginner,
he fed on hope's deceptive dinner;
the world's éclat, its thunder-roll,
still captivated his young soul.
He sweetened up with fancy's icing
the uncertainties within his heart;
for him, the objective on life's chart
was still mysterious and enticing—
something to rack his brains about,
suspecting wonders would come out.

Figure 3.11 (page 103)

Figure 4.2 (page 117)

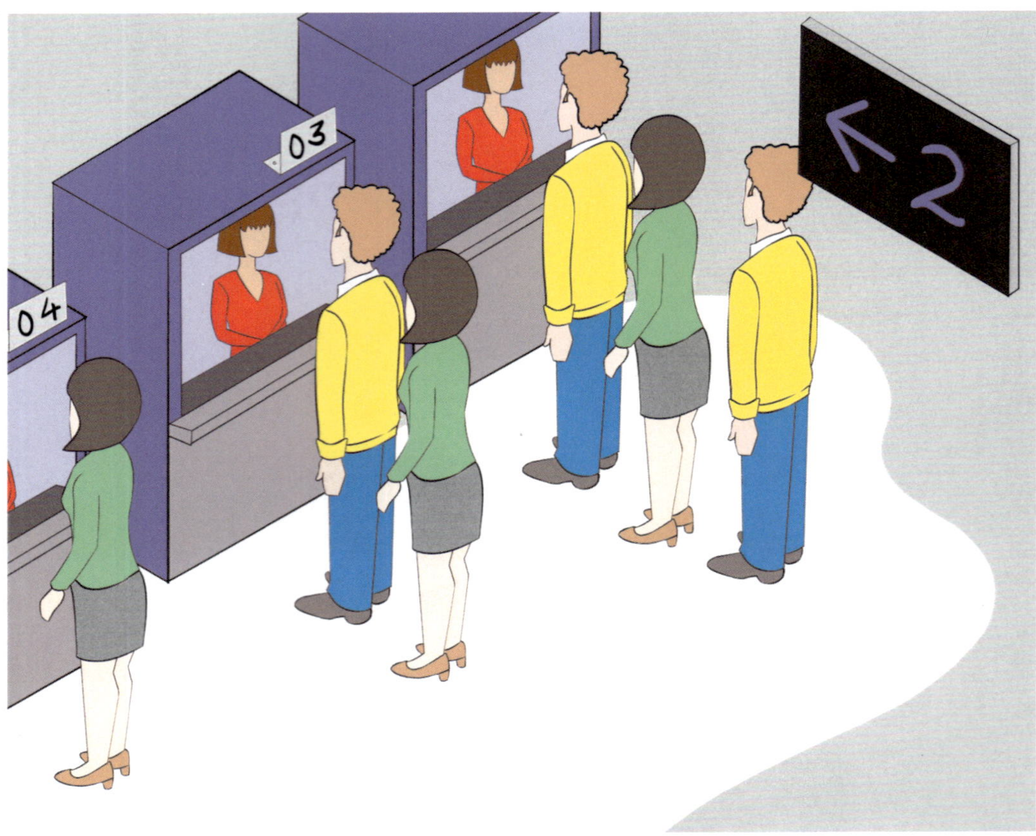

Figure 4.3 (page 120)

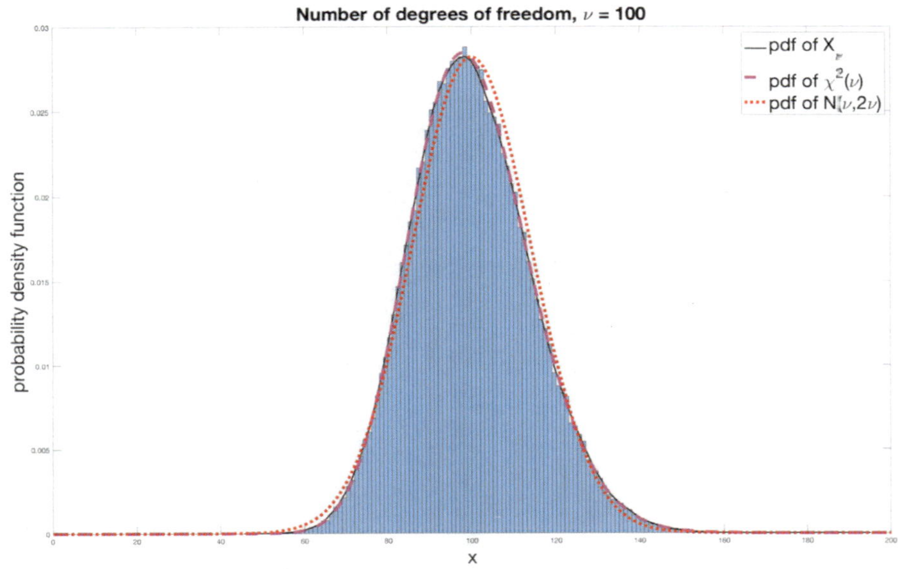

Figure 5.5 (page 148)

Figure 5.10 (page 155)

Figure 5.13 (page 158)

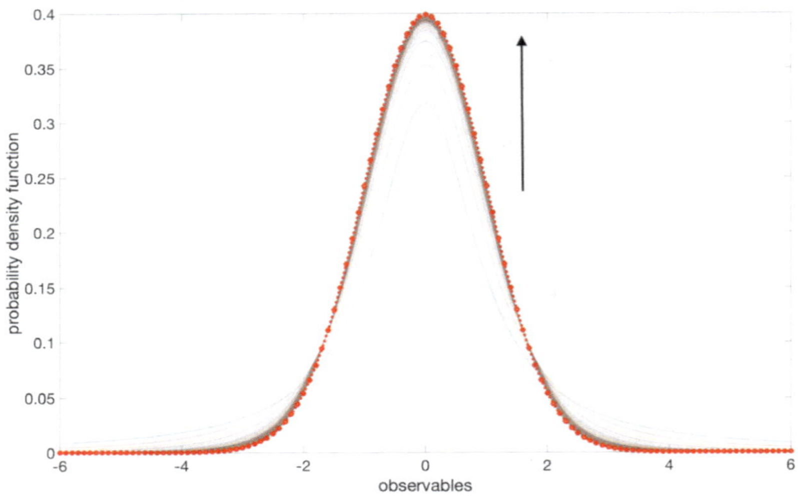

Figure 7.2 (page 216)

Figure 7.6 (page 225)